Introduction to Pseudodifferential and Fourier Integral Operators

Volume 1

Pseudodifferential Operators

THE UNIVERSITY SERIES IN MATHEMATICS

Series Editor: **Joseph H. Kohn**
Princeton University

INTRODUCTION TO PSEUDODIFFERENTIAL AND FOURIER INTEGRAL OPERATORS
François Treves
VOLUME 1: PSEUDODIFFERENTIAL OPERATORS
VOLUME 2: FOURIER INTEGRAL OPERATORS

A SCRAPBOOK OF COMPLEX CURVE THEORY
C. Herbert Clemens

Introduction to Pseudodifferential and Fourier Integral Operators

Volume 1
Pseudodifferential Operators

François Treves
Rutgers University
New Brunswick, New Jersey

Springer Science+Business Media, LLC

Library of Congress Cataloging in Publication Data

Treves, François.
Introduction to pseudodifferential and fourier integral operators.

(The University series in mathematics)
Bibliography: p.
Includes index.
1. Pseudodifferential operators. 2. Integral operators. I. Title.
QA329.7.T74 515'.72 79-25996

ISBN 978-1-4684-8782-4 ISBN 978-1-4684-8780-0 (eBook)
DOI 10.1007/978-1-4684-8780-0

Originally published by Plenum Press, New York 1980

Preface

I have tried in this book to describe those aspects of pseudodifferential and Fourier integral operator theory whose usefulness seems proven and which, from the viewpoint of organization and "presentability," appear to have stabilized. Since, in my opinion, the main justification for studying these operators is pragmatic, much attention has been paid to explaining their handling and to giving examples of their use. Thus the theoretical chapters usually begin with a section in which the construction of special solutions of linear partial differential equations is carried out, constructions from which the subsequent theory has emerged and which continue to motivate it: parametrices of elliptic equations in Chapter I (introducing pseudodifferential operators of type 1, 0, which here are called standard), of hypoelliptic equations in Chapter IV (devoted to pseudodifferential operators of type ρ, δ), fundamental solutions of strongly hyperbolic Cauchy problems in Chapter VI (which introduces, from a "naive" standpoint, Fourier integral operators), and of certain nonhyperbolic forward Cauchy problems in Chapter X (Fourier integral operators with complex phase).

Several chapters—II, III, IX, XI, and XII—are devoted entirely to applications.

Chapter II provides all the facts about pseudodifferential operators needed in the proof of the Atiyah–Singer index theorem, then goes on to present part of the results of A. Calderon on uniqueness in the Cauchy problem, and ends with a new proof (due to J. J. Kohn) of the celebrated sum-of-squares theorem of L. Hörmander, a proof that beautifully demonstrates the advantages of using pseudodifferential operators.

The subject of Chapter III is boundary problems for elliptic equations. It is perhaps the only place in the book where I have departed somewhat from standard procedure. The overall approach is the one made familiar by the works of A. Calderon, R. Seeley, and others: to transfer the problem from the domain where it was originally posed onto the boundary of that domain, where it becomes an "interior" but in general pseudodifferential rather than differential problem. The main difference is that regardless of the nature of the boundary conditions, I construct from the start the operator that effects the transfer to the boundary and show it to be a standard pseudodifferential operator (with respect to the tangential variables, depending smoothly on the variable normal to the boundary), a kind of exponential to which all the results of Chapter I are applicable. I then show that the testing of the essential properties (regularity of the solutions up to the boundary, Fredholm character, etc.) can be done for the interior problem on the boundary, which concerns the *Calderon operator* of the boundary problem. For instance, the Calderon operator of a boundary problem of the Lopatinski–Shapiro type, called *coercive* in this book, is elliptic. As a consequence the regularity up to the boundary of the solutions is an immediate corollary of the property that pseudodifferential operators are pseudolocal. It suffices to apply it to the "exponential" which effects the transfer to the boundary. Analogous results are discussed for problems of principal type, especially the subelliptic ones, for example certain oblique derivative problems, and for the $\bar{\partial}$-Neumann problem when the conditions (on the number of positive or negative eigenvalues of the Levi matrix) for hypoellipticity with loss of one order of differentiation are satisfied.

The text goes back to elliptic boundary problems at the end of Chapter V to discuss the question of analyticity up to the boundary, under the right circumstances, by exploiting the theory of analytic pseudodifferential operators, which makes up the contents of Chapter V.

Applications of Fourier integral operators are sprinkled throughout Volume 2. Chapter IX describes in great detail the reduction of suitable systems of pseudodifferential equations to the main "standard forms." An example is the microlocal transformation of systems of the induced $\bar{\partial}$ type to systems of Mizohata equations, under the hypothesis that the Levi matrix is nondegenerate. Chapter XI presents applications of Fourier integral operators with complex phase, in particular to operators that can be transformed microlocally into Mizohata's, and to establishing subelliptic estimates. It is shown that the latter can be used to refine the Carleman estimates that lead to uniqueness in the Cauchy problem (and thus improve the result in Chapter II).

Chapter XII presents three applications to the study of the spectrum of the Laplace–Beltrami operator $-\Delta$ on a compact Riemannian manifold: (1) the classical estimate, due to V. G. Akumovic, of the number $N(\lambda)$ of eigenvalues not exceeding $\lambda \sim +\infty$; (2) the generalization by J. Chazarain of the Poisson formula, relating the lengths of the closed geodesics to the singularities of the distribution on the real line, $\mathrm{Tr}(\exp(it\sqrt{-\Delta}))$; (3) the derivation of the existence of certain sequences of eigenvalues from that of Lagrangian submanifolds of the cotangent bundle on which the Riemannian length of covectors is constant and which satisfy Maslov's quantization condition. This last section of the book follows very closely the presentation of A. Weinstein [1].

With the exception of the elliptic boundary problems in Chapter III, the applications are never studied in their own right, with the pretense of describing them fully, but only as examples of what can be achieved by using pseudodifferential or Fourier integral operators. This is why I have refrained from embarking on the study of other major areas of application of the theory: solvability of linear PDEs, diffraction, well-posedness of the Cauchy problem. On the latter the reader is referred to the works of Ivrii [1–3], Ivrii–Petkov [1], and Hörmander [17].

To complete this brief outline of the contents of the book: the global theory of Fourier integral operators is described in Chapter VIII, following the laying out of the symplectic geometry background in Chapter VII. Clean phases, rather than nondegenerate ones, are used in the microlocal representations of the operators. This simplifies composition in Section 6 of Chapter VIII, and pays off nicely in the applications to Riemannian geometry in Chapter XII.

There are important aspects of pseudodifferential and Fourier integral operator theory that this book does not discuss. First, this book is totally L^2 oriented. Not a word is said about pseudodifferential action on L^p spaces for $p \neq 2$. I felt I was not qualified to go into this area. Besides, there is great advantage in restricting one's outlook to L^2, for one thus can exploit Fourier transforms to the full. This is also why the book does not deal in any depth with the kernels $K(x, y)$ associated with the operators. After all, perhaps the main thrust of pseudodifferential operators is to substitute, as often as possible, the calculus of symbols for that of kernels. Symbolic calculus has been traditionally based on Fourier (or Laplace) transforms, whose natural framework is L^2 or the Schwartz space $\mathscr{S}'$ of tempered distributions. This is of course not to deny that certain applications, such as continuity between L^p spaces and even between spaces of Hölder continuous functions, require less coarse treatment. But such questions and many others are beyond the scope of this book.

Closer to its contents are the classes of pseudodifferential operators introduced in the last few years by various authors, most notably by Beals and Fefferman [1], Boutet de Monvel [3], Hörmander [19], and Unterberger [1]. For a systematic study, see Beals [1]. In this connection my feeling has been that this is more advanced mathematics, which the reader should not have too much difficulty in learning once he has digested some of the material in this book. The same applies to the global theory of pseudodifferential operators in Euclidean space (see Kumano-Go [1]) and to various extensions of Fourier integral operators, such as the one in Guillemin [2] or those based on the *Airy function*, which turn up naturally in the study of certain problems where the characteristics are double (as occurs, for example, in geometrical optics; see, for instance, Taylor [2] and Egorov [2]). One important item related to Fourier integral operators, and which is missing from this book but undoubtedly should have been in it, is the *metaplectic representation*. On this subject I must content myself with referring the reader to other texts, for instance Leray [1] and Weinstein [1].

The prerequisites for a serious study of the material in the book vary from chapter to chapter. Most of the time they are the standard requirements in real and complex analysis and in functional analysis, with a smattering of distribution theory, whose essential concepts and notation are recalled in the section on notation and background. Manifolds, their tangent and cotangent bundles, and more general vector bundles are defined in Chapter I. Complements of "basic" differential geometry are provided in Chapter VII, following a section devoted to symplectic linear algebra and preceding one devoted to symplectic differential geometry. I hope that some chapters will be useful to anybody eager to learn the fundamental aspects of pseudodifferential and Fourier integral operator theory, or willing to teach it for the first time—I am thinking mainly of Chapters I–III and VI–VIII. Other chapters (Chapters V and IX–XI) are intended more for reference or specialized study and use. Still other chapters fall in between these two categories.

The book is rather informally written—to some this will seem an understatement—due mainly to my inclinations, lack of time, and a certain sense of urgency, the sense that a book with more or less these contents is overdue. I have not hesitated to borrow from the available literature, especially from the original article of Hörmander [11], from the lecture notes of Duistermaat [1], and from the article by Melin and Sjöstrand [1].

In matters of terminology I have tried to be as much of a conformist as I could. But one notation I could not resign myself to adopting is $L^m(\ )$ for the spaces of pseudodifferential operators. L is overused in mathematics: Lebesgue spaces, sets of linear transformations, linear partial differential operators, Lagrangian manifolds are all called L this or that. On the other hand, capital psi, Ψ, is underused, and very naturally associated with pseudo, so I write $\Psi^m(\quad)$ in the place of $L^m(\quad)$. Perhaps the only other novelty is the term *microdistribution*, which seems to me the natural analogue, in the context of distributions, of the name *microfunction* introduced by M. Sato in hyperfunction theory (see Sato [1]).

On the other hand, I have stuck to the name Fourier integral operator, although I tend to agree that it is not the most felicitous and that it may have been more equitable to use Maslov operator instead, as many Russian authors do. But Fourier integral operator is the term that people everywhere outside the Soviet Union use and are used to, and it might be too late to reverse the trend. I do not quite understand J. Dieudonné when he contends in his beautiful treatise [1] on calculus on manifolds that the name distorts the purpose of Fourier integral operators, which have succeeded, according to Dieudonné, in "eliminating" the Fourier transform. I believe rather that their purpose, and their effect, is to extend the applications of the Fourier transform from Euclidean spaces to manifolds.

François Treves

Contents of Volume 1

Pseudodifferential Operators

Contents of Volume 2

Fourier Integral Operators

Notation and Background

1. Euclidean Spaces

$\mathbb{R}^n$: n-dimensional (real) Euclidean space

$\mathbb{R}_n$: dual of $\mathbb{R}^n$

$x = (x^1, \ldots, x^n)$, $y = (y^1, \ldots, y^n)$, also $x = (x_1, \ldots, x_n)$, $y = (y_1, \ldots, y_n)$: variables and coordinates in $\mathbb{R}^n$

$\xi = (\xi_1, \ldots, \xi_n)$, $\eta = (\eta_1, \ldots, \eta_n)$: variables and coordinates in $\mathbb{R}_n$

$\mathbb{C}^n$, $\mathbb{C}_n$: n-dimensional complex space and its dual

$z = (z^1, \ldots, z^n)$, also $z = (z_1, \ldots, z_n)$: variables and coordinates in $\mathbb{C}^n$

$x \cdot \xi = x^1\xi_1 + \cdots + x^n\xi_n$: scalar product between $x \in \mathbb{R}^n$ and $\xi \in \mathbb{R}_n$

$|x| = [(x^1)^2 + \cdots + (x^n)^2]^{1/2}$, $|\xi| = [\xi_1^2 + \cdots + \xi_n^2]^{1/2}$: Euclidean norms in $\mathbb{R}^n$ and in $\mathbb{R}_n$

$\bar{z} = (\bar{z}^1, \ldots, \bar{z}^n)$: the complex conjugate of z

$z \cdot \bar{z}' = z^1\bar{z}'^1 + \cdots + z^n\bar{z}'^n$: the *hermitian product* in $\mathbb{C}^n$

$|z| = [|z^1|^2 + \cdots + |z^n|^2]^{1/2} = [z \cdot \bar{z}]^{1/2}$: Euclidean norm in $\mathbb{C}^n$

Ω: *open* subset of a Euclidean space

$\Omega \backslash S$: *complement* of a subset S in Ω

$S \subset\subset \Omega$: means that the *closure* of S is a *compact* subset of Ω (then S is said to be *relatively compact* in Ω)

2. The Multi-Index Notation

$\mathbb{Z}$: set of integers >0 or ≤ 0

$\mathbb{Z}_+$: set of integers ≥ 0

$\mathbb{Z}_+^n$: set of *n-tuples* $\alpha = (\alpha_1, \ldots, \alpha_n)$ with $\alpha_j \in \mathbb{Z}_+$ for each $j = 1, \ldots, n$

$\beta \leq \alpha$: means $\beta_j \leq \alpha_j$ for every $j = 1, \ldots, n$ $(\alpha, \beta \in \mathbb{Z}_+^n)$

$|\alpha| = \alpha_1 + \cdots + \alpha_n$: *length* of $\alpha \in \mathbb{Z}_+^n$

$\alpha! = \alpha_1! \cdots \alpha_n!$, $\binom{\alpha}{\beta} = \binom{\alpha_1}{\beta_1} \cdots \binom{\alpha_n}{\beta_n} = \dfrac{\alpha!}{\beta!(\alpha-\beta)!}$ if $\alpha, \beta \in \mathbb{Z}_+^n$ and $\beta \le \alpha$

$x^\alpha = (x^1)^{\alpha_1} \cdots (x^n)^{\alpha_n}$ if $x \in \mathbb{R}^n$, $\alpha \in \mathbb{Z}_+^n$

$\xi^\alpha = \xi_1^{\alpha_1} \cdots \xi_n^{\alpha_n}$ if $\xi \in \mathbb{R}_n$

$\partial_x^\alpha = (\partial/\partial x^1)^{\alpha_1} \cdots (\partial/\partial x^n)^{\alpha_n}$ (also denoted by ∂^α)

$\partial_\xi^\alpha = (\partial/\partial \xi_1)^{\alpha_1} \cdots (\partial/\partial \xi_n)^{\alpha_n}$

$D_x^\alpha = \left(\dfrac{1}{\sqrt{-1}} \dfrac{\partial}{\partial x^1}\right)^{\alpha_1} \cdots \left(\dfrac{1}{\sqrt{-1}} \dfrac{\partial}{\partial x^n}\right)^{\alpha_n}$ (also denoted by D^α)

$D_\xi^\alpha = \left(\dfrac{1}{\sqrt{-1}} \dfrac{\partial}{\partial \xi_1}\right)^{\alpha_1} \cdots \left(\dfrac{1}{\sqrt{-1}} \dfrac{\partial}{\partial \xi_n}\right)^{\alpha_n}$

Also, if u is a C^∞ function of x:

$u^{(\alpha)} = \partial_x^\alpha u$

$\partial_x u = u_x = \operatorname{grad} u = (\partial u/\partial x^1, \ldots, \partial u/\partial x^n)$

$D_j u = \dfrac{1}{\sqrt{-1}} \dfrac{\partial u}{\partial x^j}$, $j = 1, \ldots, n$

Taylor Expansion

(0.1)
$$u(x) = \sum_{\alpha \in \mathbb{Z}_+^n} (1/\alpha!)(x-y)^\alpha\, \partial^\alpha u(y)$$

Leibniz Formula

(0.2)
$$\partial^\alpha(uv) = \sum_{\beta \le \alpha} \binom{\alpha}{\beta} \partial^{\alpha-\beta} u\, \partial^\beta v$$

Transposed Leibniz Formula

(0.3)
$$v\, \partial^\alpha u = \sum (-1)^{|\beta|} \binom{\alpha}{\beta} \partial^{\alpha-\beta}[u\, \partial^\beta v]$$

[To prove (0.3) multiply the left-hand side by a C^∞ function w vanishing outside a compact set and integrate by parts $\int wv\, \partial^\alpha u\, dx = (-1)^{|\alpha|} \int \partial^\alpha(wv) u\, dx$, apply (0.2) and integrate by parts "back".]

Differential Operators in Ω

Linear partial differential operators are polynomials in $D = (D_1, \ldots, D_n)$ with *coefficients* belonging to $C^\infty(\Omega)$, such as

$$P(x, D) = \sum_{|\alpha| \le m} c_\alpha(x) D^\alpha.$$

If, for some α of length m, c_α does not vanish identically in Ω, m is called the *order* of $P(x, D)$. When the coefficients c_α are constant, we write $P(D)$.

${}^tP(x, D)$: *transpose* of $P(x, D)$, defined by

$$ {}^tP(x, D)u(x) = \sum_{|\alpha| \leq m} (-1)^{|\alpha|} D^\alpha [c_\alpha(x)u(x)] $$

$P(x, D)^*$: *adjoint* of $P(x, D)$, $P(x, D)^* = \overline{{}^tP(x, D)}$. The bar means that the coefficients have been replaced by their complex conjugates.

$\Delta = (\partial/\partial x^1)^2 + \cdots + (\partial/\partial x^n)^2$: the *Laplace operator* in $\mathbb{R}^n$;

$(\partial/\partial \bar{z}^j) = \frac{1}{2}(\partial/\partial x^j + \sqrt{-1}\partial/\partial y^j)$, $j = 1, \ldots, n$: the *Cauchy–Riemann operators* in $\mathbb{C}^n$.

3. Functions and Function Spaces

supp f: the *support* of the function f, i.e., the closure of the set of points at which f does *not* vanish

$C^m(\Omega)$: space of m times continuously differentiable *complex-valued* functions in Ω ($m \in \mathbb{Z}_+$ or $m = +\infty$)

$C_c^\infty(\Omega)$: space of C^∞ complex functions in Ω having compact support; the elements of $C_c^\infty(\Omega)$ are often called *test functions* in Ω

$C_c^\infty(K)$: space of C^∞ complex functions in $\mathbb{R}^n$ which vanish identically outside the compact set K

The C^∞ Topology

Let K be any compact subset of Ω, m any integer ≥ 0. For any $\phi \in C^\infty(\Omega)$, set

$$ p_{m,K}(\phi) = \operatorname{Max}_{x \in K} \sum_{|\alpha| \leq m} |\partial^\alpha \phi(x)|. $$

Then, as K and m vary in all possible manners, the $p_{m,K}$ form a basis of continuous seminorms on $C^\infty(\Omega)$. Actually, it suffices to let K range over an *exhausting sequence of compact subsets* of Ω, $\{K_\nu\}_{\nu=0,1,\ldots}$; this means that K_ν is contained in the interior of $K_{\nu+1}$ and that every compact subset of Ω is contained in some K_ν. Set $p_m = p_{m,K_m}$; the seminorms p_m define the topology of $C^\infty(\Omega)$. Every neighborhood of a "point" ϕ_o of $C^\infty(\Omega)$ contains a neighborhood

$$ V_{m,\varepsilon} = \{\phi \in C^\infty(\Omega); p_m(\phi - \phi_o) \leq \varepsilon\} $$

for a suitable choice of $m \in \mathbb{Z}_+$ and $\varepsilon > 0$. A sequence of C^∞ functions in Ω, ϕ_j $(j = 1, 2, \ldots)$, converges to ϕ_o in $C^\infty(\Omega)$ if and only if, for every $\alpha \in \mathbb{Z}^n_+$, $\partial^\alpha \phi_j$ converges to $\partial^\alpha \phi_o$ uniformly on every compact subset of Ω.

The topology of $C^\infty(\Omega)$ can be defined by a metric such as

$$\operatorname{dist}(\phi, \psi) = \sum_{m=0}^{\infty} 2^{-m} \inf(1, p_m(\phi - \psi)).$$

All such metrics are equivalent, and turn $C^\infty(\Omega)$ into a complete metric space. Equipped with its natural (i.e., the C^∞) topology $C^\infty(\Omega)$ is a Fréchet space, i.e., a locally convex topological vector space that is metrizable and complete. In the C^∞ topology every *bounded and closed* set is *compact.* (A subset of $C^\infty(\Omega)$ is bounded if every seminorm $p_{m,K}$ is bounded on it.) This property follows easily from the Ascoli–Arzela theorem.

The Natural Topology of $C_c^\infty(\Omega)$

For any compact subset K of Ω, $C_c^\infty(K)$ is a *closed* linear subspace of $C^\infty(\Omega)$ and is equipped with the induced (or relative) topology. Set-theoretically,

$$C_c^\infty(\Omega) = \bigcup_{K \subset\subset \Omega} C_c^\infty(K). \tag{0.4}$$

Then a *convex* subset of $C_c^\infty(\Omega)$ is open if and only if its intersection with every subspace $C_c^\infty(K)$ is open in the latter.

The topology of $C_c^\infty(\Omega)$ is used only through the following properties:

(i) A *sequence* converges in $C_c^\infty(\Omega)$ if and only if it is contained in $C_c^\infty(K)$ for some compact subset K of Ω and converges in $C_c^\infty(K)$.
(ii) A subset B of $C_c^\infty(\Omega)$ is *bounded* if and only if it is contained and bounded in *some* $C_c^\infty(K)$.
(iii) A linear map of $C_c^\infty(\Omega)$ into an arbitrary locally convex space E is *continuous* if and only if its restriction to every subspace $C_c^\infty(K)$, $K \subset\subset \Omega$, is *continuous* (i.e., if the image of every convergent sequence is a convergent sequence).

From (ii) it follows that *every bounded and closed subset of $C_c^\infty(\Omega)$ is compact.*

$L^p(\Omega)$: Lebesgue space of (equivalence classes) of pth power integrable functions in Ω $(1 \le p < +\infty)$
$\|u\|_{L^p(\Omega)} = (\int_\Omega |u(x)|^p \, dx)^{1/p}$, norm in $L^p(\Omega)$
$L^2(0, T; E)$: space of L^2 functions in the interval $[0, T]$ valued in the Hilbert space E

$L^\infty(\Omega)$: Lebesgue space of (equivalence classes) of essentially bounded functions in Ω

$\|u\|_{L^\infty(\Omega)}$: the norm in $L^\infty(\Omega)$

$L^p_{\text{loc}}(\Omega)$: space of *locally* L^p-functions in Ω, i.e., of the functions f such that given any test function ϕ in Ω, $\phi f \in L^p(\Omega)$

$L^p_c(\Omega)$: subspace of $L^p(\Omega)$ consisting of the functions $f \in L^p(\Omega)$ such that $\operatorname{supp} f \subset\subset \Omega$

$(u, v)_{L^2(\Omega)} = \int u(x)\overline{v(x)}\,dx$: the scalar or hermitian product in $L^2(\Omega)$, often also denoted by $(u, v)_0$

C^∞, C^∞_c, L^p, L^p_{loc}, L^p_c: the spaces when $\Omega = \mathbb{R}^n$

$\mathcal{S}$ or $\mathcal{S}(\mathbb{R}^n)$: the Schwartz space of C^∞ functions ϕ in $\mathbb{R}^n$ *rapidly decaying at infinity*, which means that, given any pair of integers m, $M \geq 0$,

$$q_{m,M}(\phi) = \sup_{x \in \mathbb{R}^n} \left[(1 + |x|)^M \sum_{|\alpha| \leq m} |\partial^\alpha \phi(x)| \right] < +\infty$$

Topology of $\mathcal{S}$

It is defined by the seminorms $q_{m,M}$ as (m, M) ranges over $\mathbb{Z}^2_+$; $\mathcal{S}$ is a Fréchet space and every bounded and closed subset of $\mathcal{S}$ is compact (a subset of $\mathcal{S}$ is bounded if all the seminorms $q_{m,M}$ are bounded on it).

The following inclusions are all continuous and have dense image:

(0.5) $$C^\infty_c(\Omega) \hookrightarrow \begin{matrix} L^p(\Omega) \\ C^\infty(\Omega) \end{matrix} \hookrightarrow L^p_{\text{loc}}(\Omega) \qquad (1 \leq p < +\infty);$$

(0.6) $$C^\infty_c \hookrightarrow \mathcal{S} \hookrightarrow L^p \qquad (1 \leq p < +\infty).$$

Let E be a Banach space, or more generally a locally convex space. We denote by $C^m(\Omega; E)$, $C^\infty_c(\Omega; E)$, $C^\infty_c(K; E)$, $L^p(\Omega; E)$, $L^p_{\text{loc}}(\Omega; E)$, $L^p_c(\Omega; E)$, $\mathcal{S}(\mathbb{R}^n; E)$ the analogues of the preceding spaces but relative to functions valued in E. The definitions are the same except that the absolute value in $\mathbb{C}$ (where the functions were valued) must be replaced by the norm, or the continuous seminorms, in the space E (see Treves [3], Section 39).

4. Distributions and Distribution Spaces

$\mathcal{D}'(\Omega)$: the space of distributions in Ω, i.e., of the continuous linear maps $C^\infty_c(\Omega) \to \mathbb{C}$, i.e., the *dual* of $C^\infty_c(\Omega)$

$\mathcal{E}'(\Omega)$: the space of *compactly supported* distributions in Ω, by definition the dual of $C^\infty(\Omega)$

supp T: the *support* of the distribution T, i.e., the intersection of all closed subsets in whose complement T vanishes identically

$\langle T, \phi\rangle = T(\phi) = \int T(x)\phi(x)\,dx$: the *duality bracket* between test functions and distributions. Thus $T \in \mathscr{D}'(\Omega)$ and $\phi \in C_c^\infty(\Omega)$, but we can also take $T \in \mathscr{E}'(\Omega)$ and $\phi \in C^\infty(\Omega)$.

Convergence of Distributions

It is the uniform convergence on the bounded subsets of $C_c^\infty(\Omega)$. For *sequences* it is the same as the weak convergence: $T_j \to T_o$ $(j = 1, 2, \ldots)$ if and only if $\langle T_j, \phi\rangle \to \langle T_o, \phi\rangle$ for each test function ϕ.

Bounded Sets of Distributions

A set B of distributions is bounded if and only if for *each* $\phi \in C_c^\infty(\Omega)$,

$$\sup_{T\in B} |\langle T, \phi\rangle| < +\infty.$$

Sets that are bounded and closed in $\mathscr{D}'(\Omega)$ (or in $\mathscr{E}'(\Omega)$) are compact.

Differential Operators Acting on Distributions

If $P(x, D)$ is a differential operator in Ω its action on $T \in \mathscr{D}'(\Omega)$ is defined by the integration-by-parts formula

$$\langle P(x, D)T, \phi\rangle = \langle T, {}^tP(x, D)\phi\rangle, \qquad \phi \in C_c^\infty(\Omega). \tag{0.7}$$

It is clear that $P(x, D)$ defines a continuous linear map of $\mathscr{D}'(\Omega)$ (resp., of $\mathscr{E}'(\Omega)$) into itself. A particular case is that of a differential operator of order *zero*, that is, *multiplication* by a C^∞ function ψ: $\langle \psi T, \phi\rangle = \langle T, \phi\psi\rangle$, which defines a continuous endomorphism of $\mathscr{D}'(\Omega)$ (resp., $\mathscr{E}'(\Omega)$). Note that $\operatorname{supp} P(x, D)T \subset \operatorname{supp} T$: *differential operators decrease the support.*

Local Structure of a Distribution

Given any $T \in \mathscr{D}'(\Omega)$ and any open set $\Omega' \subset\subset \Omega$, there is a finite set of continuous functions f_j and of differential operators P_j in Ω $(j = 1, \ldots, N)$ such that

$$T = \sum_{j=1}^{N} P_j f_j \qquad \textit{in } \Omega', \tag{0.8}$$

that is,

$$\langle T, \phi \rangle = \sum_{j=1}^{N} \int f_j(x)\, {}^t P_j \phi(x)\, dx, \qquad \phi \in C_c^\infty(\Omega').$$

If T has compact support, the finite-sum representation (0.8) may be taken to be valid in Ω itself, and the continuous functions f_j can be taken to vanish outside an arbitrary neighborhood of supp T.

Distributions That Are Functions

A distribution T in Ω is said to be a function if there is $f \in L^1_{\text{loc}}(\Omega)$ such that

$$\langle T, \phi \rangle = \int f(x)\phi(x)\, dx, \qquad \phi \in C_c^\infty(\Omega). \tag{0.9}$$

For now write T_f instead of T if (0.9) holds. Then $f \mapsto T_f$ is a continuous (linear) injection of $L^1_{\text{loc}}(\Omega)$ into $\mathscr{D}'(\Omega)$. In turn it defines the continuous injections into $\mathscr{D}'(\Omega)$ of

$$C^m(\Omega) \quad (0 \le m \le +\infty), \qquad C_c^\infty(\Omega), \qquad L^p_{\text{loc}}(\Omega) \quad (1 \le p \le +\infty).$$

We also have the continuous injections into $\mathscr{E}'(\Omega)$ of

$$C_c^m(\Omega) \quad (0 \le m \le +\infty), \quad L_c^p(\Omega) \quad (1 \le p \le +\infty).$$

These injections all have a *dense* image. This can most quickly be seen as follows: since all bounded subsets of $C_c^\infty(\Omega)$, or of $C^\infty(\Omega)$, have compact closure, these spaces are reflexive: $C_c^\infty(\Omega)$ is the dual of $\mathscr{D}'(\Omega)$, $C^\infty(\Omega)$ that of $\mathscr{E}'(\Omega)$. To prove that a subspace M of $\mathscr{D}'(\Omega)$ (resp., $\mathscr{E}'(\Omega)$) is dense, it suffices to show that any function $\phi \in C_c^\infty(\Omega)$ (resp., $C^\infty(\Omega)$) such that $\langle T, \phi \rangle = 0$ for all $T \in M$ must be identically zero. Take $M = C_c^\infty(\Omega)$. Let $\psi \in C^\infty(\Omega)$ be such that $\int \psi\phi\, dx = 0$ for all $\phi \in M$. Choose $\phi = \chi\bar{\psi}$, with $\chi \in C_c^\infty(\Omega)$, $\chi \ge 0$ arbitrary. We must have $\int |\psi|^2 \chi\, dx = 0$ for all such χ, hence $\psi \equiv 0$. □

If $\mathcal{O}$ is any open subset of Ω, one can say that T is a (locally L^1) function in $\mathcal{O}$, if this is true of the restriction of T to $\mathcal{O}$ (i.e., to $C_c^\infty(\mathcal{O})$). Then one can further specify the kind of function that T is, for instance a C^∞ function.

sing supp T: the *singular support* of T, i.e., the smallest closed set in the complement of which T is a C^∞ function

If $P(x, D)$ is a differential operator in Ω and T is a C^∞ function in $\mathcal{O}$, so is $P(x, D)T$: *differential operators decrease the singular support.*

δ: *Dirac measure* at the origin in $\mathbb{R}^n$; this is the distribution $\phi \mapsto \phi(0)$ $(\phi \in C_c^\infty(\mathbb{R}^n))$

δ_{x_o}, or $\delta(x - x_o)$: Dirac measure at the point x_o

$\delta^{(\alpha)} = \partial_x^\alpha \delta$: αth derivative of the Dirac measure.

These distributions are *not* functions, for $\alpha \neq 0$ they are not even *Radon measures*, that is, continuous linear functionals on the space C^0 of continuous functions in $\mathbb{R}^n$ (equipped with the topology of uniform convergence on compact subsets of $\mathbb{R}^n$).

$\mathcal{S}' = \mathcal{S}'(\mathbb{R}^n)$, the dual of $\mathcal{S}$: $\mathcal{S}'$ is the space of tempered (or slowly growing at infinity) distributions in $\mathbb{R}^n$.

Structure of a Tempered Distribution

Given any $T \in \mathcal{S}'$ there is a continuous function f in $\mathbb{R}^n$ such that, for suitable integers M, $m \geq 0$,

$$\sup_{x \in \mathbb{R}^n} (1 + |x|)^{-M} |f(x)| < +\infty,$$

and

$$T = (1 - \Delta)^m f. \tag{0.10}$$

In (0.10), Δ is the Laplace operator.

The dual of $\mathcal{S}'$ is $\mathcal{S}$; $\mathcal{S}$ is continuously embedded and dense in $\mathcal{S}'$ (see the preceding argument), and thus this is also true of C_c^∞.

We often write $\mathcal{D}'$, $\mathcal{E}'$ instead of $\mathcal{D}'(\mathbb{R}^n)$, $\mathcal{E}'(\mathbb{R}^n)$ respectively.

If E is a Banach space or, more generally, a locally convex topological vector space, we use the following notation:

$\mathcal{D}'(\Omega; E)$: the space of continuous linear maps $C_c^\infty(\Omega) \to E$, equipped with the topology of uniform convergence on the bounded subsets of $C_c^\infty(\Omega)$

$\mathcal{E}'(\Omega; E)$: the subspace of $\mathcal{D}'(\Omega; E)$ consisting of the compactly supported E-valued distributions in Ω

5. Convolution and Fourier Transform of Distributions

Until otherwise specified all functions and distributions in this subsection are defined in the whole of $\mathbb{R}^n$.

Convolution of Functions

$f * g$: the *convolution* of two functions f, g:

$$(0.11)\qquad (f * g)(x) = \int_{\mathbb{R}^n} f(x - y)g(y)\, dy = \int_{\mathbb{R}^n} f(y)g(x - y)\, dy.$$

One may assume that $f \in L^1_{\text{loc}}$, $g \in L^1_c$, or that both f, g belong to L^1, or one may make other assumptions such as

$$f \in L^p, \quad g \in L^q, \qquad 1 \le p, q \le +\infty, \qquad 1/p + 1/q - 1 \ge 0.$$

Then $f * g \in L^r$ with $1/r = 1/p + 1/q - 1$, and we have the *Hölder inequalities*

$$(0.12)\qquad \|f * g\|_{L^r} \le \|f\|_{L^p}\|g\|_{L^q}.$$

In particular we may take $p = 1$, $1 \le q \le +\infty$ to be arbitrary, and we get

$$(0.13)\qquad \|f * g\|_{L^q} \le \|f\|_{L^1}\|g\|_{L^q}.$$

Thus L^q is a convolution L^1-module, and L^1 is a Banach algebra.

Convolution of Distributions with Functions

$\check{\phi}$: if ϕ is a function in $\mathbb{R}^n$, $\check{\phi}(x) = \phi(-x)$.
Let $\phi \in C^\infty_c$, $T \in \mathscr{D}'$ or, alternatively, $\phi \in C^\infty$, $T \in \mathscr{E}'$.
$T * \phi$: convolution of T with ϕ, written also $\phi * T$:

$$(0.14)\qquad \langle T * \phi, \psi\rangle = \langle T, \check{\phi} * \psi\rangle, \qquad \psi \in C^\infty_c.$$

Observe that $\psi \mapsto \check{\phi} * \psi$ is a continuous endomorphism (i.e., linear map into itself) of C^∞_c, or of C^∞. Then

$$(\phi, T) \mapsto T * \phi$$

is a separately continuous bilinear map of $C^\infty_c \times \mathscr{D}'$, or of $C^\infty \times \mathscr{E}'$, into C^∞.

Convolution among Distributions

$\check{T}$: the distribution in $\mathbb{R}^n$ defined by

$$(0.15)\qquad \langle \check{T}, \phi\rangle = \langle T, \check{\phi}\rangle, \qquad \phi \in C^\infty_c.$$

$S * T$, also denoted $\int S(y)T(x - y)\, dy$ or $\int S(x - y)T(y)\, dy$: convolution of $S \in \mathscr{E}'$, $T \in \mathscr{D}'$

$$(0.16)\qquad \langle S * T, \phi\rangle = \langle S, \check{T} * \phi\rangle, \qquad \phi \in C^\infty_c.$$

By what precedes, $\phi \mapsto \check{T} * \phi$ is a continuous linear map of C_c^∞ into C^∞, and therefore (0.16) is a good definition: $S * T \in \mathscr{D}'$, and

$$(S, T) \mapsto S * T$$

is a separately continuous bilinear map of $\mathscr{E}' \times \mathscr{D}'$ into $\mathscr{D}'$. We have

$$\operatorname{supp}(S * T) \subset \operatorname{supp} S + \operatorname{supp} T, \tag{0.17}$$

where the right-hand side is the set of the *vector sums* $x + y$ of any element x of supp S with any element $y \in$ supp T.

By virtue of (0.17), $\mathscr{E}'$ is a convolution algebra. In $\mathscr{E}'$ convolution is commutative; the Dirac measure δ at the origin is the identity.

Convolution of $m + 1$ distributions, of which m have compact support, makes sense, and it is associative. If $P(D)$ is a differential operator in $\mathbb{R}^n$, then for any $S \in \mathscr{E}'$, $T \in \mathscr{D}'$, we have

$$P(D)(S * T) = [P(D)S] * T = S * [P(D)T]. \tag{0.18}$$

Since, for all $T \in \mathscr{D}'$,

$$\delta * T = T, \tag{0.19}$$

we also have

$$P(D)T = [P(D)\delta] * T. \tag{0.20}$$

Fourier Transforms of Functions

If $u \in \mathscr{S}$, its Fourier transform is

$$\hat{u}(\xi) = \int_{\mathbb{R}^n} e^{-ix\cdot\xi} u(x)\, dx. \tag{0.21}$$

The Fourier transformation $u \mapsto \hat{u}$ *defines an isomorphism of* $\mathscr{S}(\mathbb{R}^n)$ *onto* $\mathscr{S}(\mathbb{R}_n)$. The inverse of this isomorphism is given by the *Fourier inversion formula*:

$$u(x) = (2\pi)^{-n} \int_{\mathbb{R}^n} e^{ix\cdot\xi} \hat{u}(\xi)\, d\xi. \tag{0.22}$$

(Other authors follow slightly different conventions.) The Fourier transformation extends as an isomorphism of $L^2(\mathbb{R}^n)$ onto $L^2(\mathbb{R}_n)$, and we have the *Plancherel–Parseval formulas*:

$$\int |u|^2\, dx = (2\pi)^{-n} \int |\hat{u}|^2\, d\xi, \tag{0.23}$$

$$\int u\bar{v}\,dx = (2\pi)^{-n}\int \hat{u}\bar{\hat{v}}\,d\xi \qquad (u, v \in L^2(\mathbb{R}^n)). \tag{0.24}$$

Also worth mentioning is the *Lebesgue theorem*, which states that *the Fourier transform of a function $f \in L^1(\mathbb{R}^n)$ is a continuous function in $\mathbb{R}_n$ converging to zero at infinity.*

Fourier Transforms of Distributions

$\hat{T}$: Fourier transform of the *tempered* distribution T, defined as follows:

$$(2\pi)^{-n}\int \hat{T}(\xi)\overline{\hat{\phi}(\xi)}\,d\xi = \int T(x)\overline{\phi(x)}\,dx, \qquad \phi \in \mathcal{S}(\mathbb{R}^n).$$

The Fourier transformation $T \mapsto \hat{T}$ is an isomorphism of $\mathcal{S}'(\mathbb{R}^n)$ onto $\mathcal{S}'(\mathbb{R}_n)$. It extends the Fourier transformation on $L^2(\mathbb{R}^n)$.

Theorems of Paley–Wiener (–Schwartz)

In order for a tempered distribution u on $\mathbb{R}_n$ to be the Fourier transform of a compactly supported distribution (resp., C^∞ function), it is necessary and sufficient for u to be a C^∞ function slowly growing at infinity (resp., rapidly decaying at infinity, i.e., belonging to $\mathcal{S}$) extendable to $\mathbb{C}_n$ as an *entire function $u(z)$ of exponential type*, that is, satisfying everywhere the *Cauchy–Riemann equations*,

$$\partial u/\partial \bar{z}_j = 0, \qquad j = 1, \dots, n, \tag{0.25}$$

and such that, for suitable constants A, $B > 0$,

$$|u(z)| \le A\, e^{B|z|}, \qquad z \in \mathbb{C}_n. \tag{0.26}$$

When $T \in \mathcal{E}'$ we have

$$\hat{T}(\xi) = \int e^{-ix\cdot\xi}T(x)\,dx. \tag{0.27}$$

In particular, the Fourier transform of the Dirac measure is the constant function 1,

$$\hat{\delta} = 1, \tag{0.28}$$

and if $P(D) = \sum_{|\alpha|\le m} c_\alpha D^\alpha$ is any differential operator with constant coefficients in $\mathbb{R}^n$, then

$$\widehat{P(D)\delta} = P(\xi) \qquad \Big(= \sum_{|\alpha|\le m} c_\alpha \xi^\alpha\Big). \tag{0.29}$$

Fourier Transform of a Convolution

If $S \in \mathscr{E}'$, $T \in \mathscr{S}'$, we can form $S * T$ and compute its Fourier transform. We have

$$\widehat{S * T} = \hat{S}\hat{T}. \tag{0.30}$$

Since $\hat{S} \in C^\infty$, the right-hand side is well defined; since all derivatives of $\hat{S}$ are slowly growing at infinity, it is a tempered distribution (as expected).

Combining (0.20), (0.29), and (0.30) gives

$$\widehat{P(D)T} = P(\xi)\hat{T}; \tag{0.31}$$

in particular,

$$\widehat{D^\alpha T} = \xi^\alpha \hat{T}, \tag{0.32}$$

$$\widehat{\Delta T} = -|\xi|^2 \hat{T}. \tag{0.33}$$

Sobolev Spaces

In the following definitions s denotes an arbitrary *real* number.

$H^s = H^s(\mathbb{R}^n)$: the space of tempered distributions u in $\mathbb{R}^n$ whose Fourier transform $\hat{u}$ is a square-integrable function in $\mathbb{R}_n$ for the measure $(1 + |\xi|^2)^s \, d\xi$

$(u, v)_s$: the *inner product* in H^s,

$$(u, v)_s = (2\pi)^{-n} \int \hat{u}(\xi)\overline{\hat{v}(\xi)}(1 + |\xi|^2)^s \, d\xi$$

$\|u\|_s = [(u, u)_s]^{1/2}$: the *norm* in H^s, which is a Hilbert space when equipped with the inner product $(\ ,\ )_s$

$H^s_c(K)$: the subspace of H^s consisting of the distributions having their support in the compact set K; $H^s_c(K)$ is a *closed* linear subspace of H^s

$H^s_c(\Omega)$: the union of the spaces $H^s_c(K)$ for K ranging over the collection of all compact subsets of Ω

$H^s_{\text{loc}}(\Omega)$: the space of distributions u in Ω such that $\phi u \in H^s$ for any $\phi \in C^\infty_c(\Omega)$.

The topology of $H^s_{\text{loc}}(\Omega)$ is that defined by the seminorms $u \mapsto \|\phi u\|_s$, $\phi \in C^\infty_c(\Omega)$. It suffices to take ϕ ranging over a sequence $\{\phi_\nu\}$ such that the compact sets $K_\nu = \{x \in \Omega; \phi_\nu(x) = 1\}$ exhaust Ω (see definition of the C^∞ topology), and $\phi_\nu \leq \phi_{\nu+1}$, $\nu = 0, 1, \ldots$. Thus $H^s_{\text{loc}}(\Omega)$ is easily seen to be a (reflexive) Fréchet space.

The topology of $H_c^s(\Omega)$ is defined as follows: for each $K \subset\subset \Omega$, $H_c^s(K)$ is equipped with the Hilbert space structure induced by H^s. Then a *convex* set in $H_c^s(\Omega)$ is open if and only if its intersection with every $H_c^s(K)$ is open.

We have the following continuous linear injections with dense images

$$\mathscr{S} \hookrightarrow H^s \hookrightarrow H^{s'} \hookrightarrow \mathscr{S}' \qquad (s' \leq s),$$

$$C_c^\infty(\Omega) \hookrightarrow H_c^s(\Omega) \hookrightarrow H_{\text{loc}}^s(\Omega) \hookrightarrow \mathscr{D}'(\Omega).$$

We have the *set-theoretical* equalities (the first one is topological);

$$(0.34) \qquad C^\infty(\Omega) = \bigcap_s H_{\text{loc}}^s(\Omega), \qquad C_c^\infty(\Omega) = \bigcap_s H_c^s(\Omega),$$

$$(0.35) \qquad \mathscr{E}'(\Omega) = \bigcup_s H_c^s(\Omega), \qquad \mathscr{D}'^F(\Omega) = \bigcup_s H_{\text{loc}}^s(\Omega),$$

where $\mathscr{D}'^F(\Omega)$ stands for the space of *distributions of finite order* in Ω (i.e., distributions having a finite-sum representation (0.8) in the whole of Ω, not just in sets $\Omega' \subset\subset \Omega$).

$(1-\Delta)^s$: the *convolution operator*

$$u(x) \mapsto (2\pi)^{-n} \int e^{ix\cdot\xi}(1+|\xi|^2)^s \hat{u}(\xi)\, d\xi.$$

As s varies over $\mathbb{R}$, $(1-\Delta)^s$ forms a group of automorphisms of $\mathscr{S}(\mathbb{R}^n)$, or of $\mathscr{S}'(\mathbb{R}^n)$. Given any $t \in \mathbb{R}$, $(1-\Delta)^s$ is an *isometry* of H^t onto H^{t-2s}, in particular, of H^s onto H^{-s}.

We have $H^0 = L^2$; the equality applies also to the Hilbert space structures. The dense image injection $\mathscr{S} \hookrightarrow H^s$ transposes into the injection $(H^s)' \hookrightarrow \mathscr{S}'$ whose image is equal to H^{-s}. Thus the following pairs of spaces can be naturally regarded as dual pairs:

$$H^s \text{ and } H^{-s}, \qquad H_{\text{loc}}^s(\Omega) \text{ and } H_c^{-s}(\Omega),$$

and $(1-\Delta)^s$ as the natural *linear* isometry of H^s onto its *antidual*, H^{-s} (*antiduality* is defined by the bracket $\langle u, \bar{v}\rangle$, whereas duality is defined by $\langle u, v\rangle$).

I

Standard Pseudodifferential Operators

This chapter is the basic one in the book and, I hope, the most elementary. Its contents are essentially the definitions and fundamental properties of what are called here *standard* pseudodifferential operators, often called operators of type $(1, 0)$, to contrast them with operators of type (ρ, δ), studied in Chapter IV. The presentation follows the line now widely adopted of first defining the operators by means of *amplitudes*, then proving (easily) the five basic theorems (pseudolocal character, continuity, transposition, composition, invariance under diffeomorphism), and only afterwards going to the *symbolic calculus*, by then a purely "algebraic" affair.

The symbolic calculus opens the road to *microlocalization*, that is, localization in the cotangent bundle; this is perhaps the most important step forward in our understanding of linear partial differential equations since distributions. Microlocalization is attained through conic cutoffs, "smoothness" in the cotangent bundle, of distributions in the base—and its complementary aspect, singular microsupport or, in Hörmander's terminology (which we follow), *wave-front set*. Here the right objects to use and to study are *microdistributions*, truly the singularities "upstairs" of distributions "downstairs." These are the analogues of what hyperfunction theorists call microfunctions, a terminology to resist in the C^∞ category, I believe. At such an early stage in the book we content ourselves with defining microdistributions by a sheaf in the cotangent bundle, as is customary, almost in passing and use them virtually not at all. But as the book progresses, their presence and importance tends to expand irresistibly, simply because they are the correct concept. The same applies to "germs" of pseudodifferential operators and related notions.

Much of the chapter is devoted to recalling some common definitions—C^∞ manifold, its cotangent bundle, vector-valued distributions, vector bundles on a manifold, and distribution sections of such a bundle.

Section 1 describes in some detail the construction of a parametrix for an elliptic linear partial differential equation. It is a procedure known since the middle 1950s (at least). It is safe to say that pseudodifferential operators are born of it.

1. Parametrices of Elliptic Equations

Consider a linear partial differential equation with constant coefficients,

$$P(D)u = f, \tag{1.1}$$

where the right-hand side f is, say, a C^∞ function with compact support in $\mathbb{R}^n$, i.e., $f \in C_c^\infty(\mathbb{R}^n)$. We have used the standard notation

$$D = (D_1, \ldots, D_n), \qquad D_j = -\sqrt{-1}\,\partial/\partial x^j, \qquad j = 1, \ldots, n;$$

$P(\xi)$ is a polynomial with complex coefficients in n (real) variables $\xi_1, \ldots, \xi_n$. The method of solving (1.1) that first comes to mind is to use the Fourier transformation to transform the differential problem of solving (1.1) into the division problem of solving the multiplicative equation

$$P(\xi)\hat{u} = \hat{f}. \tag{1.2}$$

The upper hats $\hat{}$ denote the Fourier transforms:

$$\hat{u}(\xi) = \int_{\mathbb{R}^n} e^{-ix\cdot\xi} u(x)\, dx, \tag{1.3}$$

where $x \cdot \xi = x^1\xi_1 + \cdots + x^n\xi_n$, $dx = dx^1 \cdots dx^n$. We recall the Fourier inversion formula, which will be of vital importance to us in the sequel:

$$u(x) = (2\pi)^{-n} \int_{\mathbb{R}_n} e^{ix\cdot\xi} \hat{u}(\xi)\, d\xi, \tag{1.4}$$

where $d\xi = d\xi_1 \cdots d\xi_n$. We would like to take advantage of (1.4) by writing

$$u(x) = (2\pi)^{-n} \int e^{ix\cdot\xi} \frac{\hat{f}(\xi)}{P(\xi)}\, d\xi, \tag{1.5}$$

which should be a solution of (1.1) in view of (1.2). Unfortunately the integral on the right side of (1.5) does not, in general, make sense because of the zeros of the polynomial P in the denominator of the integrand. There are cases, however, for which a slight modification of formula (1.5) still yields an approximate solution. Perhaps the most important of these cases is that in which the operator $P(D)$, or equivalently the polynomial $P(\xi)$, is *elliptic*.

Suppose that the degree of $P(\xi)$ is m, and write

$$P(\xi) = P_m(\xi) + Q(\xi),$$

where the degree of $Q(\xi)$ is at most $m - 1$. It is customary to call $P_m(\xi)$ (resp. $P_m(D)$) the *principal symbol* (resp. the *principal part*) of $P(D)$.

DEFINITION 1.1. *The differential operator $P(D)$ (resp. the polynomial $P(\xi)$) is said to be elliptic if $P_m(\xi) \neq 0$ for all $\xi \in \mathbb{R}_n$, $\xi \neq 0$.*

In the rather uninteresting case $m = 0$, i.e., $P(\xi)$ is a constant function on $\mathbb{R}_n$, the ellipticity of P simply means that its value is not zero. When $m \geq 1$, the origin is always a zero of the polynomial $P_m(\xi)$. In general, since P_m is homogeneous of degree m, the set of its zeros forms a cone C_{P_m}, called the *characteristic cone*. To say that P is elliptic is to say that this cone consists of a single point, the origin (its vertex).

When $n = 1$, all differential operators with constant coefficients are elliptic. When $n > 1$, important examples of elliptic operators are the Laplace operator $\Delta = (\partial/\partial x^1)^2 + \cdots + (\partial/\partial x^n)^2$ and, when $n = 2$, the Cauchy–Riemann operator

$$\frac{\partial}{\partial \bar{z}} = \frac{1}{2}\left(\frac{\partial}{\partial x} + \sqrt{-1}\,\frac{\partial}{\partial y}\right).$$

As usual, we have denoted by (x, y) the variable in $\mathbb{R}^2$ and set $z = x + iy$, $\bar{z} = x - iy$. The principal symbol of the Laplace operators is

$$-|\xi|^2 = -\xi_1^2 - \cdots - \xi_n^2.$$

The symbol of the Cauchy–Riemann operator is $\frac{1}{2}i\zeta$ where $\zeta = \xi + i\eta$.

The property of elliptic polynomials that is important to our purpose here is partly expressed in the following lemma.

LEMMA 1.1. *If P is elliptic, the set of zeros of the polynomial $P(\xi)$ in $\mathbb{R}_n$ is compact.*

PROOF. Let $V_P = \{\xi \in \mathbb{R}_n;\, P(\xi) = 0\}$. If P is elliptic, $P_m(\xi)$ does not vanish on the unit sphere of $\mathbb{R}_n$; therefore $|P_m(\xi)| \geq c > 0$ on that sphere. By homogeneity we see that

$$|P_m(\xi)| \geq c|\xi|^m \qquad \text{for all } \xi \in \mathbb{R}_n. \tag{1.6}$$

On the other hand, we have $|Q(\xi)| < C|\xi|^{m-1}$ for all $\xi \in \mathbb{R}_n$, $|\xi| > 1$. Thus if

$\xi \in V_P$, $|\xi| > 1$, then we have

$$c|\xi|^m \leq |P_m(\xi)| = |Q(\xi)| \leq C|\xi|^{m-1},$$

whence $|\xi| \leq C/c$. □

REMARK 1.1. It is not true that the elliptic polynomials are the only ones with compact sets of zeros in $\mathbb{R}_n : \xi_1 + i$ is not elliptic in $\mathbb{R}_2$ and has no zeros there.

We continue to assume that P is elliptic and let ρ be a positive number such that the real zeros of $P(\xi)$ are contained in the open ball of center the origin and of radius ρ. We have now the right to consider the integral

$$v(x) = (2\pi)^{-n} \int e^{ix\cdot\xi} \frac{\hat{f}(\xi)}{P(\xi)} \chi(\xi)\, d\xi, \tag{1.7}$$

where $\chi \in C^\infty(\mathbb{R}_n)$, $\chi(\xi) = 0$ if $|\xi| < \rho$, $\chi(\xi) = 1$ if $|\xi| > \rho' > \rho$. Because of the presence of the cutoff function χ, the function v cannot be an exact solution of equation (1.1), but as we shall now see it does not differ substantially from a solution. We have

$$P(D)v(x) = (2\pi)^{-n} \int e^{ix\cdot\xi} \hat{f}(\xi)\chi(\xi)\, d\xi = f(x) - Rf(x), \tag{1.8}$$

where we have set

$$Rf(x) = (2\pi)^{-n} \int e^{ix\cdot\xi} \hat{f}(\xi)\{1 - \chi(\xi)\}\, d\xi. \tag{1.9}$$

These formulas call for several interesting remarks. First, the function $P^{-1}\chi$ is C^∞ and bounded in $\mathbb{R}_n$. Indeed,

$$|P(\xi)| \geq |P_m(\xi)| - |Q(\xi)| \geq (c|\xi| - C)|\xi|^{m-1} > 1$$

as soon as $|\xi|$ is large enough. Consequently, $P^{-1}\chi$ defines a *tempered* distribution in $\mathbb{R}_n$ which is the Fourier transform of a tempered distribution K in $\mathbb{R}^n$. We have

$$v = K * f \qquad (\textit{convolution}). \tag{1.10}$$

On the other hand

$$Rf = h * f, \tag{1.11}$$

where h is the inverse Fourier transform of $1 - \chi$. The latter is a C^∞ function with compact support in $\mathbb{R}_n$. Hence, by the easy part of the Paley–Wiener theorem, h is extendable to $\mathbb{C}^n$ as an entire analytic function of exponential type. Its restriction to $\mathbb{R}^n$ belongs to the Schwartz space $\mathcal{S}$ of the C^∞ functions rapidly decaying at infinity. In both formulas (1.10) and (1.11), f need not be a function; it may be a distribution. But it must have *compact* support. We observe that even when f is a distribution with compact support, Rf is a C^∞ function (extendable to $\mathbb{C}^n$ as an entire function of exponential type). We may rewrite (1.8) in the following way:

$$P(D)K = \delta - h, \qquad \delta = \textit{Dirac distribution}, \tag{1.12}$$

or equivalently (denoting by K the convolution operator $K*$):

$$P(D)K = I - R, \qquad I = \textit{identity mapping}, \tag{1.13}$$

where R is a continuous linear operator of $\mathcal{E}'$, the space of distributions with compact support, into C^∞. A distribution such as K (or an operator such as K) is called a *parametrix* of $P(D)$. In the study of elliptic equations a parametrix can serve many purposes, as we shall see later. In fact, we may show right away that it can immediately yield an exact solution of (1.1) by yielding a fundamental solution of $P(D)$, that is, a solution of the equation

$$P(D)E = \delta. \tag{1.14}$$

(It suffices then to take $u = E * f$.) Indeed, it can be shown either directly by functional analysis or by the Cauchy–Kovalevska theorem (see Treves [3], Chapter 10), that the equation

$$P(D)w = h \tag{1.15}$$

always has a solution w that is an entire function in $\mathbb{C}^n$; w can even be taken to have exponential type. Then $E = K + w$ satisfies (1.14).

The next question that we want to investigate is of paramount importance for the sequel: Can we extend some of the preceding techniques to elliptic equations with variable coefficients? First we must define precisely what this means.

We shall denote by Ω an open subset of $\mathbb{R}^n$. A (linear partial) *differential operator* in Ω will be an operator of the form

$$P(x, D) = \sum_{|\alpha| \leq m} c_\alpha(x) D^\alpha, \tag{1.16}$$

where the coefficients c_α are complex-valued C^∞ functions in Ω. We have used the standard multi-index notation $\alpha = (\alpha_1, \ldots, \alpha_n)$, $D^\alpha = D_1^{\alpha_1} \cdots D_n^{\alpha_n}$, $|\alpha| = \alpha_1 + \cdots + \alpha_n$. We shall also assume that m is the effective order of $P(x, D)$, that is, that there is at least one coefficient c_α, with

$|\alpha| = m$, that does not vanish identically in Ω. The principal symbol of $P(x, D)$ is the polynomial with respect to ξ, with coefficients in $C^\infty(\Omega)$,

$$P_m(x, \xi) = \sum_{|\alpha|=m} c_\alpha(x)\xi^\alpha. \tag{1.17}$$

For every $x \in \Omega$, we set

$$C_{P_m}(x) = \{\xi \in \mathbb{R}_n; P_m(x, \xi) = 0\}. \tag{1.18}$$

Since $P_m(x, \xi)$ is homogeneous of degree m with respect to ξ, $C_{P_m}(x)$ is a cone in $\mathbb{R}_n$, called the *characteristic cone* of P at x. We shall often write P as shorthand for $P(x, D)$. The differential operator $P_m(x, D)$ is called the *principal part* (sometimes also the *leading term*) of $P(x, D)$.

DEFINITION 1.2. *The differential operator $P(x, D)$ is said to be elliptic in Ω if, for every x, the characteristic cone $C_{P_m}(x)$ contains at most one point, the origin. When $m = 0$ this means that $C_{P_m}(x)$ is empty for all $x \in \Omega$.*

Definition 1.2 says that a differential operator with variable coefficients, $P(x, D)$, is elliptic if every constant coefficient operator obtained by "freezing" the coefficients at a point is elliptic, in the sense of Definition 1.1.

We shall try to construct an approximate solution of the equation

$$P(x, D)u = f \tag{1.19}$$

by modifying formula (1.7) to obtain

$$P(x, D)v = f - Rf, \tag{1.20}$$

where, as before,

$$R : \mathscr{E}'(\Omega) \to C^\infty(\Omega). \tag{1.21}$$

As usual, the arrow means a *continuous linear mapping*. We shall try a formula of the following kind, generalizing (1.7):

$$v(x) = Kf(x) = (2\pi)^{-n} \int e^{ix\cdot\xi}k(x, \xi)\hat{f}(\xi)\, d\xi. \tag{1.22}$$

We shall first make a formal determination of the symbol $k(x, \xi)$ so as to satisfy

$$P(x, D)K = I \tag{1.23}$$

and then modify it so that the integral on the right in (1.22) makes sense. This modification leads to a solution, not of equation (1.23), but of

$$P(x, D)K = I - R, \tag{1.24}$$

where R satisfies (1.21).

We have

$$P(x, D)v(x) = (2\pi)^{-n} \int e^{ix\cdot\xi} P(x, D_x + \xi)k(x, \xi)\hat{f}(\xi)\, d\xi,$$

since

$$(1.25) \qquad P(x, D_x)(e^{ix\cdot\xi}w(x)) = e^{ix\cdot\xi}P(x, D_x + \xi)w(x).$$

By virtue of Fourier's inversion formula (1.4) it suffices to solve the equation

$$(1.26) \qquad P(x, D_x + \xi)k(x, \xi) = 1.$$

In equation (1.26), ξ plays the role of a parameter (varying in $\mathbb{R}_n$). We may view $P(x, D_x + \xi)$ as a polynomial with respect to ξ whose coefficients are differential operators in Ω (in the x variables). We have

$$(1.27) \qquad P(x, D_x + \xi) = P_m(x, \xi) + \sum_{j=1}^{m} P_j(x, \xi, D_x)$$

where $P_j(x, \xi, D_x)$ is a differential operator with respect to x (in Ω) of order j whose coefficients are *homogeneous* polynomials with respect to ξ of degree $m - j$. The idea is then to take the symbol $k(x, \xi)$ as a sum of functions of (x, ξ) homogeneous with respect to ξ. Because of the use of the Fourier transformation and of our requirement that the operator K act on distributions, the symbol $k(x, \xi)$ must be *tempered* in the ξ variables, and the homogeneity degrees of the various components, therefore, has to remain bounded. We shall need an infinite series of such terms and their homogeneity degrees will be negative integers, tending to $-\infty$. We write

$$(1.28) \qquad k(x, \xi) = \sum_{j=0}^{+\infty} k_j(x, \xi),$$

where $k_j(x, \xi)$ is homogeneous with respect to ξ, of degree $d_j \to -\infty$. We shall then try to determine the successive terms k_j by identifying the terms with the same homogeneity degree with respect to ξ on the two sides of (1.26). According to (1.27) the equation for the terms of highest homogeneity degree in ξ is simply

$$(1.29) \qquad P_m(x, \xi)k_o(x, \xi) = 1.$$

Thus we see that $d_o = -m$, since $k_o = 1/P_m$. Equating the terms of degrees <0 in (1.26) yields at once

$$(1.30) \qquad P_m(x, \xi)k_j(x, \xi) = - \sum_{\substack{j'=0;\\ (j' \geq j-m)}}^{j-1} P_{j-j'}(x, \xi, D_x)k_{j'}(x, \xi), \qquad j > 0.$$

We see that the successive determination of the terms k_j is possible: the right-hand side of (1.30) depends only on the $k_{j'}$ with $j' < j$ (which have already been determined). We also note that the homogeneity degree d_j of k_j with respect to ξ is equal to $-(m + j)$; thus the d_j do indeed form a strictly decreasing sequence of negative integers.

However, it is apparent that we are going to run into trouble, for two reasons. First, every term $k_j(x, \xi)$ can be represented as a rational function whose denominator is of the form $P_m(x, \xi)^{\varpi_j}$ with $\varpi_j > 0$. As a consequence the zeros of $P_m(x, \xi)$ engender some difficulty. The zeros in question are points of the kind $(x, 0)$, since we assume the differential operator $P(x, D)$ to be elliptic. We may hope to take care of this difficulty by introducing a cutoff function χ as in (1.7). But now we encounter a new kind of difficulty, due to the fact that the series (1.28) is infinite and that the question of its convergence (in some suitable sense) arises. This will also have to be fixed. In fact, it can be fixed by using infinitely many cutoff functions, each one multiplying one of the homogeneous terms $k_j(x, \xi)$. Let us indicate rapidly how this is done. Let $\chi(t)$ be a C^∞ function on the real line, vanishing for $t < \frac{1}{2}$ and equal to one for $t > 1$. For each $j = 0, 1, \ldots,$ we set

$$\chi_j(\xi) = \chi(\rho_j^{-1}|\xi|), \tag{1.31}$$

where the ρ_j form a strictly increasing sequence of positive numbers, tending to $+\infty$. Let the k_j be defined by (1.29) and (1.30) as before, but now define $k(x, \xi)$ not by (1.28) but by the following equation:

$$k(x, \xi) = \sum_{j=0}^{+\infty} \chi_j(\xi) k_j(x, \xi). \tag{1.32}$$

We know that each k_j is a C^∞ function of (x, ξ) in $\Omega \times (\mathbb{R}_n \backslash \{0\})$, homogeneous of degree $-(m + j)$ with respect to ξ. Consequently, if $\mathcal{K}'$ is an arbitrary compact subset of Ω and α, β are any two n-tuples, then there is a constant $C_{\alpha,\beta}^{(j)}(\mathcal{K}') > 0$ such that

$$|D_\xi^\alpha D_x^\beta k_j(x, \xi)| < C_{\alpha,\beta}^{(j)}(\mathcal{K}')|\xi|^{-(m+j+|\alpha|)}, \qquad \forall x \in \mathcal{K}', 0 \neq \xi \in \mathbb{R}_n. \tag{1.33}$$

We then select an *exhausting sequence of compact subsets* $\mathcal{K}_\nu$ $(\nu = 1, 2, \ldots)$ of Ω. This always means, as far as we are concerned, that the union of the $\mathcal{K}_\nu$ is equal to Ω and that for each ν, $\mathcal{K}_\nu$ is contained in the interior of $\mathcal{K}_{\nu+1}$, although the latter precision is often not needed. We avail ourselves of (1.33) and of the following two facts; on the support of χ_j we have $|\xi| \geq \rho_j/2$, on that of grad χ_j, $|\xi| \leq \rho_j$. After suitably increasing the constants $C_{\alpha,\beta}^{(j)}(\mathcal{K}_\nu)$ we derive from this

$$\left|D_\xi^\alpha D_x^\beta \sum_{j=0}^{+\infty} \chi_j(\xi) k_j(x, \xi)\right| \leq |\xi|^{-(m+|\alpha|)} \sum_{j=0}^{+\infty} C_{\alpha,\beta}^{(j)}(\mathcal{K}_\nu)\rho_j^{-j}.$$

It suffices then to require

$$\rho_j \geq 2 \sup_{\nu \leq j, |\alpha+\beta| \leq j} C^{(j)}_{\alpha,\beta}(\mathcal{K}_\nu)^{1/j}, \tag{1.34}$$

in order to reach the conclusion that the series (1.32) converges in $C^\infty(\Omega \times \mathbb{R}_n)$ and that its sum satisfies inequalities similar to (1.33).

Now we ask whether $k(x, \xi)$, defined by (1.32), satisfies (1.26). The answer is clearly negative, since we have cut off whole pieces (in neighborhoods of $\xi = 0$) of the homogeneous terms $k_j(x, \xi)$. What we have is this:

$$P(x, D_x + \xi)k(x, \xi) = 1 - r(x, \xi). \tag{1.35}$$

It is not difficult to compute, by exploiting (1.29) and (1.30), the exact expression of the symbol $r(x, \xi)$. We find

$$r(x, \xi) = 1 - \chi_o(\xi) + \sum_{j=1}^{\infty} r_j(x, \xi), \tag{1.36}$$

where

$$r_j(x, \xi) = \sum_{\substack{k=1 \\ (k \leq m)}}^{j} [\chi_{j-k}(\xi) - \chi_j(\xi)]P_k(x, \xi, D_x)k_{j-k}(x, \xi). \tag{1.37}$$

We note that the support of each symbol $r_j(x, \xi)$ is contained in a compact subset of the ξ-space (specifically, in the ball $|\xi| \leq \rho_j$). Using this fact, together with the properties of series (1.32) and equation (1.35), one can prove that the operator R defined by

$$Rf(x) = (2\pi)^{-n} \int e^{ix\cdot\xi} r(x, \xi)\hat{f}(\xi)\, d\xi \tag{1.38}$$

satisfies (1.21). Since one of the purposes of pseudodifferential operators theory is precisely to formalize this type of argument, we shall not give the full details here. For the time being, we shall content ourselves with pointing out this property and the fact that (1.24) is satisfied, which is an immediate consequence of (1.35).

The operator K that we have defined is called a *parametrix* of the differential operator $P(x, D)$. Although it is not a right inverse of $P(x, D)$, that is, it does not satisfy (1.23), it can serve many purposes that a right inverse would serve. And by an easy argument, as we shall see in Chapter II (Proposition 1.4), it yields at least *locally* (i.e., when acting on distributions with sufficiently small support) a right inverse of $P(x, D)$.

2. Definition and Continuity of the "Standard" Pseudodifferential Operators in an Open Subset of Euclidean Space. Pseudodifferential Operators are Pseudolocal

Pseudodifferential operators may be viewed as a generalization of operators of the kind (1.22) and of the differential operators. These two types of operators can be represented by formulas of the following kind:

$$Au(x) = (2\pi)^{-n} \int e^{ix\cdot\xi} a(x, \xi)\hat{u}(\xi)\, d\xi. \tag{2.1}$$

That is precisely how we constructed K in Section 1, and it is quite evident for a differential operator $P(x, D)$. It suffices to apply $P(x, D)$ to $u(x)$ given by the Fourier inversion formula (1.4); the symbol $a(x, \xi)$ is nothing else, in this case, but the polynomial $P(x, \xi)$ obtained by substituting the variable ξ_j for the partial differentiations

$$D_j = \frac{1}{\sqrt{-1}} \frac{\partial}{\partial x_j}, \qquad 1 \le j \le n.$$

Parametrices of elliptic equations and linear partial differential operators have several interesting properties in common. If we assume that they are both defined in an open set Ω, they define continuous linear mappings of $C_c^\infty(\Omega)$ into $C^\infty(\Omega)$ and of $\mathscr{E}'(\Omega)$ into $\mathscr{D}'(\Omega)$. They are *pseudolocal.* An operator $A : \mathscr{E}'(\Omega) \to \mathscr{D}'(\Omega)$ is called pseudolocal if, given any $u \in \mathscr{E}'(\Omega)$, Au is a C^∞ function in every open set where this is true of u.†

Before pursuing the study of these properties, it is convenient to recall a few generalities about continuous linear operators acting on distributions, and first of all, the *Schwartz kernels theorem: To any continuous linear map* $C_c^\infty(\Omega) \to \mathscr{D}'(\Omega)$, K, *there corresponds a unique distribution* $K(x, y)$ *in* $\Omega \times \Omega$ *such that, for all* $u \in C_c^\infty(\Omega)$,

$$Ku(x) = \int K(x, y)u(y)\, dy. \tag{2.2}$$

Here we are using the physicists notation: the "integral" at the right in (2.2) stands for the duality bracket between test functions and distributions in Ω, both with respect to the variable y. We shall refer systematically to $K(x, y)$ as the *distribution kernel* associated with (or defining) the operator K, and to K as the operator associated with the kernel $K(x, y)$.

We say that $K(x, y)$ is *separately regular in* x *and* y if it is a C^∞ function with respect to each one of these variables, with values distributions with respect to the other. This is exactly equivalent to saying that both K and its

† Differential operators are also local: if $u = 0$ in some open subset of Ω, the same is true of $P(x, D)u$. Parametrices of elliptic equations of order ≥ 1 do not have the local property. In fact, it characterizes the differential operators.

transpose ${}^tK$ (which is defined by the kernel $K(y, x)$) map $C_c^\infty(\Omega)$ into $C^\infty(\Omega)$. It is also equivalent to saying that K maps $C_c^\infty(\Omega)$ into $C^\infty(\Omega)$ and extends as a mapping of $\mathscr{E}'(\Omega)$ into $\mathscr{D}'(\Omega)$. All mappings considered in this context are linear and continuous.

We say that $K(x, y)$ is *very regular* if it is separately regular and if, furthermore, it is a C^∞ function in the complement of the diagonal of $\Omega \times \Omega$. This terminology of separately regular or very regular will be applied also to the linear operator K, whenever it applies to its associated kernel.

One of the main properties of pseudodifferential operators is that their associated kernels are very regular. This is important in view of the following result.

LEMMA 2.1. *If* $K(x, y)$ *is very regular, the associated operator* K *is pseudolocal* (i.e., given any compactly supported distribution u in Ω, sing supp $Ku \subset$ sing supp u).

PROOF. Let U and V be open subsets of Ω such that $V \subset\subset U$. Let $\phi \in C_c^\infty(U)$, $\phi = 1$ in a neighborhood of the closure of V. Let $u \in \mathscr{E}'(\Omega)$ be such that $u \in C^\infty(U)$. Then $Ku = K(\phi u) + K[(1-\phi)u]$. Since $\phi u \in C_c^\infty(\Omega)$ and since K is separately regular, we have $K(\phi u) \in C^\infty(\Omega)$. On the other hand,

$$K[(1-\phi)u](x) = \int K(x, y)[1-\phi(y)]u(y)\, dy.$$

Notice that if x belongs to V and y to the support of $1-\phi$, then (x, y) stays in the complement of a suitable neighborhood of the diagonal of $\Omega \times \Omega$, in which $K(x, y)$ is a C^∞ function. That $K[(1-\phi)u]$ is C^∞ in V follows by differentiation under the integral sign. □

We shall say that the operator (2.2) is *regularizing* if it extends as a continuous linear map of $\mathscr{E}'(\Omega)$ into $C^\infty(\Omega)$. In order for this to be the case, it is necessary and sufficient for the associated kernel $K(x, y)$ to be C^∞ in $\Omega \times \Omega$. Pseudodifferential operators are going to be defined modulo regularizing operators; as a result, they are best adapted to situations in which one can factor out the contributions of smoothing operators. The operator R in Section 1 (see (1.37)) was regularizing.

Let us briefly describe a situation, generalizing that of Section 1, which points to the relevance of the concepts just defined.

LEMMA 2.2. *Let* P *denote a differential operator in* Ω. *Suppose that there is a very regular operator* $K : \mathscr{E}'(\Omega) \to \mathscr{D}'(\Omega)$ *such that* $KP - I$ *is regularizing* *(which is sometimes expressed by saying that K is a left parametrix of P). Then P is hypoelliptic, i.e., it has the following property: Given any open set U of Ω, then every distribution u in U such that $Pu \in C^\infty(U)$ is a C^∞ function in U.*

PROOF. Let V and ϕ be as in the proof of Lemma 2.1. We have

$$P(\phi u) = \phi Pu + w,$$

where $w \in \mathscr{E}'(\Omega)$, $w = 0$ in V. Set $R = KP - I$. We have, in V,

$$\begin{aligned} u = \phi u &= KP(\phi u) - R(\phi u) \\ &= K(\phi Pu) + Kw - R(\phi u). \end{aligned}$$

Since R is regularizing, $R(\phi u) \in C^\infty(\Omega)$; since K is separately regular and $\phi Pu \in C_c^\infty(\Omega)$, $K(\phi Pu) \in C^\infty(\Omega)$. Since $w = 0$ in V and K is pseudolocal, $Kw \in C^\infty(V)$. □

A common feature of differential operators and of parametrices such as K of Section 1 is that the corresponding symbols $a(x, \xi)$ can, in both cases, be represented by series of terms that are homogeneous of decreasing degree with respect to ξ (for $|\xi|$ large). The series is finite in the case of differential operators; $P(x, \xi)$ is simply a polynomial in ξ. It is infinite in the case of the parametrix K. It has been recognized that the inequalities such as (1.33), which accompany those series representations, are the key to many of the useful properties of the operators in question. For this reason it has been agreed to adopt an estimate of the kind (1.33) as the starting point for the definition and the theory of pseudodifferential operators, at least of those we shall call the *standard* pseudodifferential operators. (More general classes of pseudodifferential operators are often needed and their definition stems from generalizations of estimate (1.33).) Thus pseudodifferential operators are often defined as operators of the kind given by (2.1), where the symbol $a(x, \xi)$ is a C^∞ function in $\Omega \times \mathbb{R}_n$ such that to every compact subset $\mathscr{K}$ of Ω and to every pair $\alpha, \beta \in \mathbb{Z}_+^n$ there is a constant $C_{\alpha,\beta}(\mathscr{K}) > 0$ such that

$$|D_\xi^\alpha D_x^\beta a(x, \xi)| \leq C_{\alpha,\beta}(\mathscr{K})(1 + |\xi|)^{m-|\alpha|}, \qquad \forall x \in \mathscr{K}, \xi \in \mathbb{R}_n.$$

The accepted terminology is to say that $a(x, \xi)$ is a symbol of order $\leq m$; here m is any real number.

However, as the theory has matured, it has been realized that a slight modification of this definition has considerable expository advantages. The modified version is essentially equivalent to the unmodified one, as we shall see. The modification is based on the observation that by virtue of the

Fourier inversion formula (1.3), the expression (2.1) is equivalent to the following one:

$$Au(x) = (2\pi)^{-n} \iint e^{i(x-y)\cdot\xi} a(x, \xi) u(y)\, dy\, d\xi. \tag{2.3}$$

If we assume that $u \in C_c^\infty$, the integrations must be performed in the indicated order. An alternative, which is preferable, is that Fourier transform must be interpreted in the distribution sense, as we shall do below. Once we have the expression in the form (2.3), there does not appear to be any obvious reason for handling only symbols $a(x, \xi)$ independent of y. This modivates the following definition.

DEFINITION 2.1. *Let m be any real number. We shall denote by $S^m(\Omega, \Omega)$ the linear space of C^∞ functions in $\Omega \times \Omega \times \mathbb{R}_n$, $a(x, y, \xi)$, which have the following property:*

(2.4) *To every compact subset $\mathcal{K}$ of $\Omega \times \Omega$ and to every triplet of n-tuples α, β, γ, there is a constant $C_{\alpha,\beta,\gamma}(\mathcal{K}) > 0$ such that*

$$|D_\xi^\alpha D_x^\beta D_y^\gamma a(x, y, \xi)| \le C_{\alpha,\beta,\gamma}(\mathcal{K})(1 + |\xi|)^{m-|\alpha|}, \qquad \forall (x, y) \in \mathcal{K},\ \xi \in \mathbb{R}_n. \tag{2.5}$$

We are going to call the elements of $S^m(\Omega, \Omega)$ *amplitudes* of degree $\le m$ (in $\Omega \times \Omega$). What we shall later introduce as *symbols* will be (classes of) amplitudes that are independent of y. A polynomial with respect to ξ of degree m, now an integer ≥ 0, with coefficients in $C^\infty(\Omega)$ is of course an amplitude (and, in fact, a symbol) of degree m. The intersection of the sets $S^m(\Omega, \Omega)$ as m ranges over $\mathbb{R}$ will be denoted by $S^{-\infty}(\Omega, \Omega)$.

The space $S^m(\Omega, \Omega)$ is endowed with a natural locally convex topology: denote by $\mathfrak{p}_{\mathcal{K};\alpha,\beta,\gamma}(a)$ the infimum of the constants $C_{\alpha,\beta,\gamma}(\mathcal{K})$ such that (2.5) is true; it is seen at once that $\mathfrak{p}_{\mathcal{K};\alpha,\beta,\gamma}$ is a *seminorm* on $S^m(\Omega, \Omega)$ and defines the topology of this space when $\mathcal{K}$ ranges over the collection of all compact subsets of Ω and α, β, γ over that of all n-tuples. Thus topologized, $S^m(\Omega, \Omega)$ is a Fréchet space.

From (2.4) it follows at once that $a(x, y, \xi)$ is a C^∞ function of (x, y) in $\Omega \times \Omega$ valued in the space $\mathcal{S}'_\xi$ of tempered distributions of $\xi \in \mathbb{R}_n$. Hence it can be regarded as the inverse Fourier transform, from the variable ξ to a new variable z (in $\mathbb{R}^n$), of a C^∞ function of (x, y) in $\Omega \times \Omega$, valued in $\mathcal{S}'_z$, $A^\#(x, y, z)$. We may then define the following distribution on $\Omega \times \Omega$:

$$A(x, y) = A^\#(x, y, x - y). \tag{2.6}$$

A remark for the reader familiar with *topological tensor products* (see Treves

[3], Part III): $A^{\#}(x, y, z)$ is an element of $C^\infty_{x,y}(\Omega \times \Omega) \hat{\otimes} \mathcal{S}'_z$ and can be represented (in infinitely many manners) as a convergent series $\sum_j a_j(x)b_j(y)T_j(z)$, with a_j and b_j belonging to $C^\infty(\Omega)$ and $T_j \in \mathcal{S}'(\mathbb{R}^n)$. The distribution $A(x, y)$ can then be defined as the series $\sum_j a_j(x)b_j(y)T_j(x - y)$ in a trivial manner when the sum is finite, and from there by continuity when the series is infinite. Of course, in such an approach, one must also show that the definition of $A(x, y)$ is independent of the choice of the series representation of $A^{\#}(x, y, z)$. In turn, the distribution kernel $A(x, y)$ defines a continuous linear map $C_c^\infty(\Omega) \to \mathcal{D}'(\Omega)$ by the formula

$$Au(x) = \int A(x, y)u(y)\, dy. \tag{2.7}$$

We have

$$A(x, y) = (2\pi)^{-n} \int e^{i(x-y)\cdot\xi} a(x, y, \xi)\, d\xi. \tag{2.8}$$

When the distribution $A^{\#}(x, y, z)$ is represented by a series $\sum_j a_j(x)b_j(y)T_j(z)$, the linear operator A may be equated to the sum, possibly infinite, of operators, $u \mapsto \sum_j a_j[T_j * (b_j u)]$.

Actually it is more within the spirit of this book to regard (2.8) as an *oscillatory integral*, in fact as the simplest kind of such integrals. The assertion that $A(x, y)$ is a distribution in $\Omega \times \Omega$ can be directly checked by integration by parts and by availing oneself of the property (2.4) of the amplitude $a(x, y, \xi)$. Indeed, we may write

$$(2\pi)^n A(x, y) = \int (1 - \Delta_x)^N (e^{i(x-y)\cdot\xi}) \frac{a(x, y, \xi)}{(1 + |\xi|^2)^N}\, d\xi, \tag{2.9}$$

where N is an integer as large as we wish. We use then the transposed Leibniz formula (0.3) of Notation and Background:

(2.10)
$$a(x, y, \xi)(1 - \Delta_x)^N (e^{i(x-y)\cdot\xi}) = \sum_{|\alpha+\beta|\le 2N} D_x^\alpha [c_{\alpha,\beta}\, e^{i(x-y)\cdot\xi} D_x^\beta a(x, y, \xi)],$$

for a suitable choice of coefficients $c_{\alpha,\beta}$ (which, incidentally, depend only on the integer N). Thus, by putting (2.10) into (2.9), we get

$$A(x, y) = \sum_{|\alpha|\le 2N} D_x^\alpha A_{(\alpha)}(x, y), \tag{2.11}$$

where

(2.12)

$$(2\pi)^n A_{(\alpha)}(x, y) = \sum_{|\beta| \leq 2N - |\alpha|} c_{\alpha,\beta} \int e^{i(x-y)\cdot\xi} D_x^\beta a(x, y, \xi)(1 + |\xi|^2)^{-N} \, d\xi.$$

If we then require

$$2N > m + n + 1, \tag{2.13}$$

where m is the real number in (2.5), we see that the integrands in the integrals at the right in (2.12) are dominated by $g(\xi) \in L^1$, and therefore, by Lebesgue's dominated convergence theorem, $A_{(\alpha)}$ is a continuous function of (x, y) in $\Omega \times \Omega$. This means that (2.11) is the usual representation of a distribution as a finite sum of derivatives of continuous functions. (Note that the representation here is global: it is valid in the whole of $\Omega \times \Omega$, not only in relatively compact subsets of $\Omega \times \Omega$.)

In fact, let k be any integer ≥ 0 and take

$$2N > m + n + k + 1. \tag{2.14}$$

In this case we can differentiate k times with respect to (x, y) under the integral sign in (2.12) and still get integrands that are suitably L^1-dominated. This means that in the representation (2.11), provided we take N large enough, we can achieve that $A_{(\alpha)} \in C^k(\Omega \times \Omega)$ for every α, $|\alpha| \leq 2N$. This shows that $A(x, y)$ is a C^∞ function of y in Ω valued in $\mathscr{D}'_x(\Omega)$. But obviously the definition (2.8) of $A(x, y)$ is "symmetric" with respect to x and to y, and therefore it is also a C^∞ function of x in Ω valued in $\mathscr{D}'_y(\Omega)$. In other words, the kernel $A(x, y)$ is separately regular in x and in y. Later we shall again obtain this property as a corollary to a precise continuity result (Theorem 2.1).

In summary, to view $A(x, y)$ given by (2.8) as an oscillatory integral is to view it as a distribution in $\Omega \times \Omega$ given by (2.11), with continuous functions $A_{(\alpha)}$ given by (2.12), where N satisfies (2.13). It is worth repeating, however, that the earlier definition via $A^{\#}(x, y, z)$ is the more direct and intrinsic. We may also content ourselves with the interpretation analogous to that of (2.3):

$$Au(x) = (2\pi)^{-n} \iint e^{i(x-y)\cdot\xi} a(x, y, \xi) u(y) \, dy \, d\xi, \tag{2.15}$$

where it is understood that the integration with respect to y is effected first and the one with respect to ξ last.

Definition 2.2. *Let $a \in S^m(\Omega, \Omega)$. We shall denote by* Op a *the linear operator $C_c^\infty(\Omega) \to \mathscr{D}'(\Omega)$ defined by* (2.15).

PROPOSITION 2.1. *Any regularizing operator* $A: \mathscr{E}'(\Omega) \to C^\infty(\Omega)$ *is of the form* Op a, *with* $a \in S^{-\infty}(\Omega, \Omega)$.

PROOF. The kernel $A(x, y)$ of A belongs to $C^\infty(\Omega \times \Omega)$. It suffices then to take

$$a(x, y, \xi) = e^{-i(x-y)\cdot\xi} A(x, y)\chi(\xi).$$

with $\chi \in C_c^\infty(\mathbb{R}_n)$ such that $\int \chi \, d\xi = 1$. □

DEFINITION 2.3. *A linear operator* $A: \mathscr{E}'(\Omega) \to \mathscr{D}'(\Omega)$ *is called a standard pseudodifferential operator of order m if there is an amplitude* $a \in S^m(\Omega, \Omega)$ *such that* $A =$ Op a.

The set of standard pseudodifferential operators of order m will be denoted by $\Psi^m(\Omega)$.

The union of the spaces $\Psi^m(\Omega)$, m real, will be denoted by $\Psi(\Omega)$, their intersection by $\Psi^{-\infty}(\Omega)$. The elements of $\Psi(\Omega)$ will of course be the standard pseudodifferential operators in Ω. In the remainder of this chapter, since no confusion is likely to arise, we shall usually omit the adjective "standard."

As we shall see in Corollary 2.2, the elements of $\Psi^{-\infty}(\Omega)$ are the regularizing operators in Ω.

If $P(x, \xi)$ is a polynomial with respect to ξ of degree $m \in \mathbb{Z}_+$ with coefficients in $C^\infty(\Omega)$, then the differential operator $P(x, D)$ is equal to Op a with $a(x, y, \xi) = P(x, \xi)$. Of course it belongs to $\Psi^m(\Omega)$.

Let $A \in \Psi^m(\Omega)$. Any element $a(x, y, \xi)$ of $S^m(\Omega, \Omega)$ such that $A =$ Op a will be called an *amplitude* for (or of) A. As we shall see, a pseudodifferential operator has (infinitely) many amplitudes; in contrast, it will be shown to have a unique symbol.

Pseudodifferential operators are not just continuous from $C_c^\infty(\Omega)$ into $\mathscr{D}'(\Omega)$ but actually map $C_c^\infty(\Omega)$ into $C^\infty(\Omega)$ and can be extended as continuous linear mappings of $\mathscr{E}'(\Omega)$ into $\mathscr{D}'(\Omega)$. (Their associated kernels are separately regular in x and y.) We shall now state and prove a much more precise property. The added precision is based on use of the Sobolev spaces H^s. For the basic definitions and notation connected with these spaces we refer to Section 5 of Notation and Background. For more details on this subject we refer to the abundant literature (for instance, Hörmander [2], Chapter II; Lions and Magenes [1], Chapter 1; Treves [3], Section 25].

THEOREM 2.1. *Let A be a pseudodifferential operator in* Ω, *of order* $\leq m$. *Given any real number s*, $u \mapsto Au$ *can be extended as a continuous linear mapping of* $H_c^s(\Omega)$ *into* $H_{\mathrm{loc}}^{s-m}(\Omega)$.

PROOF. We must prove that for every function $g \in C_c^\infty(\Omega)$ and for every compact subset $\mathcal{K}$ of Ω, there is a constant $C > 0$ such that

$$\|gAu\|_{s-m} \leq C\|u\|_s \qquad \textit{for all } u \in C_c^\infty(\mathcal{K}). \tag{2.16}$$

It is clear that nothing is changed if we replace gA by gAh with $h \in C_c^\infty(\Omega)$ equal to one in some neighborhood of $\mathcal{K}$. Observe that an amplitude for gAh is given by

$$g(x)a(x, y, \xi)h(y), \tag{2.17}$$

assuming that $a(x, y, \xi)$ is an amplitude of A. Since (2.17) has the property that the (x, y)-projection of its support is contained in a compact subset of $\Omega \times \Omega$, we may as well assume that $a(x, y, \xi)$ possesses this property and omit mention of the cutoff functions g and h. Our other assumption is that to every triplet of n-tuples α, β, γ, there is a constant $C_{\alpha,\beta,\gamma} > 0$ such that

$$|D_\xi^\alpha D_x^\beta D_y^\gamma a(x, y, \xi)| \leq C_{\alpha,\beta,\gamma}(1 + |\xi|)^{m-|\alpha|}, \qquad \forall x, y \in \mathbb{R}^n, \xi \in \mathbb{R}_n. \tag{2.18}$$

Let us denote by $\hat{a}(\xi', \eta', \xi)$ the Fourier transform of $a(x, y, \xi)$ with respect to (x, y), which has a meaning since $a(x, y, \xi) \in (C_c^\infty)_{x,y}$. From (2.18) we derive at once that to every n-tuple α and to every pair of integers $k, l \geq 0$ there is a constant $C_{\alpha;k,l} > 0$ such that

$$|D_\xi^\alpha \hat{a}(\xi', \eta', \xi)| \leq C_{\alpha;k,l}(1 + |\xi'|)^{-k}(1 + |\eta'|)^{-l}(1 + |\xi|)^{m-|\alpha|} \tag{2.19}$$

for every $\xi, \xi', \eta' \in \mathbb{R}_n$.

By virtue of the Fourier inversion formula we have

$$\widehat{Au}(\xi') = (2\pi)^{-2n} \iiint e^{-iy\cdot(\xi-\eta')}\tilde{a}(\xi' - \xi, y, \xi)\hat{u}(\eta')\, dy\, d\xi\, d\eta'$$

where $\tilde{a}(\xi', y, \xi)$ denotes the Fourier transform of $a(x, y, \xi)$ with respect to x. In the preceding integral we have the right to perform the integration with respect to y. It yields

$$\widehat{Au}(\xi') = (2\pi)^{-2n} \iint \hat{a}(\xi' - \xi, \xi - \eta', \xi)\hat{u}(\eta')\, d\xi\, d\eta'. \tag{2.20}$$

We multiply both sides of (2.20) by $(1 + |\xi'|^2)^{(s-m)/2}$. We shall use the following simple inequality (sometimes called *Peetre's inequality*).

(2.21) *For all real numbers s and for all vectors θ, θ' in $\mathbb{R}_n$,*

$$(1 + |\theta|^2)^s \leq 2^{|s|}(1 + |\theta - \theta'|^2)^{|s|}(1 + |\theta'|^2)^s.$$

PROOF. When $s > 0$, the assertion is evident after taking the sth root of both sides. When $s < 0$, one exchanges θ and θ' and applies the result for $-s$.

We apply (2.21) with $\theta = \xi'$, $\theta' = \xi$ and with $\frac{1}{2}(s - m)$ in the place of s. We get

$$(1 + |\xi'|^2)^{(s-m)/2}|\widehat{Au}(\xi')| \leq$$
$$C_s \iint (1 + |\xi - \xi'|^2)^{|s-m|/2}|\hat{a}(\xi' - \xi, \xi - \eta', \xi)|\,(1 + |\xi|^2)^{(s-m)/2}|\hat{u}(\eta')|\, d\xi\, d\eta'.$$

Next we apply (2.21) with $\theta = \xi$, $\theta' = \eta'$ and $s/2$ substituted for s. We obtain

$$(2.22)\quad (1 + |\xi'|^2)^{(s-m)/2}|\widehat{Au}(\xi')| \leq C_s \iint (1 + |\xi - \xi'|)^{|s-m|}(1 + |\xi - \eta'|)^{|s|}(1 + |\xi|)^{-m}|\hat{a}(\xi' - \xi, \xi - \eta', \xi)| \times (1 + |\eta'|^2)^{s/2}|\hat{u}(\eta')|\, d\xi\, d\eta'.$$

At this point we apply (2.19) with $\alpha = 0$, $k = |s - m| + n + 1$, $l = |s| + n + 1$, $\xi' - \xi$ substituted for ξ' and $\xi - \eta'$ for η'. We obtain at once

$$(2.23)\quad (1 + |\xi'|^2)^{(s-m)/2}|\widehat{Au}(\xi')| \leq C'_s \iint (1 + |\xi - \xi'|)^{-n-1}(1 + |\xi - \eta'|)^{-n-1}(1 + |\eta'|^2)^{s/2}|\hat{u}(\eta')|\, d\xi\, d\eta'.$$

We may regard the right-hand side as a double convolution and apply the classical Hölder's inequality, in iterated form:

$$(2.24)\qquad \|f * g * h\|_{L^2} \leq \|f\|_{L^1}\|g\|_{L^1}\|h\|_{L^2}.$$

We obtain at once

$$(2.25)\qquad \|Au\|_{s-m} \leq C''_s\|u\|_s,$$

which is precisely what we wanted to prove. □

COROLLARY 2.1. *Every pseudodifferential operator A in Ω defines a continuous linear mapping of $C_c^\infty(\Omega)$ into $C^\infty(\Omega)$ which extends as a continuous linear mapping of $\mathscr{E}'(\Omega)$ into $\mathscr{D}'(\Omega)$.*

PROOF. From the set-theoretical relations (0.34) and (0.35) and from Theorem 2.1 we deduce that A defines a linear map of $C_c^\infty(\Omega)$ into $C^\infty(\Omega)$ and one of $\mathscr{E}'(\Omega)$ into $\mathscr{D}'(\Omega)$. To prove the continuity of these maps it suffices to prove the continuity of their restrictions to the subspaces $C_c^\infty(\mathscr{K})$ and

$\mathscr{E}'(\mathscr{K})$ consisting of the functions or distributions supported in a fixed but arbitrary compact subset $\mathscr{K}$ of Ω. But $C_c^\infty(\mathscr{K})$ is equal to the intersection of the $H^s(\mathscr{K})$, as a *topological* vector space. And $\mathscr{E}'(\mathscr{K})$ is the inductive limit of the $H^s(\mathscr{K})$ (all this as s ranges over $\mathbb{R}$), which implies that a linear map defined in $\mathscr{E}'(\mathscr{K})$ is continuous if and only if its restriction to each $H^s(\mathscr{K})$ is continuous. On the other hand the first equality (0.34) is also valid topologically, and of course every $H^s_{\text{loc}}(\Omega)$ is continuously embedded in $\mathscr{D}'(\Omega)$. All this, together with Theorem 2.1, implies the result. Details can be found in Treves [2, 3]. □

Corollary 2.1 may be restated by saying that the kernel distribution $A(x, y)$ associated with the pseudodifferential operator A (see (2.7)) is separately regular in x and in y, or in the notation of topological tensor products (see Treves [2], Part III) by writing

$$A(x, y) \in \{C_x^\infty(\Omega) \hat{\otimes} \mathscr{D}'_y(\Omega)\} \cap \{C_y^\infty(\Omega) \hat{\otimes} \mathscr{D}'_x(\Omega)\}. \tag{2.26}$$

COROLLARY 2.2. *Every pseudodifferential operator A of order $-\infty$ in Ω is regularizing, i.e., maps $\mathscr{E}'(\Omega)$ into $C^\infty(\Omega)$.*

PROOF. Evident by Theorem 2.1. □

COROLLARY 2.3. *The kernel $A(x, y)$ of a pseudodifferential operator A of order $-\infty$ in Ω is a C^∞ function in $\Omega \times \Omega$.*

PROOF. Follows from Corollary 2.2 and from the Schwartz kernels theorem which asserts that the kernel of any continuous linear map $\mathscr{E}'(\Omega) \to C^\infty(\Omega)$ belongs to $C^\infty(\Omega \times \Omega)$. □

We recall that by Proposition 2.1 every regularizing operator in Ω belongs to $\Psi^{-\infty}(\Omega)$.

Theorem 2.1 is the generalization of the following well-known property of differential operators:

COROLLARY 2.4. *Any (linear partial) differential operator of order m (with C^∞ coefficients) in Ω defines continuous linear maps $H_c^s(\Omega) \to H_c^{s-m}(\Omega)$ and $H^s_{\text{loc}}(\Omega) \to H^{s-m}_{\text{loc}}(\Omega)$ for all real s.*

We have taken advantage of the fact that differential operators are local. In turn, Corollary 2.4 is a generalization of the well-known property that multiplication by a C^∞ function $a(x)$ defines a continuous endomorphism of $H_c^s(\Omega)$ (resp., $H_{\mathrm{loc}}^s(\Omega)$). It is easy to show that in the case $\Omega = \mathbb{R}^n$ and when all derivatives of $a(x)$ are bounded in the whole of $\mathbb{R}^n$, multiplication by $a(x)$ induces a continuous endomorphism of H^s. (Direct proof when s is an integer ≥ 0; by transposition when s is an integer <0; by interpolation when s is not an integer).

Actually we can extract from the proof of Theorem 2.1 slightly more than we have stated. The space $L(H_c^s(\Omega); H_{\mathrm{loc}}^{s-m}(\Omega))$ of continuous linear operators $T: H_c^s(\Omega) \to H_{\mathrm{loc}}^{s-m}(\Omega)$ can be endowed with a natural structure of locally convex Hausdorff topological vector space, namely that defined by the seminorms $\mathfrak{p}_{\mathscr{K},\phi}$, where $\mathscr{K}$ is any compact subset of Ω, ϕ any function belonging to $C_c^\infty(\Omega)$. Then

$$\mathfrak{p}_{\mathscr{K},\phi}(T) = \sup \|\phi T u\|_{s-m}/\|u\|_s,$$

where the supremum is taken over all C^∞ functions u with support in $\mathscr{K}$. Inequality (2.22) and the subsequent observations imply the following:

PROPOSITION 2.2. *The assignment $a \mapsto \mathrm{Op}\, a$ defines a continuous linear map $S^m(\Omega, \Omega) \to L(H_c^s(\Omega); H_{\mathrm{loc}}^{s-m}(\Omega))$ for all real numbers m, s.*

Next we establish the property announced at the beginning of this section, namely that all pseudodifferential operators are pseudolocal:

THEOREM 2.2. *The kernel $A(x, y)$ of a pseudodifferential operator A in Ω is a C^∞ function off the diagonal in $\Omega \times \Omega$, and A is pseudolocal.*

PROOF. In order to prove the first part of the statement it suffices to show that $g(x)A(x, y)h(y) \in C^\infty(\Omega \times \Omega)$ for all $g, h \in C_c^\infty(\Omega)$ with disjoint supports. Let $a(x, y, \xi) \in S^m(\Omega, \Omega)$ be an amplitude of A. Whatever $u \in C^\infty(\Omega)$ we have

$$\begin{aligned} g(x)\{A(hu)(x)\} &= (2\pi)^{-n} \iint |x-y|^{2k} e^{i(x-y)\cdot\xi} \frac{a^{\#}(x, y, \xi)}{|x-y|^{2k}} u(y)\, dy\, d\xi \\ &= (2\pi)^{-n} \iint \{(-\Delta_\xi)^k e^{i(x-y)\cdot\xi}\} \frac{a^{\#}(x, y, \xi)}{|x-y|^{2k}} u(y)\, dy\, d\xi \\ &= (2\pi)^{-n} \iint e^{i(x-y)\cdot\xi} \left\{(-\Delta_\xi)^k \frac{a^{\#}(x, y, \xi)}{|x-y|^{2k}}\right\} u(y)\, dy\, d\xi, \end{aligned}$$

where $a^{\#}(x, y, \xi) = g(x)a(x, y, \xi)h(y)$. This, together with (2.5), shows that the operator $u \mapsto gA(hu)$ has an amplitude of degree $\leq m - 2k$ and therefore, since k is arbitrarily large, that it is of order $-\infty$. Corollary 2.3 then implies our assertion.

If we combine the first part of the statement of Theorem 2.2 with Corollary 2.1, we see that the kernel distribution $A(x, y)$ is very regular. Lemma 2.1 then implies that A is pseudolocal. □

3. Transposition, Composition, Transformation under Diffeomorphisms of Pseudodifferential Operators

Let A be a pseudodifferential operator in the open set Ω. We know that A defines a continuous linear map of $C_c^\infty(\Omega)$ into $C^\infty(\Omega)$ and of $\mathscr{E}'(\Omega)$ into $\mathscr{D}'(\Omega)$. Therefore the same is true of its *transpose* ${}^tA$, which can be defined by

$$\langle {}^tAu, v\rangle = \langle u, Av\rangle, \qquad u, v \in C_c^\infty(\Omega). \tag{3.1}$$

As a matter of fact, we have

THEOREM 3.1. *The transpose of a pseudodifferential operator A of order $\leq m$ in Ω, ${}^tA$, is a pseudodifferential operator of order $\leq m$ in Ω.*

PROOF. Let $a(x, y, \xi)$ be an amplitude of A, belonging to $S^m(\Omega, \Omega)$. Let u and v be test functions in Ω. By (3.1) we have

$$\langle {}^tAu, v\rangle = (2\pi)^{-n} \iiint e^{i(x-y)\cdot\xi} a(x, y, \xi) u(x) v(y)\, dx\, dy\, d\xi,$$

which shows that

$$ {}^tAu(x) = (2\pi)^{-n} \iint e^{i(x-y)\cdot\xi} a(y, x, -\xi) u(y)\, dy\, d\xi, \tag{3.2}$$

and therefore that ${}^tA = \operatorname{Op} b$, where $b(x, y, \xi) = a(y, x, -\xi)$. □

REMARK 3.1. Let us emphasize the meaning of formula (3.2). If $a(x, y, \xi)$ is an amplitude of $A \in \Psi^m(\Omega)$, then $a(y, x, -\xi)$ is one of ${}^tA$.

REMARK 3.2. We shall always distinguish the *adjoint* A^* of A from its transpose, ${}^tA$. A^* is the *complex conjugate* of ${}^tA$:

$$A^*u = \overline{{}^tA\bar{u}}, \qquad u \in \mathscr{E}'(\Omega). \tag{3.3}$$

It follows at once from Theorem 3.1 that

$$A \in \Psi^m(\Omega) \Rightarrow A^* \in \Psi^m(\Omega);$$

and from (3.2) it follows that if $a(y, x, \xi)$ is an amplitude of A, then

$$\overline{a(y, x, \xi)} \tag{3.4}$$

is one of A^*.

We want now to define the *compose* $A \circ B$ of two pseudodifferential operators (of orders $\leq m, m'$ respectively) in Ω. This is not always possible. Indeed, in general, A acts only on distributions with compact support in Ω, whereas the image (or range) of B consists of distributions in Ω that do not necessarily have compact support. We must therefore impose the following restriction on B:

(3.5) *B maps $\mathscr{E}'(\Omega)$ into itself.*

It is not difficult to characterize the kernel distributions associated with pseudodifferential operators having property (3.5): if $B(x, y)$ is the kernel of B, then (3.5) is equivalent to saying that given any test function h in Ω, the support of $B(x, y)h(y)$ is a compact subset of $\Omega \times \Omega$. Let us also point out that (3.5) is equivalent to the following property:

(3.6) *B maps $C_c^\infty(\Omega)$ into itself.*

Indeed, if (3.5) holds, B maps $C_c^\infty(\Omega)$ into $C^\infty(\Omega) \cap \mathscr{E}'(\Omega)$. Conversely, suppose that (3.6) holds. Then the transpose ${}^tB$ of B maps $\mathscr{D}'(\Omega)$ into itself and hence (being pseudolocal, by Theorem 2.2) maps $C^\infty(\Omega)$ into itself. Consequently, the transpose of ${}^tB$, which is B, must satisfy (3.5).

About the amplitudes of the pseudodifferential operators that satisfy (3.5) the following can be said. Let B satisfy (3.5) and let $b(x, y, \xi) \in S^m(\Omega, \Omega)$ be one of its amplitudes. Let $\{h_j\}$, $j = 1, 2, \ldots$, be an arbitrary locally finite partition of unity in $C_c^\infty(\Omega)$. By (3.5) we know that, to every j, there is a compact subset K_j of Ω such that

(3.7) *Given any distribution u in Ω, $B(h_j u)$ vanishes in $\Omega \backslash K_j$.*

Then let $g_j \in C_c^\infty(\Omega)$ be equal to one in a neighborhood of K_j and set

$$b^\#(x, y, \xi) = \sum_j g_j(x)h_j(y)b(x, y, \xi). \tag{3.8}$$

It is checkeed at once that $b^\#(x, y, \xi) \in S^m(\Omega, \Omega)$ is an amplitude of B. Of course, $b^\#$ has the following property:

(3.9) *Given any $h \in C_c^\infty(\Omega)$, the x-projection of the support of $h(y)b^\#(x, y, \xi)$ is compact.*

In what now follows we assume that B has property (3.5) and that $b = b^{\#}$ is an amplitude of B endowed with property (3.9). Let u be an arbitrary test function in Ω. Then we have

$$A(Bu)(x) = (2\pi)^{-2n} \iiiint e^{i(x-y)\cdot\xi + i(y-z)\cdot\eta} a(x, y, \xi) b(y, z, \eta) u(z)\, dy\, dz\, d\xi\, d\eta,$$

and after exchanging y and z it follows that

$$A(Bu)(x) = (2\pi)^{-n} \iint e^{i(x-y)\cdot\xi} k(x, y, \xi) u(y)\, dy\, d\xi, \tag{3.10}$$

where

$$k(x, y, \xi) = (2\pi)^{-n} \iint e^{i(y-z)\cdot(\xi-\eta)} a(x, z, \xi) b(z, y, \eta)\, dz\, d\eta. \tag{3.11}$$

The integration in (3.11) can be understood as follows. Let y remain in a compact subset of Ω, K'. Then by (3.9) $b = b^{\#}$ vanishes for $z \notin K''$, another compact subset of Ω. Consequently, the integration with respect to z yields a function of $\xi - \eta$ which belongs to the space $\mathcal{S}$ with respect to $\xi - \eta$. If ξ remains fixed, or in a bounded set of $\mathbb{R}_n$, this function is integrable with respect to η, even after multiplication by powers of $|\eta|$.

By using the expressions (3.10) and (3.11) we may now prove

THEOREM 3.2. *Let $A \in \Psi^m(\Omega)$, $B \in \Psi^{m'}(\Omega)$. Suppose that B maps $\mathscr{E}'(\Omega)$ into itself. Then $A \circ B \in \Psi^{m+m'}(\Omega)$.*

PROOF. We are going to prove that $k(x, y, \xi)$, defined in (3.11), belongs to $S^{m+m'}(\Omega, \Omega)$. Formula (3.10) shows then that it is an amplitude of $A \circ B$.

We have, by the Leibniz formula, and in the multi-index notation (see Notation and Background)

$$\begin{aligned}
&(2\pi)^n D_\xi^\alpha D_x^\beta D_y^\gamma k(x, y, \xi) \\
&\quad = \sum_{\alpha' \leq \alpha} \sum_{\gamma' \leq \gamma} \binom{\alpha}{\alpha'} \binom{\gamma}{\gamma'} \iint \{D_\xi^{\alpha-\alpha'} D_y^{\gamma-\gamma'} (e^{i(y-z)\cdot(\xi-\eta)})\} \\
&\qquad \times D_\xi^{\alpha'} D_x^\beta a(x, z, \xi) D_y^{\gamma'} b(z, y, \eta)\, dz\, d\eta \\
&\quad = \sum_{\alpha' \leq \alpha} \sum_{\gamma' \leq \gamma} \binom{\alpha}{\alpha'} \binom{\gamma}{\gamma'} \iint \{(-D_\eta)^{\alpha-\alpha'} (-D_z)^{\gamma-\gamma'} (e^{i(y-z)\cdot(\xi-\eta)})\} \\
&\qquad \times D_\xi^{\alpha'} D_x^\beta a(x, z, \xi) D_y^{\gamma'} b(z, y, \eta)\, dz\, d\eta
\end{aligned}$$

(cont.)

$$= \sum_{\alpha' \le \alpha} \sum_{\gamma' \le \gamma} \binom{\alpha}{\alpha'}\binom{\gamma}{\gamma'} \iint e^{i(y-z)\cdot(\xi-\eta)}$$
$$\times D_z^{\gamma-\gamma'}\left(D_\xi^{\alpha'} D_x^\beta a(x, z, \xi) D_\eta^{\alpha-\alpha'} D_y^{\gamma'} b(z, y, \eta)\right) dz\, d\eta$$
$$= \sum_{\alpha' \le \alpha} \sum_{\gamma' \le \gamma} \binom{\alpha}{\alpha'}\binom{\gamma}{\gamma'} \iint e^{i(y-z)\cdot(\xi-\eta)} (1 + |\xi - \eta|^2)^{-M}$$
$$\times (1 - \Delta_z)^M D_z^{\gamma-\gamma'}\left(D_\xi^{\alpha'} D_x^\beta a(x, z, \xi) D_\eta^{\alpha-\alpha'} D_y^{\gamma'} b(z, y, \eta)\right) dz\, d\eta,$$

where we have repeatedly integrated by parts with respect to z and η. As for M it is a large positive integer which we are soon going to choose. We see that $D_\xi^\alpha D_x^\beta D_y^\gamma k(x, y, \xi)$ is a linear combination of terms of the kind

$$\iint \exp\{i(y - z)\cdot(\xi - \eta)\}(1 + |\xi - \eta|^2)^{-M} D_\xi^{\alpha'} D_x^\beta D_z^\lambda a(x, z, \xi)$$
$$\times \{D_\eta^{\alpha-\alpha'} D_z^\mu D_y^{\gamma'} b(z, y, \eta)\}\, dz\, d\eta.$$

It is time now to avail ourselves of the fact that $a \in S^m(\Omega, \Omega)$, $b \in S^{m'}(\Omega, \Omega)$. We obtain at once, for (x, y) in a compact subset $\mathscr{K}$ of $\Omega \times \Omega$,

$$|D_\xi^\alpha D_x^\beta D_y^\gamma k(x, y, \xi)| \le$$
$$\text{const} \sum_{\alpha' \le \alpha} \int (1 + |\xi - \eta|)^{-2M} (1 + |\xi|)^{m-|\alpha'|} (1 + |\eta|)^{m'-|\alpha-\alpha'|}\, d\eta.$$

We have tacitly used the fact that if y remains in a compact subset of Ω (the y-projection of $\mathscr{K}$) and (z, y, η) remains in the support of $b(z, y, \eta)$ (which we recall satisfies (3.9)), then z remains in a compact subset of Ω and consequently the integration with respect to z can be achieved without further ado. It now suffices to observe (cf. (2.21)) that

$$(1 + |\eta|)^{m'-|\alpha-\alpha'|} \le (1 + |\xi|)^{m'-|\alpha-\alpha'|} (1 + |\xi - \eta|)^{|m'|+|\alpha-\alpha'|}.$$

By interchanging η and $\xi - \eta$ in the preceding integral, we obtain

$$|D_\xi^\alpha D_x^\beta D_y^\gamma k(x, y, \xi)| \le \text{const}(1 + |\xi|)^{m+m'-|\alpha|} \int (1 + |\eta|)^{-2M+|m'|+|\alpha|}\, d\eta,$$

and it is enough to choose $2M \ge n + 1 + |m'| + |\alpha|$, to see that $k(x, y, \xi)$ indeed belongs to $S^{m+m'}(\Omega, \Omega)$. □

If we require that both B and its transpose ${}^tB$ have property (3.5), then we obtain an interesting class of pseudodifferential operators.

Definition 3.1. *A pseudodifferential operator A in Ω will be said to be properly supported if both A and ${}^tA$ map $\mathscr{E}'(\Omega)$ into itself.*

PROPOSITION 3.1. *A pseudodifferential operator A in Ω is properly supported if and only if its associated kernel $A(x, y) \in \mathscr{D}'(\Omega \times \Omega)$ has the following property.*

(3.12) *Given any function $g \in C_c^\infty(\Omega)$, both $g(x)A(x, y)$ and $A(x, y)g(y)$ have compact support (contained in $\Omega \times \Omega$).*

The proof of this property is virtually evident.

PROPOSITION 3.2. *If $A \in \Psi(\Omega)$ is properly supported, it can be extended as a continuous linear map of $C^\infty(\Omega)$ into itself and of $\mathscr{D}'(\Omega)$ into itself. Moreover, it maps continuously $C_c^\infty(\Omega)$ into itself.*

PROOF. The last part in the statement is obvious, since A maps $C_c^\infty(\Omega)$ into $C^\infty(\Omega)$ (being a pseudodifferential operator) and into $\mathscr{E}'(\Omega)$ (being properly supported). Its transpose ${}^tA$ has the same property. Consequently, the transpose of ${}^tA$, which is A, maps the dual of $C_c^\infty(\Omega)$ into itself, and also the dual of $\mathscr{E}'(\Omega)$ into itself. □

PROPOSITION 3.3. *If A is properly supported, so are ${}^tA$ and A^*. If B is another properly supported pseudodifferential operator in Ω, so is the compose $A \circ B$.*

These assertions are easy to prove: we leave their proof to the reader.

We see that the properly supported pseudodifferential operators in Ω form a (noncommutative) algebra, with two involutions.

PROPOSITION 3.4. *Let $A \in \Psi^m(\Omega)$. There is a properly supported pseudodifferential operator $A^\# \in \Psi^m(\Omega)$ such that $A - A^\# \in \Psi^{-\infty}(\Omega)$.*

PROOF. Let $a(x, y, \xi) \in S^m(\Omega, \Omega)$ be an amplitude of A. Let $\phi(x, y) \in C^\infty(\Omega \times \Omega)$ have the property (3.12), namely that $g(x)\phi(x, y)$ and $\phi(x, y)g(y)$ have compact support in $\Omega \times \Omega$, whatever $g \in C_c^\infty(\Omega)$.† We further require that ϕ be identically equal to one in a neighborhood of the diagonal in $\Omega \times \Omega$. Set

$$a^\#(x, y, \xi) = \phi(x, y)a(x, y, \xi).$$

By using the argument in the proof of Theorem 2.2 we can show that

$$\{1 - \phi(x, y)\}a(x, y, \xi) \in S^{-\infty}(\Omega, \Omega).$$

† A function or distribution in $\Omega \times \Omega$ with this property is said to be *properly supported.*

Consequently, $a^{\#}(x, y, \xi)$ is an amplitude of a pseudodifferential operator $A^{\#}$ in Ω which has the required properties and which is obviously properly supported. □

The meaning of Proposition 3.4 may be roughly expressed as follows: modulo $\Psi^{-\infty}(\Omega)$ the pseudodifferential operators in Ω correspond to kernel distributions concentrated along the diagonal of $\Omega \times \Omega$.

The simplest properly supported pseudodifferential operators are the differential operators.

REMARK 3.3. Every kernel distribution $A(x, y)$ that possesses property (3.12) defines a continuous linear map A having a representation (2.1), where the symbol $a(x, \xi)$, however, does not necessarily satisfy an estimate of the kind (2.2). Indeed, for x restricted to any relatively compact open subset of Ω, the y-projection of the support of $A(x, y)$ is relatively compact, and we may therefore form the kernel distribution in $\Omega \times \mathbb{R}_n$:

$$a(x, \xi) = \int \exp\{-i(x - y)\cdot\xi\}A(x, y)\, dy$$

(in physicists' notation). It is checked at once that $a(x, \xi)$ is a function of ξ that is slowly growing at infinity, C^∞, valued in $\mathscr{D}'_x(\Omega)$, and that we have

$$Au(x) = \iint A(x, y)u(y)\, dy = (2\pi)^{-n} \int e^{ix\cdot\xi} a(x, \xi)\hat{u}(\xi)\, d\xi$$

for every $u \in C_c^\infty(\Omega)$. Of course, these symbols $a(x, \xi)$ are much too general for a useful theory to systematize their handling.

Finally we shall see what happens to a pseudodifferential operator when a change of variables is performed, or rather when one performs a diffeomorphism $x \mapsto y = \phi(x)$ of the open set Ω onto another open subset of $\mathbb{R}^n$, Ω'. We recall that $x \mapsto y$ is a bijective C^∞ mapping of Ω onto Ω' whose Jacobian matrix J_ϕ is invertible at every point. We denote by $y \mapsto x = \overset{-1}{\phi}(y)$ the inverse diffeomorphism. Also it will be convenient to denote by $\mathscr{J}(y)$ the Jacobian matrix of $\overset{-1}{\phi}$ viewed as a function of y.

The diffeomorphism ϕ induces an isomorphism $\phi^*: C_c^\infty(\Omega') \to C_c^\infty(\Omega)$ by the formula

$$(\phi^* u)(x) = u(\phi(x)), \qquad u \in C_c^\infty(\Omega').$$

From there, by the transposition formula

$$\langle \phi_* F, u\rangle = \langle F, \phi^* u\rangle, \qquad u \in C_c^\infty(\Omega'),\ F \in \mathscr{D}'(\Omega),$$

it defines an isomorphism $\phi_*:\mathscr{D}'(\Omega)\to\mathscr{D}'(\Omega')$. The restriction of ϕ_* to $C_c^\infty(\Omega)$ does not coincide with $\overset{-1}{\phi}{}^*$: when a distribution in Ω is defined by a locally integrable function $f(y)$, its ϕ_*-image is also defined by a locally integrable function, namely, by

$$(3.13)\qquad (\phi_* f)(y) = f(\overset{-1}{\phi}(y))\,|\det \mathscr{J}(y)|, \qquad y\in\Omega'.$$

The restriction of ϕ_* to $\mathscr{E}'(\Omega)$ is an isomorphism of this space onto $\mathscr{E}'(\Omega')$.

We shall regard pseudodifferential operators as mapping distributions with compact supports into distributions. Then given any $A\in\Psi^m(\Omega)$, we set

$$(3.14)\qquad A^\phi = \phi_*\circ A\circ(\overset{-1}{\phi_*}):\mathscr{E}'(\Omega')\to\mathscr{D}'(\Omega').$$

We shall often refer to A^ϕ as the *transfer of* A *to* Ω' *via* ϕ.

THEOREM 3.3. *Let ϕ be a diffeomorphism of Ω onto Ω'. For every real number m, $A\to A^\phi$ is a bijection of $\Psi^m(\Omega)$ onto $\Psi^m(\Omega')$.*

PROOF. It suffices to prove that if $A\in\Psi^m(\Omega)$ then $A^\phi\in\Psi^m(\Omega')$, for there is an obvious inverse mapping $B\to B^\psi$ with $\psi=\overset{-1}{\phi}$. Actually we shall prove a more precise result, namely that if $A=\operatorname{Op} a$ with $a\in S^m(\Omega,\Omega)$, then $A^\phi=\operatorname{Op} a^\phi$ with $a^\phi\in S^m(\Omega',\Omega')$ constructed explicitly from a.

We shall denote by x, x' (resp. y, y') the variables on Ω (resp. in Ω'). We shall need the following lemma.

LEMMA 3.1. *There is an open covering $\{\mathscr{W}_j\}$ $(j\in J)$ of $\Omega'\times\Omega'$ such that for each j there is a C^∞ function $\mathscr{J}_j(y,y')$ valued in the linear space of $n\times n$ matrices, invertible and such that*

$$(3.15)\qquad \overset{-1}{\phi}(y)-\overset{-1}{\phi}(y') = \mathscr{J}_j(y,y')(y-y')$$

for every (y,y') in $\mathscr{W}_j$.

PROOF OF LEMMA 3.1. If y and y' lie in a convex open subset of Ω' we may write

$$(3.16)\qquad \overset{-1}{\phi}(y)-\overset{-1}{\phi}(y') = \int_0^1 \mathscr{J}(ty+(1-t)y')(y-y')\,dt$$

where $\mathscr{J}(z)$ is the Jacobian matrix of $\overset{-1}{\phi}$ at the point $y=z$. If we note that

$$\mathscr{J}_o(y,y') = \int_0^1 \mathscr{J}(ty+(1-t)y')\,dt = \mathscr{J}(y')+O(|y-y'|),$$

we see that $\mathcal{J}_o(y, y')$ is obviously a C^∞ matrix-valued function, everywhere invertible in some neighborhood $\mathcal{W}_o$ of the diagonal in $\Omega' \times \Omega'$.

Consider now an arbitrary pair (y_o, y'_o) such that $\mathbf{v}_o = y_0 - y'_o \neq 0$. There is an open neighborhood $U(\mathbf{v}_o)$ in $\mathbb{R}^n$ and a C^∞ matrix-valued function $G(\mathbf{v})$, everywhere invertible in $U(\mathbf{v}_o)$, such that

$$\mathbf{v} = G(\mathbf{v})\mathbf{v}_o \qquad \textit{for every } \mathbf{v} \in U(\mathbf{v}_o). \tag{3.17}$$

We leave the (elementary) proof of this assertion to the reader. Similarly, since $\mathbf{u}_o = \overset{-1}{\phi}(y_o) - \overset{-1}{\phi}(y'_o) \neq 0$, there is an open neighborhood $U(\mathbf{u}_o)$ in $\mathbb{R}^n$ such that

$$\mathbf{u} = F(\mathbf{u})\mathbf{u}_o \qquad \textit{for every } \mathbf{u} \in U(\mathbf{u}_o), \tag{3.18}$$

where F is an invertible matrix with C^∞ entries in $U(\mathbf{u}_o)$. Let us then select arbitrarily an invertible matrix M_o such that $\mathbf{u}_o = M_o\mathbf{v}_o$, and set

$$M(\mathbf{u}, \mathbf{v}) = F(\mathbf{u})M_oG(\mathbf{v})^{-1}.$$

We have

$$\mathbf{u} = M(\mathbf{u}, \mathbf{v})\mathbf{v} \qquad \textit{for all } (\mathbf{u}, \mathbf{v}) \in U(\mathbf{u}_o) \times U(\mathbf{v}_o), \tag{3.19}$$

and clearly $M(\mathbf{u}, \mathbf{v})$ is a C^∞ function with values in the nth linear group, in the set $U(\mathbf{u}_o) \times U(\mathbf{v}_o)$. We may form

$$\mathcal{J}(y, y'; y_o, y'_o) = M(\overset{-1}{\phi}(y) - \overset{-1}{\phi}(y'), y - y'),$$

and it is clear that there is an open neighborhood $\mathcal{W}(y_o, y'_o)$ of (y_o, y'_o) in which the matrix $\mathcal{J}(y, y'; y_o, y'_o)$ is C^∞ and everywhere invertible. □

We return to the proof of Theorem 3.3. We may assume the covering $\{\mathcal{W}_j\}_{(j\in J)}$ in Lemma 3.1 to be locally finite, and we may form a C^∞ partition of unity $h_j(y, y')$ in $\Omega' \times \Omega'$ subordinate to it. Let then $a(x, x', \xi)$ be any amplitude of A and v be any test function in Ω', regarded as a distribution (with compact support) in Ω'. We have

$$\begin{aligned} A^\phi v(y) &= \phi_*(A\{(v \circ \phi)|\det \mathcal{J}|^{-1}\}) = \\ &(2\pi)^{-n}\phi_* \iint e^{i(x-x')\cdot\xi}a(x, x', \xi)v(\phi(x'))|\det \mathcal{J}(x')|\, dx'\, d\xi = \\ &(2\pi)^{-n}|\det \mathcal{J}(y)| \iint \exp\{i(\overset{-1}{\phi}(y) - \overset{-1}{\phi}(y'))\xi\}a(\overset{-1}{\phi}(y), \overset{-1}{\phi}(y'), \xi)v(y')\, dy'\, d\xi. \end{aligned}$$

Let us set

$$a_j^{\#}(y, y', \xi) = h_j(y, y')a(\overset{-1}{\phi}(y), \overset{-1}{\phi}(y'), \xi).$$

By Lemma 3.1 we may write, for (y, y') in the support of $a_j^{\#}$,

$$\overset{-1}{\phi}(y) - \overset{-1}{\phi}(y') = \mathscr{J}_j(y, y')(y - y'),$$

whence

$$(2\pi)^n |\det \mathscr{J}(y)|^{-1} A^{\phi}v(y) =$$

$$\sum_j \iint \exp\{i(y - y') \cdot {}^t\mathscr{J}_j(y, y')\xi\} a_j^{\#}(y, y', \xi)v(y')\, dy'\, d\xi =$$

$$\sum_j \iint e^{i(y-y')\cdot\eta} a_j^{\#}\big(y, y', {}^t\mathscr{J}_j^{-1}(y, y')\eta\big) |\det \mathscr{J}_j(y, y')|^{-1} v(y')\, dy'\, d\eta.$$

Let us therefore set

(3.20) $\quad a^{\phi}(y, y', \eta) =$

$$|\det \mathscr{J}(y)| \sum_j h_j(y, y') |\det \mathscr{J}_j(y, y')|^{-1} a\big(\overset{-1}{\phi}(y), \overset{-1}{\phi}(y'), {}^t\mathscr{J}_j^{-1}(y, y')\eta\big).$$

We have proved that

$$A^{\phi}v(y) = (2\pi)^{-n} \iint e^{i(y-y')\cdot\eta} a^{\phi}(y, y', \eta)v(y')\, dy'\, d\eta. \tag{3.21}$$

It only remains to show that $a^{\phi} \in S^m(\Omega', \Omega')$, which is a consequence of (2.5) and an easy exercise in differentiation. The key fact is that although differentiation with respect to y and y' brings out, as factors, powers of η, via differentiation of $a(\overset{-1}{\phi}(y), \overset{-1}{\phi}(y'), \xi)$ with respect to $\xi = {}^t\mathscr{J}_j^{-1}(y, y')\eta$, each such differentiation also lowers the degree with respect to ξ, by virtue of (2.5), and therefore compensates the effect of the factor η_i, $i = 1, \ldots, n$, which appears because of it. □

REMARK 3.4. It is important to recall that one of the elements (denoted $\mathscr{W}_o$ in the proof of Lemma 3.1) of the covering $\{\mathscr{W}_j\}$ can be taken to be an open neighborhood of the diagonal in $\Omega' \times \Omega'$ and the corresponding matrix to be

$$\mathscr{J}_o(y, y') = \int_0^1 \mathscr{J}(ty + (1 - t)y')\, dt, \tag{3.22}$$

where $\mathscr{J}(y)$ is the Jacobian matrix of $\overset{-1}{\phi}(y)$.

Theorem 3.3 enables us to prove the so-called invariance of the Sobolev spaces:

THEOREM 3.4. *Let ϕ be as in Theorem 3.3. For all real s the "direct image" map ϕ_* induces an isomorphism of $H_c^s(\Omega)$ onto $H_c^s(\Omega')$ (resp. of $H_{\text{loc}}^s(\Omega)$ onto $H_{\text{loc}}^s(\Omega')$).*

PROOF. It suffices to prove the statement when the subscript is c and use duality when it is loc. If $u \in H_c^s(\Omega)$ and $A \in \Psi^{-s}(\Omega)$, we have (Theorem 2.1) $Au \in L^2_{\text{loc}}(\Omega)$. Conversely suppose that $u \in \mathscr{E}'(\Omega)$ and that $(1-\Delta)^{-s/2}u \in L^2_{\text{loc}}(\Omega)$ (see (2.13)); then $u \in H^s$. In other words, $u \in H_c^s(\Omega)$ if and only if $Au \in L^2_{\text{loc}}(\Omega)$ for all $A \in \Psi^{-s}(\Omega)$. But $\phi_*(Au) = A^\phi(\phi_* u)$; and $A^\phi \in \Psi^{-s}(\Omega')$. Therefore it suffices to prove the statement when $s = 0$, in which case it is well known. (Follows at once from the formula of changes of variable in the Lebesgue integral.) □

4. The Symbolic Calculus of Pseudodifferential Operators

For the purposes of application it is advisable, at this stage, to move back, closer to our original models: the parametrices of elliptic equations and the differential operators, where the amplitudes are independent of the second variable y. The introduction of amplitudes depending on y has led to very straightforward and relatively simple proofs of the basic properties of pseudodifferential operators. But from the viewpoint of computation, the handling of such amplitudes has two important drawbacks. One has already been pointed out: to a given pseudodifferential operator there correspond several different amplitudes. This will become clearer in a moment. The second disadvantage is that the computation rules on pseudodifferential operators are not truly of the symbolic calculus type, where, for instance, one expects the composition of operators to translate into the multiplication of their symbols. The "true" symbols that we shall introduce and study in this section come closer to this goal than do the amplitudes. It suffices to recall formula (3.11) to see that amplitudes of the compose of two operators are obtained by an integration with respect to one of the "base" variables and thus partake more of the composition (sometimes called Volterra product) of the associated distribution kernels. As we are going to see, the multiplication of the true symbols will be more—but not quite—like a usual multiplication.

DEFINITION 4.1. *Let m be any real number. We shall denote by* $S^m(\Omega)$ *the subspace of* $S^m(\Omega, \Omega)$ *consisting of the amplitudes* $a(x, \xi)$ *independent of* y.

We shall denote by $S^{-\infty}(\Omega)$ *the intersection of the spaces* $S^m(\Omega)$ *as* m *ranges over the real line.*

Needless to say, any element $a(x, \xi)$ of $S^m(\Omega)$ will be regarded as a C^∞ function in $\Omega \times \mathbb{R}_n$ rather than in $\Omega \times \Omega \times \mathbb{R}_n$. Such a function satisfies inequalities of the kind (2.5). The topology of $S^m(\Omega)$ is that induced by $S^m(\Omega, \Omega)$ of which $S^m(\Omega)$ is a closed linear subspace; hence $S^m(\Omega)$ is a Fréchet space.

DEFINITION 4.2. *We denote by* $\dot{S}^m(\Omega)$ *the quotient vector space* $S^m(\Omega)/S^{-\infty}(\Omega)$. *The elements of* $\dot{S}^m(\Omega)$ *will be called the symbols of degree* $\leq m$ *in* Ω.

We denote by $\dot{S}(\Omega)$ *the union of the* $\dot{S}^m(\Omega)$.

As is customary when dealing with quotient sets, we shall often represent an equivalence class by one of its representatives. But we shall also represent it by certain formal infinite series, which we now introduce and which are related to the infinite series representing the parametrix K in Section 1 (see (1.28) and (1.32)).

Indeed it is often convenient to deal with formal symbols rather than true ones. By a *formal symbol* we mean a sequence of symbols $a_j \in S^{m_j}(\Omega)$ whose orders m_j are strictly decreasing and converging to $-\infty$. It is standard to represent it by the formal series

$$\sum_{j=0}^{+\infty} a_j(x, \xi). \tag{4.1}$$

Now, from such a formal symbol one can build true symbols, elements of $S^{m_0}(\Omega)$ in the present case, which all belong to the same class modulo $S^{-\infty}(\Omega)$. We shall represent this class by the notation (4.1). One may proceed as follows:

First, one may state that a symbol $a(x, \xi)$ belongs to the class (4.1) if, given any large positive number M, there is an integer $J \geq 0$ such that

$$a(x, \xi) - \sum_{j=0}^{J} a_j(x, \xi) \in S^{-M}(\Omega).$$

Second, one can construct such a symbol $a(x, \xi)$ as the sum of a series

$$\sum_{j=0}^{+\infty} \chi_j(\xi) a_j(x, \xi), \tag{4.2}$$

where $\chi_j(\xi) = \chi(\rho_j^{-1}|\xi|)$ as in (1.31). It is possible to select the numbers $\rho_j > 0$ (in general, $\nearrow +\infty$) in such a way that the series (4.2) converges in the Fréchet space $S^{m_0}(\Omega)$. We leave this as an exercise to the reader (cf. the argument at the end of Section 1). Of course, one may choose to use different kinds of cutoff functions. The class of (4.2) mod $S^{-\infty}(\Omega)$ will at any rate be independent of this choice.

Since we are forced to use cutoff functions such as the χ_j, we may even dispense with the requirement that $a_j(x, \xi)$ be "true" elements of $S^{m_j}(\Omega)$. This enables us to deal with terms $a_j(x, \xi)$ that are not C^∞ functions of ξ near the origin or are not even defined in neighborhoods of $\xi = 0$, neighborhoods that may depend on the variable x. Thus, for instance, we may deal with "symbols" $a_j(x, \xi)$ defined for $x \in \mathcal{K}, \xi \in \mathbb{R}_n$, with $|\xi| > \rho_{\mathcal{K}}$, a positive number depending on $\mathcal{K}$; the latter ranges over the family of all compact subsets of Ω. In dealing with such $a_j(x, \xi)$ we ought to use cutoff functions $\chi_j(x, \xi)$ that also depend on x.

EXAMPLE 4.1. Perhaps the most important examples of formal symbols (4.1) with terms $a_j(x, \xi)$ which are not C^∞ functions of ξ at the origin are the *classical symbols*: this will always mean for us that each term $a_j(x, \xi)$ is a *positive-homogeneous*† function of degree m_j of ξ and also that the degrees m_j differ from one another by an integer: $m_j - m_{j+1} \in \mathbb{Z}_+$. Observe that this is the general form of the symbols both of the differential operators and of the parametrices of elliptic equations, as shown in (1.28)–(1.30).

The motivation for the preceding definitions lies in the following result, which provides the foundation for the symbolic calculus of pseudodifferential operators:

THEOREM 4.1. *The mapping* $a(x, \xi) \mapsto \mathrm{Op}\, a$ *from* $S^m(\Omega)$ *into* $\Psi^m(\Omega)$, *defined by*

$$(\mathrm{Op}\, a)u(x) = (2\pi)^{-n} \int e^{ix\cdot\xi} a(x, \xi)\hat{u}(\xi)\, d\xi, \tag{4.3}$$

induces a linear bijection of $\dot{S}^m(\Omega)$ *onto* $\Psi^m(\Omega)/\Psi^{-\infty}(\Omega)$.

PROOF. Let A be an arbitrary element of $\Psi^m(\Omega)$ and $a(x, y, \xi)$ any one of its amplitudes. We can find a neighborhood $\mathcal{U}$ of the diagonal in $\Omega \times \Omega$

† A function $f(\xi)$ is said to be positive-homogeneous of degree d with respect to ξ in $\mathbb{R}_n\backslash\{0\}$ if $f(\rho\xi) = \rho^d f(\xi)$ for every $\rho > 0$ (but not necessarily for every real ρ). Thus the Heaviside function on $\mathbb{R}_1\backslash\{0\}$ is positive-homogeneous, but not homogeneous, of degree zero.

with the following property: *Given any $x \in \Omega$, the set of y's in Ω such that $(x, y) \in \mathcal{U}$ is an open ball centered at x.* Let then $\phi(x, y) \in C^\infty(\Omega \times \Omega)$ have its support in $\mathcal{U}$ and be equal to 1 in a subneighborhood of the diagonal. Let $A^\#$ denote a pseudodifferential operator with amplitude $\phi(x, y)a(x, y, \xi)$. We have $A - A^\# \in \Psi^{-\infty}(\Omega)$. We may therefore assume that $A = A^\#$. We shall therefore drop any mention of the cutoff $\phi(x, y)$ and simply assume that the (x, y)-projection of the support of $a(x, y, \xi)$ lies in the neighborhood $\mathcal{U}$ of the diagonal. This enables us to write the finite Taylor expansion of $a(x, y, \xi)$ with respect to y in the neighborhood of x in the following manner:

$$a(x, y, \xi) = \sum_{|\alpha| \leq N} \frac{1}{\alpha!}(y - x)^\alpha \, \partial_y^\alpha a(x, x, \xi) + R_N(x, y, \xi), \tag{4.4}$$

where $\partial_y = \sqrt{-1}\, D_y$ and where

$$\begin{aligned} R_N(x, y, \xi) = {} & \\ & \frac{1}{N!} \int_0^1 (1 - t)^N (\partial/\partial t)^{N+1} a(x, x + t(y - x), \xi)\, dt = \\ & (N + 1) \sum_{|\alpha| = N+1} \frac{1}{\alpha!}(y - x)^\alpha \int_0^1 (1 - t)^N (\partial_y^\alpha a) \\ & \qquad (x, x + t(y - x), \xi)\, dt. \end{aligned} \tag{4.5}$$

We shall now consider an amplitude of the kind

$$(y - x)^\alpha b(x, y, \xi), \tag{4.6}$$

where $b(x, y, \xi) \in S^m(\Omega, \Omega)$. Let $B_\alpha = \mathrm{Op}(4.6)$; note that we have

$$\begin{aligned} B_\alpha u(x) &= (2\pi)^{-n} \iint (-D_\xi)^\alpha \{e^{i(x-y)\cdot\xi}\} b(x, y, \xi) u(y)\, dy\, d\xi \\ &= (2\pi)^{-n} \iint e^{i(x-y)\cdot\xi} D_\xi^\alpha b(x, y, \xi) u(y)\, dy\, d\xi. \end{aligned}$$

We reach the conclusion that another amplitude of the operator B_α is

$$D_\xi^\alpha b(x, y, \xi). \tag{4.7}$$

This provides an example of an operator with at least two different amplitudes. It also shows that the pseudodifferential operator B_α is of order $\leq m - |\alpha|$.† We apply these considerations to the terms in the finite Taylor expansion (4.4) and in particular to the remainder R_N. We reach the

† This is not too surprising, at least *a posteriori*; on inspection of (4.6), we see that the operator B with amplitude $b(x, y, \xi)$ is of order $\leq m$. Its order is closely related to the singularities of the associated kernel distribution $B(x, y)$, which are concentrated along the diagonal of $\Omega \times \Omega$. Multiplication by $(y - x)^\alpha$ necessarily improves the regularity of $B(x, y)$.

following conclusion:

(4.8) *If $a(x, y, \xi) \in S^m(\Omega, \Omega)$ is an amplitude of A, another amplitude of A is given by*

$$\sum_{|\alpha| \leq N} \frac{1}{\alpha!} D_\xi^\alpha \partial_y^\alpha a(x, x, \xi) + R_N^\#(x, y, \xi), \tag{4.9}$$

where

$$R_N^\#(x, y, \xi) = (N+1) \sum_{|\alpha|=N+1} \frac{1}{\alpha!} \int_0^1 (1-t)^N D_\xi^\alpha \partial_y^\alpha a(x, x + t(y-x), \xi)\, dt. \tag{4.10}$$

Let us denote by $A_\alpha^\#$ (resp. $R_N^\#$) a pseudodifferential operator with amplitude $D_\xi^\alpha \partial_y^\alpha a(x, x, \xi)$ (resp. $R_N(x, y, \xi)$). We have just shown that

$$A \equiv \sum_{|\alpha| \leq N} \frac{1}{\alpha!} A_\alpha^\# + R_N^\# \mod \Psi^{-\infty}(\Omega). \tag{4.11}$$

It follows at once from (2.5) that

$$D_\xi^\alpha \partial_y^\alpha a(x, x, \xi) \in S^{m-|\alpha|}(\Omega), \tag{4.12}$$

$$R_N^\#(x, y, \xi) \in S^{m-(N+1)}(\Omega, \Omega). \tag{4.13}$$

Consequently,

$$A_\alpha^\# \in \Psi^{m-|\alpha|}(\Omega), \qquad R_N^\# \in \Psi^{m-(N+1)}(\Omega). \tag{4.14}$$

Another consequence of (4.12) and (4.13) is that the formal series

$$\sum_\alpha \frac{1}{\alpha!} D_\xi^\alpha \partial_y^\alpha a(x, x, \xi) \tag{4.15}$$

is exactly of the kind (4.1). This series defines a symbol $a(x, \xi) \in \dot{S}^m(\Omega)$ which, in turn, defines a class modulo $\Psi^{-\infty}(\Omega)$ of pseudodifferential operators in Ω. Formula (4.11) implies at once that this is the class of A and proves the *surjectivity* of the map under consideration.

To prove the *injectivity* of the map $\dot{S}^m(\Omega) \to \Psi^m(\Omega)/\Psi^{-\infty}(\Omega)$, we use an asymptotic formula describing the action of a pseudodifferential operator on rapidly oscillating test functions $u(x)\exp(ix \cdot \xi)$ with $u \in C_c^\infty(\Omega)$. This formula will be proved in Section 3 of Chapter VI (Theorem 3.1). We refer the reader to that section: the statements and proofs there do not require any knowledge that we do not already have. Actually we need a version "with parameters" of Theorem 3.1 of Chapter VI, a version obvious to state and easy to prove by the same arguments used in the proof of the version without parameters (allowing for differentiation with respect to the parameters).

We must prove that if $a(x, \xi) \in S^m(\Omega)$ is such that Op a is regularizing, then perforce $a \in S^{-\infty}(\Omega)$. The kernel $A(x, y)$ of Op a is a C^∞ function in $\Omega \times \Omega$; therefore the integral

$$\int A(x, y) u(y)\, e^{iy \cdot \xi}\, dy$$

defines an element of $S^{-\infty}(\Omega)$, as the reader can easily ascertain. On the other hand, by Theorem 3.1 of Chapter VI (or rather by its version with parameters) that integral is asymptotically equivalent, for $\rho = |\xi| \sim +\infty$, to

$$e^{ix \cdot \xi} \sum_\alpha \frac{1}{\alpha!} a^{(\alpha)}(x, \xi) \mathfrak{N}_\alpha(\xi, D_x) u.$$

We have denoted by $\mathfrak{N}_\alpha(\xi, D_x)$ what in the notation of Section 3, Chapter VI, would have been called $\mathcal{N}_\alpha(\phi; |\xi|, D_x)$ with $\phi(x) = x \cdot \xi / |\xi|$. Actually if we apply formulas (3.10), (3.11), and (3.14) of Chapter VI with this choice of ϕ, we find that $\mathfrak{N}_\alpha(\xi, D_x) = D_x^\alpha$. If we then take $u = 1$ in an open subset $U \subset\subset \Omega$ the preceding asymptotic series reduces, in U, to $a(x, \xi)\, e^{ix \cdot \xi}$, which is thus seen to belong to $S^{-\infty}(U)$. This implies at once what we wanted. □

REMARK 4.1. The expression (4.15) and the reasoning that completes the proof of surjectivity are important, insofar as they provide us with a systematic method to construct the symbol from an amplitude.

REMARK 4.2. Let us return to the formal symbol (4.1) and suppose there is a pseudodifferential operator A in Ω with the following property. Given any integer $N > 0$ there is another integer $J > 0$ such that

$$A - \sum_{j \leq J} \operatorname{Op} a_j \in \Psi^{-N}(\Omega). \tag{4.16}$$

Then the operator A belongs to the class in $\Psi^m(\Omega)/\Psi^{-\infty}(\Omega)$ defined by any one of the true symbols, such as (4.2), constructed from (4.1).

EXAMPLE 4.2. Let the pseudodifferential operator A in Ω have the property in Remark 4.2 but assume moreover that the formal symbol (4.1) is *classical* (Example 4.1). Then we shall say that A is a *classical pseudodifferential operator in* Ω.

From now on we shall use the notation $\dot{\Psi}^m(\Omega)$ to denote the quotient space $\Psi^m(\Omega)/\Psi^{-\infty}(\Omega)$. We often refer to the elements of $\dot{\Psi}^m(\Omega)$ as classes $(\text{mod } \Psi^{-\infty}(\Omega))$ of pseudodifferential operators in Ω, of order m. The union of the $\dot{\Psi}^m(\Omega)$ will be denoted by $\dot{\Psi}(\Omega)$.

The notions of transpose, adjoint, and compose can be transferred in a natural manner to the classes mod $\Psi^{-\infty}(\Omega)$ of pseudodifferential operators in Ω; for if A and B are congruent mod $\Psi^{-\infty}(\Omega)$, then the same is true of their transposes ${}^tA$ and ${}^tB$ and also of their adjoints A^* and B^*. The compose $\dot{A} \circ \dot{B}$ of $\dot{A} \in \dot{\Psi}^m(\Omega)$ and $\dot{B} \in \dot{\Psi}^{m'}(\Omega)$ can be defined as the class mod $\Psi^{-\infty}(\Omega)$ of two properly supported representatives of $\dot{A}$ and $\dot{B}$ (such representatives exist, by Proposition 3.4). It is clear that this definition does not depend on the choice of those representatives. One can define the action of a class mod $\Psi^{-\infty}(\Omega)$ of pseudodifferential operators $\dot{A}$ on a class mod $C^\infty(\Omega)$ of distributions in Ω, $\dot{u}$. In other words, the elements of $\dot{\Psi}(\Omega)$ define endomorphisms on the vector space $\mathscr{D}'(\Omega)/C^\infty(\Omega)$ (not topologized). It suffices to select any properly supported representative of $\dot{A}$, A, any representative u of $\dot{u}$ and take as value $\dot{A}\dot{u}$ the class mod $C^\infty(\Omega)$ of Au.

If the operator Op a in (4.3) belongs to the class $\dot{A} \in \Psi^m(\Omega)$, we say that the class of $a(x, \xi)$ mod $S^{-\infty}(\Omega)$ is the *symbol* of $\dot{A}$, or, for that matter, of any representative A of $\dot{A}$; we also say that $a(x, \xi)$ itself is the symbol of $\dot{A}$ and of A.

This said, we may state:

THEOREM 4.2. *Let $\dot{A} \in \dot{\Psi}^m(\Omega)$ with symbol $a(x, \xi) \in \dot{S}^m(\Omega)$. Then the symbol of the transpose ${}^t\dot{A}$ of $\dot{A}$ is defined by the formal symbol*

$$\sum_\alpha (-1)^{|\alpha|} \frac{1}{\alpha!} (\partial_\xi^\alpha D_x^\alpha a)(x, -\xi), \tag{4.17}$$

and the symbol of the adjoint A^*, by

$$\sum_\alpha \frac{1}{\alpha!} \partial_\xi^\alpha D_x^\alpha \overline{a(x, \xi)}. \tag{4.18}$$

PROOF. We replace $\dot{A}$ by a representative A, $a(x, \xi)$ by one of its representatives, still denoted by $a(x, \xi)$, such that (2.3) holds. We may then regard $a(x, \xi)$ as an amplitude of A, which happens not to depend on y. Then $a(y, -\xi)$ and $\overline{a(y, \xi)}$ are amplitudes of ${}^tA$ and A^* respectively. In order to get (4.17) and (4.18) it suffices to form the expression (4.15) where we replace $a(x, y, \xi)$ by these two symbols, respectively. □

The following consequence of Theorem 4.2 is useful in many applications. It can be roughly stated by saying that if the "principal symbol" of a pseudodifferential operator A (or of a class of such, $\dot{A}$) is real, then A is "essentially" self-adjoint. More precisely,

COROLLARY 4.1. *Let $\dot{A}$ be as in Theorem 4.2, but assume that there is a real symbol $a_o(x, \xi) \in \dot{S}^m(\Omega)$ such that $a(x, \xi) - a_o(x, \xi) \in \dot{S}^{m-1}(\Omega)$. Then $\dot{A} - \dot{A}^* \in \dot{\Psi}^{m-1}(\Omega)$.*

PROOF. We write $a(x, \xi) = a_o(x, \xi) + r(x, \xi)$, $r(x, \xi) \in \dot{S}^{m-1}(\Omega)$. Then, by (4.18), the symbol of A^* is equal to

$$a_o(x, \xi) + \overline{r(x, \xi)} + \sum_{|\alpha|>0} \frac{1}{\alpha!} \partial_\xi^\alpha D_x^\alpha \overline{a(x, \xi)}, \tag{4.19}$$

and therefore that of $\dot{A} - \dot{A}^*$ belongs to $\dot{S}^{m-1}(\Omega)$. □

It is convenient to introduce the following differential operator of infinite order:

$$\exp\{\partial_\xi D_x\} = \sum_\alpha \frac{1}{\alpha!} \partial_\xi^\alpha D_x^\alpha, \tag{4.20}$$

where we have set

$$\partial_\xi D_x = \partial_{\xi_1} D_{x^1} + \cdots + \partial_{\xi_n} D_{x^n}.$$

We are going to use this operator in the proof of the next result. Let us note that the expressions (4.15), (4.17), (4.18) can be conveniently rewritten by means of (4.20). Let A be a representative of $\dot{A} \in \dot{\Psi}^m(\Omega)$, $a(x, y, \xi)$ one of its amplitudes, and let $a(x, \xi)$ denote the symbol of $\dot{A}$. We have

$$a(x, \xi) = \exp\{\partial_\xi D_y\} a(x, y, \xi)|_{y=x}, \tag{4.21}$$

$$\text{symbol of } {}^t\dot{A} = \exp\{\partial_\xi D_x\} a(x, -\xi), \tag{4.22}$$

$$\text{symbol of } \dot{A}^* = \exp\{\partial_\xi D_x\} \overline{a(x, \xi)}. \tag{4.23}$$

In view of the basic estimates (2.5) defining amplitudes and of the considerations concerning formal symbols, which follow Definition 4.2, the right-hand sides of (4.21)–(4.23) are indeed such formal symbols and define elements of $\dot{S}^m(\Omega)$ in a straightforward manner.

The next result concerns the compose of two (classes of) pseudodifferential operators.

THEOREM 4.3. *Let $\dot{A} \in \dot{\Psi}^m(\Omega)$, $\dot{B} \in \dot{\Psi}^{m'}(\Omega)$ with respective symbols $a(x, \xi)$, $b(x, \xi)$. Then the symbol of $\dot{A} \circ \dot{B}$ is defined by the formal symbol*

$$(a \odot b)(x, \xi) = \sum_\alpha \frac{1}{\alpha!} \partial_\xi^\alpha a(x, \xi) D_x^\alpha b(x, \xi). \tag{4.24}$$

PROOF. We consider $a(x, \xi)$ to be the amplitude of a representative A of $\dot{A}$ given by (2.3). We introduce a cutoff function $\phi(x, y) \in C^\infty(\Omega \times \Omega)$

properly supported (see † footnote on p. 25), equal to 1 in a neighborhood of the diagonal. We regard $b(x, \xi)\phi(x, y)$ as the amplitude of a representative B of $\dot{B}$, and since B is properly supported, we may form the compose $A \circ B$, of which an amplitude is given in (3.11). In this particular case, the formula simplifies somewhat and reads

$$k(x, y, \xi) = a(x, \xi)(\tau b)(y, \xi), \tag{4.25}$$

where

$$(\tau b)(y, \xi) = (2\pi)^{-n} \iint \exp\{i(y - z)\cdot(\xi - \eta)\} b(z, \eta)\phi(z, y)\, dz\, d\eta. \tag{4.26}$$

By taking the Taylor expansion of $b(z, \eta)\phi(z, y)$ with respect to (z, η) about (y, ξ), we easily obtain that

$$(\tau b)(y, \xi) \equiv \exp\{-\partial_\xi D_z\} b(z, \xi)\phi(z, y)|_{z=y} \tag{4.27}$$

where $\equiv$ is the congruence mod $S^{-\infty}(\Omega)$, i.e., in $\dot{S}^{m'}(\Omega)$. We shall apply the following formula:

$$\exp\{\partial_w D_z\}(f(z, w)g(z)) = \sum_\alpha \frac{1}{\alpha!} D_z^\alpha g(z)\, \partial_w^\alpha(\exp(\partial_w D_z) f(z, w)). \tag{4.28}$$

The proof is an easy exercise, left to the reader. We apply it, with $w = \xi$, substituting $-z$ for z and taking $f(z, \xi) = b(-z, \xi)$, $g(z) = \phi(-z, y)$. From (4.27) we derive

$$(\tau b)(y, \xi) \equiv \exp\{-\partial_\xi D_y\} b(y, \xi), \tag{4.29}$$

since all the derivatives of $\phi(z, y)$ vanish identically in a neighborhood of the diagonal $z = y$. We may rewrite (4.25) in the form

$$k(x, y, \xi) = a(x, \xi) \exp\{-\partial_\xi D_y\} b(y, \xi), \tag{4.30}$$

and we know that the symbol of $\dot{A} \circ \dot{B}$ is

$$\exp\{\partial_\xi D_y\} k(x, y, \xi)|_{y=x}. \tag{4.31}$$

It suffices therefore to apply once more (4.28) now with $z = \xi$, $w = y$, $f(\xi, y) = \exp\{-\partial_\xi D_y\} b(y, \xi)$, $g(\xi) = a(x, \xi)$. We obtain at once (4.24). □

Theorem 4.3 has a well known and important consequence, namely that commutation of pseudodifferential operators lowers the order by one unit.

COROLLARY 4.2. *The commutator* $[\dot{A}, \dot{B}] = \dot{A}\dot{B} - \dot{B}\dot{A}$ *belongs to* $\dot{\Psi}^{m+m'-1}(\Omega)$ *and the symbol of* $[\dot{A}, \dot{B}]$ *is congruent to*

$$\frac{1}{\sqrt{-1}}\{a(x, \xi), b(x, \xi)\} \tag{4.32}$$

mod $\dot{S}^{m+m'-2}(\Omega)$, *where* $\{a, b\}$ *denotes the Poisson bracket of a and b, i.e.,*

$$\{a, b\} = \sum_{j=1}^{n} \frac{\partial a}{\partial \xi_j}\frac{\partial b}{\partial x^j} - \frac{\partial a}{\partial x^j}\frac{\partial b}{\partial \xi_j}. \tag{4.33}$$

REMARK 4.3. Formulas (4.17), (4.18), (4.24) make sense (i.e., define bona fide formal symbols) when $a(x, \xi)$ and $b(x, \xi)$ are symbols. Actually they also make sense when they themselves are formal symbols, of the kind (4.1). For instance, the terms in

$$\sum_{j=0}^{+\infty} \sum_{\alpha \in Z_+^n} \frac{1}{\alpha!} \partial_\xi^\alpha D_x^\alpha \overline{a_j(x, \xi)}$$

can be rearranged into a formal series of symbols whose degree $\searrow -\infty$, in such a way that each term depends only on finitely many a_j.

Last, we look at the symbol of the transform of a pseudodifferential operator under a diffeomorphism $\phi: \Omega \to \Omega'$. By Theorem 3.3, $A \to A^\phi$ is a bijection of $\Psi^m(\Omega)$ onto $\Psi^m(\Omega')$ and hence induces an isomorphism of $\Psi^{-\infty}(\Omega)$ onto $\Psi^{-\infty}(\Omega')$ and therefore also of the quotient space $\dot{\Psi}^m(\Omega)$ onto the quotient space $\dot{\Psi}^m(\Omega')$. This defines the transfer $\dot{A}^\phi$ of $\dot{A}$ via ϕ. Incidentally, if A is properly supported, the same is true of A^ϕ.

THEOREM 4.4. *If* $\dot{A} \in \dot{\Psi}^m(\Omega)$ *has the symbol* $a(x, \xi)$, *the symbol of* $\dot{A}^\phi$ *is congruent to*

$$a\big(\overset{-1}{\phi}(y), {}^t\mathcal{J}^{-1}(y)\eta\big) \tag{4.34}$$

mod $\dot{S}^{m-1}(\Omega')$, *where* $\mathcal{J}$ *denotes the Jacobian matrix of the diffeomorphism* $\overset{-1}{\phi}: \Omega' \to \Omega$.

PROOF. We exploit formulas (3.20) and (3.21). Observe that we may assume that an amplitude of a representative A of $\dot{A}$ used in (3.20) is

$$a^{\#}(x, x', \xi) = g\big(\phi(x), \phi(x')\big)a(x, \xi),$$

where $g(y, y') \in C^\infty(\Omega' \times \Omega')$ is equal to one in a neighborhood of the diagonal and is such that $h_o g = g$, $h_j g = 0$ for $j > 0$ ($\{h_j\}$ is the partition of unity used in (3.20)). Then (3.20) reduces to

$$(4.35)\qquad a^{\phi}(y, y', \eta) = \left|\frac{\det \mathcal{J}(y)}{\det \mathcal{J}_o(y, y')}\right| g(y, y')a(\overset{-1}{\phi}(y), {}^t\mathcal{J}_o^{-1}(y, y')\eta),$$

where the matrix $\mathcal{J}_o$ is that defined in (3.22). Since $\mathcal{J}_o(y, y) = \mathcal{J}(y)$, we obtain

$$(4.36)\qquad a^{\phi}(y, y, \eta) = a(\overset{-1}{\phi}(y), {}^t\mathcal{J}^{-1}(y)\eta).$$

It suffices, then, to write that the symbol of $\dot{A}^{\phi}$ is equal to

$$(4.37)\qquad \exp\{\partial_\eta D_{y'}\}a^{\phi}(y, y', \eta)|_{y'=y}, \qquad \square$$

Appendix: Elliptic Pseudodifferential Operators and Their Parametrices

DEFINITION 4.4. *The adjective "elliptic of order m" will be applied to the following objects:*

(I) *Any "symbol" $a(x, \xi)$ belonging to $S^m(\Omega)$ such that there exist two strictly positive continuous functions ρ, c in Ω with the property that, for every x in Ω,*

$$(4.38)\qquad c(x)|\xi|^m \leq |a(x, \xi)| \qquad \textit{if } \xi \in \mathbb{R}_n,\ |\xi| \geq \rho(x).$$

(II) *Any symbol (class) with a representative $a(x, \xi)$ having the property (I).*

(III) *Any pseudodifferential operator or any class of pseudodifferential operator* mod$\Psi^{-\infty}(\Omega)$ *with symbol as in* (II).

PROPOSITION 4.1. *Let a belong to $\dot{S}^m(\Omega)$. The following properties are equivalent:*

(i) *a is elliptic of degree m.*

(ii) *There is $b \in \dot{S}^{-m}(\Omega)$ such that $a \odot b = 1$ (see (4.24)).*

(iii) *There is $b \in \dot{S}^{-m}(\Omega)$ such that $b \odot a = 1$.*

PROOF. Let $a(x, \xi) \in S^m(\Omega)$ be a representative of a with property (4.38). Determine recursively the "symbols" b_j by means of the relations

$$(4.39)\qquad b_o(x, \xi)a(x, \xi) = 1,$$

and for $j = 1, 2, \ldots,$

$$(4.40)\qquad b_j(x, \xi)a(x, \xi) = -\sum_{1\leq|\alpha|\leq j} \frac{1}{\alpha!}\partial_\xi^\alpha a(x, \xi)D_x^\alpha b_{j-|\alpha|}(x, \xi),$$

respectively,

$$b_j(x,\xi)a(x,\xi) = -\sum_{1\le|\alpha|\le j}\frac{1}{\alpha!}\partial_x^\alpha a(x,\xi)D_\xi^\alpha b_{j-|\alpha|}(x,\xi). \tag{4.41}$$

This makes sense only for $|\xi| \ge \rho(x)$. We select a monotone increasing sequence of continuous functions $\rho_j > \rho$ and select $\chi_j \in C^\infty(\Omega \times \mathbb{R}_n)$, $\chi_j(x,\xi) = 0$ if $|\xi| \le \rho_j(x)$, $\chi_j(x,\xi) = 1$ if $|\xi| \ge 2\rho_j(x)$; variations of this type of choice are allowed. One can easily prove, by induction on j, that for each j, $\chi_j b_j \in S^{-m-j}(\Omega)$ and, furthermore, that if the ρ_j are carefully chosen (cf. (1.34)), then the series $\sum_j \chi_j b_j$ converges in $S^{-m}(\Omega)$. The class b of this symbol is an element of $\dot{S}^{-m}(\Omega)$ satisfying (ii) (resp., (iii)).

Suppose that (ii) holds. This means that there are representatives $a(x,\xi) \in S^m(\Omega)$, $b(x,\xi) \in S^{-m}(\Omega)$ of a and b respectively, such that

$$a(x,\xi)b(x,\xi) - 1 \in S^{-1}(\Omega).$$

In particular this implies that, to every compact set $\mathcal{K} \subset \Omega$, there is a constant $C_{\mathcal{K}} > 0$ such that

$$|a(x,\xi)b(x,\xi) - 1| \le C_{\mathcal{K}}(1+|\xi|)^{-1}, \qquad \forall x \in \mathcal{K},\ \xi \in \mathbb{R}_n.$$

Select $\rho_{\mathcal{K}} > 0$ large enough that $C_{\mathcal{K}}(1+|\xi|)^{-1} < \frac{1}{2}$ for $|\xi| > \rho_{\mathcal{K}}$. Then, for x in $\mathcal{K}$ and these ξ's,

$$\tfrac{1}{2} \le |a(x,\xi)b(x,\xi)| \le C'_{\mathcal{K}}|a(x,\xi)|(1+|\xi|)^{-m},$$

whence (4.38) with $c(x) > (2C'_{\mathcal{K}})^{-1}$ for x in $\mathcal{K}$. □

REMARK 4.4. Since the multiplication $\odot$ is associative, the classes b in (ii) and (iii) are the same.

COROLLARY 4.3. *Let A belong to $\Psi^m(\Omega)$. The following properties are equivalent:*

(i) *A is elliptic of order m.*

(ii) *There is $B \in \Psi^{-m}(\Omega)$ such that*

$$AB \equiv BA \equiv I \qquad \text{mod}\ \Psi^{-\infty}(\Omega). \tag{4.42}$$

The operator B in (ii) can be called a parametrix of A, in analogy with the case of elliptic differential operators (Section 1).

Formulas (4.39) and (4.40) enable us to determine the symbol of such a parametrix (cf. (1.29), (1.30)).

Corollary 4.4. *If $A \in \Psi^m(\Omega)$ is elliptic of order m, any parametrix $B \in \Psi^{-m}(\Omega)$ of A is elliptic of order $-m$.*

Corollary 4.5. *If $A \in \Psi(\Omega)$ is elliptic in Ω, it has the following property:*

(4.43) *Given any distribution $u \in \mathscr{E}'(\Omega)$ and any open subset U of Ω, if $Au \in C^\infty(U)$ then $u \in C^\infty(U)$.*

Property (4.43) is very close to hypoellipticity (cf. Lemma 2.2): the only difference is that the distributions u on which A acts are required to have compact supports.

Proof. Let B be as in Corollary 4.3(ii). Then $u = B(Au) + Ru$, with R regularizing; hence $Ru \in C^\infty(\Omega)$. Since $Au \in C^\infty(U)$ and B is pseudolocal, $BAu \in C^\infty(U)$. □

Remark 4.5. When looking at a properly supported pseudodifferential operator A, we may lift the restriction that the support of u be compact.

Let $\dot{A}$ be a class of pseudodifferential operators in Ω, modulo $\Psi^{-\infty}(\Omega)$. If $\dot{A}$ is elliptic of order m, there is $\dot{B} \in \dot{\Psi}^{-m}(\Omega)$ such that

$$\dot{A}\dot{B} = \dot{B}\dot{A} = I. \tag{4.44}$$

This follows at once from Corollary 4.3. The class $\dot{B}$ is often called the *inverse* of $\dot{A}$. It is natural to denote it by $\dot{A}^{-1}$.

It follows at once from Theorem 4.4. that if A is an elliptic pseudodifferential operator of order m in Ω, and if ϕ is a diffeomorphism of Ω onto the open subset Ω' of $\mathbb{R}^n$, the transfer A^ϕ of A to Ω' is also elliptic of order m.

Now we rapidly show how to construct *fractional powers* of an elliptic pseudodifferential operator in Ω or, more accurately, of an equivalence class modulo $\Psi^{-\infty}(\Omega)$ of such operators. We shall denote by A the operator, by $\dot{A}$ the class, by m their order, by $a(x, \xi)$ a representative of the symbol class of $\dot{A}$. We shall make the following hypothesis about $a(x, \xi)$. (It is clearly a hypothesis about its equivalence class modulo $S^{m-1}(\Omega)$.)

(4.45) *There is an angle $\theta_o \in [0, 2\pi[$ and there are constants R, $c > 0$ such that*

$$|\operatorname{Arg} a(x, \xi) - \theta_o| \geq c, \qquad \forall x \in \Omega,\ \xi \in \mathbb{R}_n,\ |\xi| > R.$$

This hypothesis enables us to use the powers $a(x, \xi)^s$ for nonintegral s. Of course there can be many branches. For the sake of simplicity let us assume that $\theta_o = \pi$, a situation that we can get into by multiplying A by $\exp[i(\pi - \theta_o)]$. We then use only the main branch of z^s, that is, the branch equal to $\exp(s \log x)$ for $z = x > 0$.

Below we denote by p, q two integers greater than zero, coprime.

PROPOSITION 4.2. *There is a unique equivalence class* $\dot{C} \in \Psi^{mp/q}(\Omega)/\Psi^{-\infty}(\Omega)$ *having the following two properties:*

$$\dot{C}^q = \dot{A}^p; \tag{4.46}$$

(4.47) *modulo* $S^{mp/q-1}(\Omega)$ *the symbol of* $\dot{C}$ *is equal to* $a(x, \xi)^{p/q}$.

The class $\dot{C}$ in Proposition 4.2 will be denoted by $\dot{A}^{p/q}$. It is of course elliptic of order mp/q.

PROOF. We make use of the representative A of $\dot{A}$. Like all other pseudodifferential operators in this proof, it is taken to be *properly supported*, so that all compositions are well defined (Definition 3.1). First let C_0 denote such an arbitrary pseudodifferential operator of order mp/q in Ω, whose symbol if equal to $a^{p/q}$ for large $|\xi|$, say $|\xi| > R + 1$. Suppose that we have determined N pseudodifferential operators C_j $(j = 0, \dots, N-1)$, respectively of order $mp/q - j$, such that the order of

$$A^p - (C_0 + \cdots + C_{N-1})^q = R_N \tag{4.48}$$

is $\leq mp - N$ $(N \geq 1)$. We then select C_N so that the order of R_{N+1} does not exceep $mp - N - 1$. It suffices to require that the order of

$$qC_0^{q-1}C_N - R_N$$

not exceed $mp - N - 1$, for instance by taking

$$C_N \equiv \frac{1}{q} C_0^{1-q} R_N \qquad \text{mod}\, \Psi^{-\infty}(\Omega). \tag{4.49}$$

We are availing ourselves of the ellipticity of order mp/q of C_0 and of Corollary 4.3. If $\dot{C}$ is the equivalence class of pseudodifferential operators defined by the formal series $C_0 + \cdots + C_N + \cdots$ (Remark 4.2), it has properties (4.46) and (4.47).

About the uniqueness of the class $\dot{C}$ with this notation let us assume that there are two pseudodifferential operators of order $mp/q - 1$ in Ω, S

and T such that

$$(C_0 + S)^q = (C_0 + T)^q.$$

This is equivalent to $(I + X)^q = (I + Y)^q$, where $X = C_0^{-1}S$, $Y = C_0^{-1}T$, in other words, to

$$q(X - Y) = -\sum_{r=2}^{q} \binom{q}{r} (X^r - Y^r), \tag{4.50}$$

where the right-hand side is equal to zero when $q = 1$. In any case, as the orders of X and Y do not exceed -1, that of

$$X^r - Y^r = (X - Y + Y)^r - Y^r$$

does not exceed the order of $X - Y$ *minus one*. But then (4.50) is possible only if the order of $X - Y$ is equal to $-\infty$, which implies $S - T \in \Psi^{-\infty}(\Omega)$. □

REMARK 4.6. Substituting $\dot{A}^{-1}$ for $\dot{A}$, we can define the *negative* fractional powers of $\dot{A}$.

We leave as an exercise to the reader the proof that whatever the rational numbers s, t,

$$\dot{A}^s \dot{A}^t = \dot{A}^{s+t}. \tag{4.51}$$

5. Pseudodifferential Operators on Manifolds

In many applications it is important to use pseudodifferential operators on C^∞ manifolds. We recall rapidly the definition of such a manifold, $\mathfrak{M}$. It is a locally compact topological space. We are given an open covering $\{\mathcal{O}_j\}_{j\in J}$ of this space and, for each index j, a homeomorphism χ_j of $\mathcal{O}_j$ onto an open subset of a Euclidean space $\mathbb{R}^n$, Ω_j; the number n does not depend on j and it is called the *dimension* of $\mathfrak{M}$. There is a kind of compatibility condition between these various homeomorphisms χ_j, through which the C^∞ structure enters the picture: suppose that $\mathcal{O}_{j,j'} = \mathcal{O}_j \cap \mathcal{O}_{j'}$ is not empty. Then $\chi_j \circ \chi_{j'}^{-1}$ induces a homeomorphism of $\chi_{j'}(\mathcal{O}_{j,j'})$ onto $\chi_j(\mathcal{O}_{j,j'})$. One requires this homeomorphism to be a diffeomorphism, i.e., to be C^∞, to be a bijection and to have nowhere-vanishing Jacobian determinant.

A function f in $\mathfrak{M}$ is C^p if $(f|_{\mathcal{O}_j}) \circ \chi_j^{-1}$ is a C^p function in Ω_j for every j $(0 \le p \le +\infty)$. This defines the spaces $C_c^p(\mathfrak{M})$, $C^p(\mathfrak{M})$. They have natural topologies which are defined by straightforward generalization of the case where $\mathfrak{M}$ is an open subset of a Euclidean space. For instance, convergence in $C^\infty(\mathfrak{M})$ is the uniform convergence of the functions and of each one of their derivatives on every compact subset of $\mathfrak{M}$. *We shall always assume that $\mathfrak{M}$ is countable at infinity*. As a consequence, $C^\infty(\mathfrak{M})$ will be a Fréchet space, and $C_c^\infty(\mathfrak{M})$ an $\mathscr{LF}$-space. Furthermore, $C_c^\infty(\mathfrak{M})$ is dense in $C^\infty(\mathfrak{M})$. And the dual of $C_c^\infty(\mathfrak{M})$ is the space $\mathscr{D}'(\mathfrak{M})$ of distributions in $\mathfrak{M}$ (strictly speaking, of the *currents* of degree n on $\mathfrak{M}$) whereas the dual of $C^\infty(\mathfrak{M})$ can be identified to the space of *compactly supported* distributions in $\mathfrak{M}$.

A map f of $\mathfrak{M}$ onto an open subset Ω of $\mathbb{R}^n$ is a *diffeomorphism* if it is a homeomorphism and if, for every index j, $f \circ \overset{-1}{\chi}_j$ is a diffeomorphism of $\mathcal{O}_j$ onto $f(\mathcal{O}_j)$, which is an open subset of Ω. Of course, there might not exist any such diffeomorphism.

It is clear that an open subset $\mathcal{O}$ of $\mathfrak{M}$ can be regarded as a C^∞ manifold. We use the open covering of $\mathcal{O}$ consisting of the intersections $\mathcal{O}_j \cap \mathcal{O}$ and the restriction of the maps χ_j to $\mathcal{O}_j \cap \mathcal{O}$. A *local chart* in $\mathfrak{M}$ is then any pair $(\mathcal{O}, \chi)$ consisting of an open subset of $\mathfrak{M}$ and of a diffeomorphism χ of $\mathcal{O}$ onto an open subset of $\mathbb{R}^n$. Let $x^1, \ldots, x^n$ be the natural coordinates in $\mathbb{R}^n$. The functions $x^j \circ \chi$ are usually simply written x^j and called the *local coordinates* in the local chart (Ω, χ). One often denotes the latter by $(\Omega, x^1, \ldots, x^n)$; it is also referred to as a *coordinates patch*. The $(\mathcal{O}_j, \chi_j)$ are local charts in $\mathfrak{M}$; but there might be others.

Suppose $\mathfrak{M}'$ is a second C^∞ manifold; set $n' = \dim \mathfrak{M}'$. A map $f\colon \mathfrak{M} \to \mathfrak{M}'$ is a C^∞ mapping if it is continuous and if given any local chart $(\mathcal{O}', \chi')$ in $\mathfrak{M}'$, $\chi' \circ f$ is a C^∞ mapping of $\overset{-1}{f}(\mathcal{O}')$ into $\mathbb{R}^{n'}$. If moreover f is a homeomorphism and if given any local chart $(\mathcal{O}, \chi)$ in $\mathfrak{M}$, every compose $\chi' \circ f \circ \overset{-1}{\chi}$ defines a diffeomorphism of $\chi(\mathcal{O} \cap \overset{-1}{f}(\mathcal{O}'))$ onto $\chi'(f(\mathcal{O}) \cap \mathcal{O}')$, then one says that f is a *diffeomorphism of $\mathfrak{M}$ onto $\mathfrak{M}'$*. Suppose instead that f has the following weaker property: each point of $\mathfrak{M}$ has an open neighborhood which is mapped diffeomorphically onto an open subset of $\mathfrak{M}'$. Then we say that f is a *local diffeomorphism*. Note that f need not be onto (i.e., surjective) and certainly not injective.

Another notion of which we shall make frequent use is that of *submanifold*. For us it will always mean a C^∞ submanifold (and sometimes an analytic one). A subset Σ of $\mathfrak{M}$ is called a C^∞ submanifold of $\mathfrak{M}$ if every point x_o of Σ is contained in a coordinates patch $(\Omega, x^1, \ldots, x^n)$ such that $\Sigma \cap \Omega$ is the subset of Ω exactly defined by the equations $x^{n-d+1} = \cdots = x^n = 0$. The number d is independent of x_o. It is the dimension of Σ. It is clear that Σ can be equipped with a differentiable manifold structure, that defined by the local charts $(\Sigma \cap \Omega, x^1, \ldots, x^d)$ derived from local charts $(\Omega, x^1, \ldots, x^n)$ like the one before. Σ need not be *closed*, for instance, the set $\Sigma =$

$\{(x, y) \in \mathbb{R}^2; x > 0, y = \sin(1/x)\}$ is a C^∞ submanifold of $\mathbb{R}^2$, obviously not closed. For a submanifold Σ of $\mathfrak{M}$ to be closed, it is necessary and sufficient for the natural injection $j: \Sigma \to \mathfrak{M}$ to be *proper*, i.e., that given any compact subset $\mathcal{K}$ of $\mathfrak{M}$, $\overset{-1}{j}(\mathcal{K} \cap j(\Sigma))$ be compact.

We shall also use the terminology *diffeomorphism into* to mean a map, say from a C^∞ manifold $\mathfrak{M}$ to another one, $\mathfrak{M}'$, which is a diffeomorphism of $\mathfrak{M}$ onto a C^∞ submanifold of $\mathfrak{M}'$. If the map is a local diffeomorphism of $\mathfrak{M}$ into $\mathfrak{M}'$, it is called an *immersion*: this means that every point of $\mathfrak{M}$ has an open neighborhood that is mapped diffeomorphically onto a C^∞ submanifold of $\mathfrak{M}'$. Its image certainly need not be closed. It need not even be a submanifold. It is often called an *immersed submanifold*. The simplest example of such a "submanifold" is a curve in the plane $\mathbb{R}^2$, defined by C^∞ parametric equations $x = f(t)$, $y = g(t)$ with $f'^2 + g'^2$ never zero. Such a curve might evidently self-intersect, e.g., a smooth "eight."

We may now generalize the argument of Section 3. Let ϕ be a diffeomorphism of $\mathfrak{M}$ onto $\mathfrak{M}'$, and let A be a continuous linear map $C_c^\infty(\mathfrak{M}) \to \mathcal{D}'(\mathfrak{M})$. We use the diagram

$$\begin{array}{ccc} C_c^\infty(\mathfrak{M}) & \overset{A}{\to} & \mathcal{D}'(\mathfrak{M}) \\ \phi^*\uparrow & & \downarrow\phi_* \\ C_c^\infty(\mathfrak{M}') & \overset{A^\phi}{\to} & \mathcal{D}'(\mathfrak{M}') \end{array} \tag{5.1}$$

where ϕ^* is the pullback of function and ϕ_* is the direct image of distributions associated with ϕ, which is the transpose of the map $f \mapsto f \circ \overset{-1}{\phi}: C_c^\infty(\mathfrak{M}') \to C_c^\infty(\mathfrak{M})$. Then the map A^ϕ is *the transfer* of A via ϕ. Now let $\mathcal{O}$ be an open subset of $\mathfrak{M}$. We call restriction of A to $\mathcal{O}$ and denote by $A_\mathcal{O}$ the compose

$$C_c^\infty(\mathcal{O}) \to C_c^\infty(\mathfrak{M}) \overset{A}{\to} \mathcal{D}'(\mathfrak{M}) \to \mathcal{D}'(\mathcal{O}), \tag{5.2}$$

where the first arrow is the natural extension map (by zero in $\mathfrak{M}\backslash\mathcal{O}$) and the last one is the natural restriction map. Suppose that χ is a diffeomorphism of $\mathcal{O}$ onto an open subset of $\mathbb{R}^n$, i.e., $(\mathcal{O}, \chi)$ is a local chart in $\mathfrak{M}$. Then we may consider the transfer $A_\mathcal{O}^\chi$ of $A_\mathcal{O}$; it is a continuous linear map $C_c^\infty(\chi(\mathcal{O})) \to \mathcal{D}'(\chi(\mathcal{O}))$. In the sequel we shall sometimes refer to it as the transfer of A to $\chi(\mathcal{O})$.

DEFINITION 5.1. *A continuous linear map* $A: C_c^\infty(\mathfrak{M}) \to \mathcal{D}'(\mathfrak{M})$ *is called a pseudodifferential operator (of order m) in* $\mathfrak{M}$ *if, given any local chart*

$(\mathcal{O}, \chi)$ *in* $\mathfrak{M}$, *the transfer* $A_{\mathcal{O}}^{\chi}$ *is a pseudodifferential operator (of order* m*) in* $\chi(\mathcal{O})$.

Definition 5.1 agrees with the earlier one when $\mathfrak{M}$ is an open subset Ω of $\mathbb{R}^n$: it is evident that the restriction $A_{\mathcal{O}}$ of $A \in \Psi^m(\Omega)$ to an open subset $\mathcal{O}$ of Ω belongs to $\Psi^m(\mathcal{O})$; and by Theorem 3.1 we know that $A_{\mathcal{O}}^{\chi} \in \Psi^m(\chi(\mathcal{O}))$ whatever the diffeomorphism χ of $\mathcal{O}$ into $\mathbb{R}^n$.

We shall denote by $\Psi^m(\mathfrak{M})$ the space of pseudodifferential operators of order m in $\mathfrak{M}$, by $\Psi(\mathfrak{M})$ (resp., $\Psi^{-\infty}(\mathfrak{M})$) the union (resp., the intersection) of the spaces $\Psi^m(\mathfrak{M})$. A pseudodifferential operator in $\mathfrak{M}$ is thus a member of $\Psi(\mathfrak{M})$.

The theorems of Sections 2 and 3 all extend to the case where the open subset Ω of $\mathbb{R}^n$ is replaced by a C^∞ manifold $\mathfrak{M}$. We shall limit ourselves to a few remarks.

The spaces $H_{\text{loc}}^s(\mathfrak{M})$ and $H_c^s(\mathfrak{M})$ are well defined whatever the real number s. Indeed we may say that a distribution u in $\mathfrak{M}$ belongs to $H_{\text{loc}}^s(\mathfrak{M})$ if given any local chart $(\mathcal{O}, \chi)$ in $\mathfrak{M}$ the transfer $\chi_*(u|_{\mathcal{O}})$ of the restriction $u|_{\mathcal{O}}$ of u to $\mathcal{O}$ belongs to $H_{\text{loc}}^s(\chi(\mathcal{O}))$. This definition agrees with the standard one when $\mathfrak{M}$ is an open subset of $\mathbb{R}^n$, as one sees by applying Theorem 3.4. Of course, $H_c^s(\mathfrak{M})$ can then be defined as the linear subspace of $H_{\text{loc}}^s(\mathfrak{M})$ consisting of the compactly supported elements. But beware: the topology of $H_c^s(\mathfrak{M})$ is always strictly finer than that induced by $H_{\text{loc}}^s(\mathfrak{M})$ unless $\mathfrak{M}$ is compact, in which case the two spaces are equal! On the other hand, $H_c^s(\mathfrak{M})$ and $H_{\text{loc}}^{-s}(\mathfrak{M})$ can be regarded as the dual of one another only after one has selected a density $d\mu$ on $\mathfrak{M}$. This allows one to define the inner products $\int_{\mathfrak{M}} u\bar{v}\, d\mu$ for locally square-integrable u, v, one of which has compact support.

Then we may state the analogue of Theorem 2.1:

THEOREM 5.1. *Let* A *belong to* $\Psi^m(\mathfrak{M})$ $(m \in \mathbb{R})$. *Given any real* s, $u \mapsto Au$ *is a continuous linear map of* $H_c^s(\mathfrak{M})$ *into* $H_{\text{loc}}^{s-m}(\mathfrak{M})$.

Theorem 2.2. generalizes in a straightforward way:

THEOREM 5.2. *Any pseudodifferential operator in* $\mathfrak{M}$ *is pseudolocal.*

PROOF. Let $\{U_j\}_{j=1,2,\ldots}$ be a (countable) locally finite open covering of $\mathfrak{M}$ such that, for each U_j, there is a diffeomorphism χ_j of U_j onto an open subset of $\mathbb{R}^n$, $\{g_j\}_{j=1,2,\ldots}$ a C^∞ partition of unity subordinate to this covering. Let $u \in \mathscr{E}'(\mathfrak{M})$ be C^∞ in an open neighborhood of a point x_o of $\mathfrak{M}$. We set

$u_j = g_j u$, so that $u = \sum_{j=1}^{\infty} u_j$. It suffices to show that each Au_j is C^∞ near x_o. For each index j we can find an open neighborhood V_j of x_o in which u_j is C^∞ and one, $W_j \subset U_j$, of supp g_j such that there exists a diffeomorphism $\chi_j^{\#}$ of $V_j \cup W_j$ onto an open subset of $\mathbb{R}^n$, equal to χ_j in W_j. Since, given any local chart $(\mathcal{O}, \chi)$ in $\mathfrak{M}$, the transfer $A_{\mathcal{O}}^{\chi}$ of A is a pseudodifferential operator in $\chi(\mathcal{O})$ we conclude that $(\chi_j^{\#})_*(Au_j|_{V_j \cup W_j})$ is C^∞ in $\chi_j^{\#}(V_j)$, which at once implies the desired result. □

Transposes and adjoints of continuous linear operators from, say, $C_c^\infty(\mathfrak{M})$ to $\mathcal{D}'(\mathfrak{M})$ raise a question of definition. There are at least two approaches to this question. One is to establish a duality between complex-valued C^∞ functions on $\mathfrak{M}$ and differential n-forms on $\mathfrak{M}$ with distributions as coefficients. The latter are called *currents* of degree n. We shall not use currents, although it might be preferable to do so in many contexts; on this subject we refer the reader to De Rham [1] or to Chapter XI of Schwartz [1]. Another approach is through the use of a positive density on $\mathfrak{M}$. This can easily be defined in any local chart $(\mathcal{O}, x^1, \dots, x^n)$; in such a local chart we are given a C^∞ function $g_{\mathcal{O}}$, everywhere strictly positive. Then we can define the integral of a smooth function $f \in C_c^\infty(\mathcal{O})$ by the formula

$$\int f(x) g_{\mathcal{O}}(x)\, dx.$$

If we use different coordinates $y^1, \dots, y^n$ in $\mathcal{O}$, $g_{\mathcal{O}}(x)$ must be replaced by

$$\tilde{g}_{\mathcal{O}}(y) = g_{\mathcal{O}}(x(y)) \left| \frac{Dx}{Dy}(y) \right|,$$

where Dx/Dy is the Jacobian determinant of the x^j with respect to the y^k. This clearly defines a positive measure on any local chart, and these measures agree on intersection. Patching them up yields a positive measure on $\mathfrak{M}$, which for simplicity we denote by $g(x)\,dx$. The standard measure theory can be applied to this measure: the notion of measurable function and of the Lebesgue spaces $L^p(\mathfrak{M}, g\,dx)$ $(1 \le p \le +\infty)$ can be defined, in the usual manner; and so also can the spaces $L^p_{\text{loc}}(\mathfrak{M}, g\,dx)$ and $L^p_c(\mathfrak{M}, g\,dx)$. However, the latter spaces, as well as the notion of measurable function, do not depend on the choice of the particular density $g\,dx$. We shall therefore denote them by $L^p_{\text{loc}}(\mathfrak{M})$ and $L^p_c(\mathfrak{M})$ respectively; they are naturally associated to the manifold $\mathfrak{M}$. There is a natural duality between $L^p_{\text{loc}}(\mathfrak{M})$ and $L^q_c(\mathfrak{M})$, with $1/p + 1/q = 1$, and this duality can be extended to $\mathcal{D}'(\mathcal{M}) \times$

$C_c^\infty(\mathfrak{M})$. In physicist's notation it can be expressed by the bracket

$$\langle T, f\rangle = \int T(x)f(x)g(x)\,dx,$$

where $T \in \mathscr{D}'(\mathfrak{M})$, $f \in C_c^\infty(\mathfrak{M})$. But then we see that the transposition of a differential operator P on $\mathfrak{M}$, and likewise of a pseudodifferential operator, must take the density $g\,dx$ into account: Thus if $P(x, D) = \sum_{|\alpha|\le m} c_\alpha(x)D^\alpha$ is the expression of the differential operator P in the local coordinates x^j, that of its transpose, acting on f, is given by

$$g^{-1}\sum_{|\alpha|\le m}(-D)^\alpha(gc_\alpha f).$$

Whether defined in this manner or in any other manner (a third approach is used starting in Chapter VII: regarding distributions and functions as half-densities), transposes and adjoints of pseudodifferential operators on the manifold $\mathfrak{M}$ are also pseudodifferential operators on $\mathfrak{M}$ (Theorems 3.1, 3.2).

Composition of two pseudodifferential operators raises no particular difficulty, provided we assume that at least one of the two is *properly supported*, a notion that is defined for a manifold in the same manner as for an open subset of $\mathbb{R}^n$ (see Definition 3.1). The transfer A^ϕ of a pseudodifferential operator of order m on $\mathfrak{M}$, A, under a diffeomorphism ϕ of $\mathfrak{M}$ onto another C^∞ manifold $\mathfrak{M}'$, is a pseudodifferential operator of the same order on $\mathfrak{M}'$ (Theorem 3.3).

The following result is useful in constructing pseudodifferential operators in a manifold.

PROPOSITION 5.1. *A linear operator* $A: \mathscr{E}'(\mathfrak{M}) \to \mathscr{D}'(\mathfrak{M})$ *is a pseudodifferential operator (of order* m*) in* $\mathfrak{M}$ *if and only if it has both of the following properties:*

(5.3) *There exists a covering of* $\mathfrak{M}$ *by local charts* $(\mathcal{O}_j, \chi_j)$ $(j \in J)$ *such that, for every* j, *the transfer* $A_j = A_{\mathcal{O}_j}^{\chi_j}$ *of* A *to* $\chi_j(\mathcal{O}_j)$ *is a pseudodifferential operator (of order* m*) there.*

(5.4) *Given any* $u \in \mathscr{E}'(\mathfrak{M})$, Au *is a* C^∞ *function in* $\mathfrak{M}\backslash\operatorname{supp} u$.

PROOF. Both conditions are necessary, by Definition 5.1 and Theorem 5.2. Let us show that they are sufficient. We may refine the covering $\{\mathcal{O}_j\}$ into a locally finite one, consisting of relatively compact open sets (remember that $\mathfrak{M}$ is countable at infinity). In fact, let us assume that $\{\mathcal{O}_j\}_{j\in J}$ has these properties, and let $\{g_j\}_{j\in J}$ be a C^∞ partition of unity subordinate to this covering.

Let now $(\mathcal{O}, \chi)$ be an arbitrary local chart in $\mathfrak{M}$. For each $j \in J$ let $h_j \in C^\infty(\mathcal{O})$ be a function with support in $\mathcal{O} \cap \mathcal{O}_j$, equal to one in some neighborhood in $\mathcal{O}$ of $\mathcal{O} \cap \operatorname{supp} g_j$. Let us set, for all $u \in \mathcal{E}'(\mathcal{O})$,

$$A^{\#}u = \sum_{j\in J} h_j A(g_j u) \quad \text{in } \mathcal{O}. \tag{5.5}$$

It is clear, by virtue of (5.4), that $(A - A^{\#})u \in C^\infty(\mathcal{O})$, i.e., $A - A^{\#}$ is regularizing in $\mathcal{O}$. On the other hand,

$$\chi_*(A^{\#}u) = \sum_j (\chi_* h_j)(\chi \circ \chi_j^{-1})_*(\chi_j)_*\{A(g_j u)\}, \tag{5.6}$$

where we view $\chi \circ \chi_j^{-1}$ as a diffeomorphism of $\chi_j(\mathcal{O} \cap \mathcal{O}_j)$ onto $\chi(\mathcal{O} \cap \mathcal{O}_j)$. Take $u = \chi^{-1}_* v$ with $v \in \mathcal{E}'(\chi(\mathcal{O}))$. Then (5.6) can be rewritten:

$$(A^{\#})^\chi v = \sum_j (\chi_* h_j) A_j^{\chi\circ\chi_j^{-1}}\{(\chi_* g_j)v\}. \tag{5.7}$$

By (5.3) and Theorem 3.3 we know that $A_j^{\chi\circ\chi_j^{-1}}$ is a pseudodifferential operator in $\chi(\mathcal{O} \cap \mathcal{O}_j)$. By our choices of g_j and h_j we conclude that $(A^{\#})^\chi$ is a pseudodifferential operator in $\chi(\mathcal{O})$. □

REMARK 5.1. Condition (5.4) cannot be substantially relaxed as the following example shows: take $\mathfrak{M} = \mathbb{R}^1$ and for $\mathcal{O}_j$ the interval $]j-1, j+1[$ (we let j range over the set $\mathbb{Z}$ of all integers); we take χ_j to be the identity map. Let A be the translation operator $u(x) \mapsto u(x-2)$. For all $u \in \mathcal{E}'(\mathcal{O}_j)$ the restriction to $\mathcal{O}_j$ of Au vanishes identically; thus each transfer A_j of A to $\mathcal{O}_j$ (via χ_j) is zero. But clearly A is not pseudolocal.

Let A be a pseudodifferential operator in $\mathfrak{M}$. Given any local chart $(\mathcal{O}, \chi)$ in $\mathfrak{M}$ we may form the pseudodifferential operator $A^\chi_{\mathcal{O}}$ in $\chi(\mathcal{O}) \subset \mathbb{R}^n$. The symbol of $A^\chi_{\mathcal{O}}$ (in $\chi(\mathcal{O})$) is called the *symbol of A in the local chart* $(\mathcal{O}, \chi)$, or in the local coordinates $x^1, \dots, x^n$ defined by this chart. All concepts and properties connected with the symbolic calculus in $\chi(\mathcal{O})$ (see Section 4) are then automatically transferred to the symbolic calculus in the local chart $(\mathcal{O}, \chi)$. For instance we shall denote by $S^m(\mathcal{O})$ or, if there is a risk of confusion, by $S^m(\mathcal{O}, \chi)$ the space of symbols of degree m in $(\mathcal{O}, \chi)$; by $\dot{S}^m(\mathcal{O})$ the quotient space $S^m(\mathcal{O})/S^{-\infty}(\mathcal{O})$, etc.

The concept of *principal symbol*, however, can be globally defined. For this we require the notion of *cotangent bundle*.

As before, let $\mathfrak{M}$ denote a C^∞ manifold, of dimension n. Let x_o be an arbitrary point of $\mathfrak{M}$ and $\mathcal{J}^0_{x_o}$ the space of germs of real-valued C^∞ functions at x_o which vanish at x_o. (A germ $\dot{f}$ of function at x_o is an equivalence class of functions, each of which is defined in some neighborhood of x_o, which might depend on this function, for the equivalence relation of being equal in some neighborhood of x_o.) By using local coordinates in an open neighborhood $\mathcal{O}$ of x_o, we may define a metric $d(x, y)$ in $\mathcal{O}$. All metrics defined in this manner are equivalent in the neighborhood of x_o, by virtue of the definition of the differentiable structure of $\mathfrak{M}$. It therefore makes sense to say that some C^∞ function f, defined in a neighborhood of x_o, *vanishes to second order at* x_o, that is, $|f(x)| \leq \text{const}\, d(x, x_o)^2$ for all x in some neighborhood of x_o. Then let $\mathcal{J}^1_{x_o}$ denote the space of germs of real-valued C^∞ functions at x_o, which vanish to second order at x_o. The quotient vector space $\mathcal{J}^0_{x_o}/\mathcal{J}^1_{x_o}$ is, by definition, the *cotangent space* to $\mathfrak{M}$ at x_o. We shall denote it by $T^*_{x_o}\mathfrak{M}$ or, shortly, when there is no risk of confusion, by $T^*_{x_o}$. Its elements are the *covectors* on $\mathfrak{M}$ at x_o. There is a very important canonical map of the space of real C^∞ (or C^1) functions in the neighborhood of x_o into $T^*_{x_o}$, called the *differential* and denoted by $f \mapsto df(x_o)$. It is the compose

$$f \mapsto \text{germ at } x \text{ of } f - f(x_o) \in \mathcal{J}^0_{x_o} \mapsto \text{canonical image in } \mathcal{J}^0_{x_o}/\mathcal{J}^1_{x_o}.$$

Let $(\mathcal{O}, x^1, \ldots, x^n)$ be a local chart containing x_o. By writing the Taylor expansion of order 1 of an arbitrary C^∞ function f in a neighborhood of x_o, we obtain that

$$df(x_o) = \sum_{i=1}^{n} \frac{\partial f}{\partial x^i}(x_o)\, dx^i(x_o),$$

which shows that the $dx^i(x_o)$ span $T^*_{x_o}$. Let us denote by x^i_o the value of x^i at x_o. A linear function of the $(x^i - x^i_o)$'s, which is obviously defined in $\mathcal{O}$, cannot vanish of order two at x_o without vanishing identically, which proves that the $dx^i(x_o)$ are linearly independent and therefore form a basis of $T^*_{x_o}$. Thus $T^*_{x_o}$ is a linear space over the real numbers, of dimension $n = \dim \mathfrak{M}$. Moreover, let $T^*\mathcal{O}$ denote the set-theoretical union of the sets $T^*_{x_o}$ as x_o ranges over $\mathcal{O}$. Any point in $T^*\mathcal{O}$ can be denoted by (x_o, ξ^o) with $\xi^o \in T^*_{x_o}$. We shall denote by $\xi^o_1, \ldots, \xi^o_n$ the coordinates of ξ^o in the basis $dx^1(x_o), \ldots, dx^n(x_o)$. The mapping

$$(x_o, \xi^0) \mapsto (x^1_o, \ldots, x^n_o, \xi^o_1, \ldots, \xi^o_n) \tag{5.8}$$

defines a bijection of $T^*\mathcal{O}$ onto $\chi(\mathcal{O}) \times \mathbb{R}_n$. We shall transport the topology of $\chi(\mathcal{O}) \times \mathbb{R}_n$ onto $T^*\mathcal{O}$ using the bijection (5.8). Suppose that we use different coordinates, $y^1, \ldots, y^n$, in $\mathcal{O}$. From the chain rule

$$\frac{\partial f}{\partial y^j} = \sum_{i=1}^{n} \frac{\partial x^i}{\partial y^j} \frac{\partial f}{\partial x^i},$$

we see that the relation of the coordinates η_j in the basis $dy^1, \ldots, dy^n$ (in each cotangent space $T^*_{x_o}$) to the coordinates ξ_i in the basis $dx^1, \ldots, dx^n$ can be expressed in the following form:

$$\eta = {}^t(\partial x/\partial y)\xi, \tag{5.9}$$

where $(\partial x/\partial y)$ denotes the Jacobian matrix of the x^i's as functions of the y^j's, and ${}^t(\partial x/\partial y)$ denotes its transpose. The relation (5.9) implies that the topology of $T^*\mathcal{O}$ does not depend on the choice of the coordinates in $\mathcal{O}$. In fact, it shows more. Call Ω' the image of $\mathcal{O}$ under the diffeomorphism $x \mapsto (y^1(x), \ldots, y^n(x))$ and consider the compose

$$(x^1, \ldots, x^n, \xi_1, \ldots, \xi_n) \mapsto (x, \xi) \mapsto (y^1, \ldots, y^n, \eta_1, \ldots, \eta_n). \tag{5.10}$$

It follows at once from (5.9) and from the definition of the differential structure on $\mathfrak{M}$, which says that the y^j are C^∞ functions of the x's, that (5.10) *is a diffeomorphism of* $\chi(\mathcal{O}) \times \mathbb{R}_n$ *onto* $\Omega' \times \mathbb{R}_n$. This fact will now be exploited.

Let us denote by $T^*\mathfrak{M}$ the disjoint union of the linear spaces $T^*_{x_o}$ as x_o ranges over the whole of $\mathfrak{M}$. Clearly, the sets $T^*\mathcal{O}$ form a covering of $T^*\mathfrak{M}$ as $(\mathcal{O}, x^1, \ldots, x^n)$ ranges over the collection of all local charts of $\mathfrak{M}$. We may then define a topology on $T^*\mathfrak{M}$ by saying that a set $\mathcal{U}$ is open in $T^*\mathfrak{M}$ if its intersection with every $T^*\mathcal{O}$ is open in $T^*\mathcal{O}$. This makes sense, in view of the independence of the topology of each $T^*\mathcal{O}$ from the choice of the local coordinates. Moreover this turns $T^*\mathfrak{M}$ into a locally compact space, locally homeomorphic to $\mathbb{R}^n \times \mathbb{R}_n$. And last, we may regard each pair consisting of an open subset $T^*\mathcal{O}$ and of the associated mapping (5.8) as a local chart of $T^*\mathfrak{M}$. The condition on intersections is satisfied, because of the remark that (5.10) constitutes a diffeomorphism. We have thus turned $T^*\mathfrak{M}$ into a C^∞ manifold. By definition, the C^∞ manifold $T^*\mathfrak{M}$ is called the *cotangent bundle* over $\mathfrak{M}$. Local coordinates in $T^*\mathfrak{M}$ are given by the $(x^1, \ldots, x^n, \xi_1, \ldots, \xi_n)$ introduced earlier. The x^i's are local coordinates in the *base* manifold $\mathfrak{M}$, and the ξ_i's are the coordinates in each cotangent space with respect to the basis $dx^1, \ldots, dx^n$. It is standard terminology to call a *section* of $T^*\mathfrak{M}$ over a subset $\mathcal{A}$ of $\mathfrak{M}$ any mapping of $\mathcal{A}$ into $T^*\mathfrak{M}$ of the form $x \mapsto (x, \xi(x))$ where $\xi(x) \in T^*_x$ for each $x \in \mathcal{A}$ (or the range of such a map). In particular the mapping $x \mapsto (x, 0)$ from $\mathfrak{M}$ into $T^*\mathfrak{M}$ is called the *zero section* (and so is its range). Also the *projection on the base* is the mapping $(x, \xi) \mapsto x$ of $T^*\mathfrak{M}$ onto $\mathfrak{M}$. The cotangent space T_x^* at x is called the *fiber* of $T^*\mathfrak{M}$ at x; the variable in T_x^* is the *fiber variable*, often denoted by ξ; the mappings $(x, \xi) \mapsto (x, \rho\xi)$, $\rho > 0$, are the *fiber dilations*, etc.

REMARK 5.2. The dual of the vector space $T_x^*\mathfrak{M}$ is called the *tangent space* to $\mathfrak{M}$ at x and denoted by $T_x\mathfrak{M}$; its elements are the *tangent vectors* to $\mathfrak{M}$ at x. As x ranges over $\mathfrak{M}$ the disjoint union of these vector spaces makes up the *tangent bundle* $T\mathfrak{M}$ *over* $\mathfrak{M}$. In a local coordinates patch $(\mathcal{O}, x^1, \ldots, x^n)$ a tangent vector at x is a linear combination of the form

$$\theta = \theta^1 \frac{\partial}{\partial x^1} + \cdots + \theta^n \frac{\partial}{\partial x^n}.$$

If $\xi = \xi\, dx^1 + \cdots + \xi_n\, dx^n$ is a covector at x, the duality bracket $\langle \theta, \xi \rangle$ is equal to $\theta^1 \xi_1 + \cdots + \theta^n \xi_n$. If f is a differentiable function in a neighborhood of x with differential $df(x)$ at x, we have

$$\theta f(x) = \langle \theta, df(x) \rangle = \sum_{j=1}^{n} \theta^j \frac{\partial f}{\partial x^j}(x). \tag{5.11}$$

This describes the action of a tangent vector on a differentiable function.

The sections of the bundle $T\mathfrak{M}$ are called *vector fields*; those of $T^*\mathfrak{M}$ are called *differential forms*. If ϕ is a C^∞ mapping of $\mathfrak{M}$ into another C^∞ manifold $\mathfrak{M}'$, it defines a linear map $\phi_*: T_x\mathfrak{M} \to T_{\phi(x)}\mathfrak{M}'$, often called the *differential of the map* ϕ and denoted in a variety of fashions; in local coordinates it is represented by the Jacobian matrix. If g is a C^∞ function in a neighborhood of $\phi(x)$ and θ an element of $T_x\mathfrak{M}$, we have

$$(\phi_*\theta)g(x) = \langle \theta, d(g \circ \phi)(x) \rangle. \tag{5.12}$$

The transpose of this map is the natural map $\phi^*: T^*_{\phi(x)}\mathfrak{M}' \to T_x^*\mathfrak{M}$ associated with ϕ. Of course, we also associate with ϕ the bundle map $(x, \theta) \mapsto (\phi(x), \phi_*\theta)$ of $T\mathfrak{M}$ into $T\mathfrak{M}'$. When ϕ is a diffeomorphism, we may also associate with it the diffeomorphism $(x, \xi) \mapsto (\phi(x), \overset{1}{\phi}{}^*(\xi))$ of $T^*\mathfrak{M}$ onto $T^*\mathfrak{M}'$, which we shall sometimes denote by ϕ_+.

Immersions are easy to characterize by means of their differentials: the C^∞ map $\phi: \mathfrak{M} \to \mathfrak{M}'$ is an immersion if and only if its differential ϕ_* is injective, as a linear map $T_x\mathfrak{M} \to T_{\phi(x)}\mathfrak{M}'$, whatever $x \in \mathfrak{M}$. This is a direct consequence of the implicit function theorem. One sometimes says that ϕ is a *submersion* if its differential ϕ_* has (locally) constant rank, say r. In this case, one can cover $\mathfrak{M}$ with local charts $(\mathcal{O}, x^1, \ldots, x^n)$ having the following property: ϕ is a diffeomorphism of the submanifold of $\mathcal{O}$, defined by the equations $x^{r+1} = \cdots = x^n = 0$, onto $\phi(\mathcal{O})$; ϕ is constant on each submanifold of $\mathcal{O}$ defined by the equations $x^1 = x_o^1, \ldots, x^r = x_o^r$. When $r = n$ we recover the concept of an immersion.

Let $\mathcal{U}$ be any subset of $T^*\mathfrak{M}$ and suppose that, for every local chart $(\mathcal{O}, x^1, \ldots, x^n)$ such that $T^*\mathcal{O} \cap \mathcal{U} \neq \emptyset$, we are given a function $f_{\mathcal{O}}$ on $T^*\mathcal{O} \cap \mathcal{U}$. By using the coordinates $(x^1, \ldots, x^n, \xi_1, \ldots, \xi_n)$ in $T^*\mathcal{O}$, we may represent it by $f_{\mathcal{O}}(x^1, \ldots, x^n, \xi_1, \ldots, \xi_n)$ or by $f_{\mathcal{O}}(x, \xi)$. Suppose then that we consider another local chart $(\mathcal{O}', y^1, \ldots, y^n)$ and the associated function $f_{\mathcal{O}'}(y^1, \ldots, y^n, \eta_1, \ldots, \eta_n)$. The fact that $f_{\mathcal{O}} = f_{\mathcal{O}'}$ in the intersection $T^*\mathcal{O} \cap T^*\mathcal{O}' \cap \mathcal{U}$ can be expressed in the form

$$f_{\mathcal{O}'}(y, \eta) = f_{\mathcal{O}}\big(x(y), {}^t(\partial x/\partial y)^{-1}\eta\big). \tag{5.13}$$

If this happens to be true for all the choices of $\mathcal{O}$ and $\mathcal{O}'$, we can patch the functions $f_{\mathcal{O}}$ together and define a function f on the whole of $\mathcal{U}$. Conversely, if the $f_{\mathcal{O}}$ are already the restrictions of a function f on $\mathcal{U}$ to the respective $T^*\mathcal{O}$, the relation (5.13) must hold.

We may define now the concept of *principal symbol*. Although this can be done for all pseudodifferential operators in $\mathfrak{M}$ through an abstract approach, we shall restrict ourselves to those that do have a principal symbol. We say that $A \in \Psi^m(\mathfrak{M})$ has a principal symbol if, given any local chart $(\mathcal{O}, \chi)$ in $\mathfrak{M}$ in which the symbol of A is $a_{\mathcal{O},\chi}$, there is a C^∞ function $\sigma_{\mathcal{O},\chi}$ in $\chi(\mathcal{O}) \times (\mathbb{R}^n\backslash\{0\})$, positive-homogeneous of degree m with respect to ξ, such that $a_{\mathcal{O},\chi} - \sigma_{\mathcal{O},\chi} \in S^{m-\varepsilon}(\mathcal{O})$ for some $\varepsilon > 0$. This number ε might depend on the local chart but for simplicity we assume it to be constant. Let $x^1, \ldots, x^n$ denote the local coordinates in $(\mathcal{O}, \chi)$, and let $y^1, \ldots, y^n$ denote another system of C^∞ coordinates in $\mathcal{O}$. Then we know, by Theorem 4.4, that the symbol $a_{\mathcal{O},\chi'}(y, \eta)$ of A in $(\mathcal{O}, \chi') = (\mathcal{O}, y^1, \ldots, y^n)$ is such that

$$a_{\mathcal{O},\chi'}(y, \eta) - a_{\mathcal{O},\chi}\left(x(y), {}^t\left(\frac{\partial x}{\partial y}\right)^{-1}\eta\right) \in S^{m-1}(\mathcal{O}, \chi').$$

From this we derive

$$\sigma_{\mathcal{O},\chi'}(y, \eta) - \sigma_{\mathcal{O},\chi}\left(x(y), {}^t\left(\frac{\partial x}{\partial y}\right)^{-1}\eta\right) \in S^{m-\varepsilon}(\mathcal{O}, \chi').$$

By homogeneity with respect to η this is possible only if

$$\sigma_{\mathcal{O},\chi'}(y, \eta) = \sigma_{\mathcal{O},\chi}\left(x(y), {}^t\left(\frac{\partial x}{\partial y}\right)^{-1}\eta\right). \tag{5.14}$$

This means precisely that $\sigma_{\mathcal{O},\chi}$ is the expression in the local chart $(\mathcal{O}, \chi)$ of a

C^∞ function σ in the complement $T^*\mathfrak{M}\backslash 0$ of the zero section in $T^*\mathfrak{M}$. This function σ is positive-homogeneous of degree m with respect to the fiber variable. It is the principal symbol of A.

If A has a principal symbol we shall often denote it by $\sigma(A)$.

The notion of principal symbol is particularly important when dealing with classical pseudodifferential operators. We shall call *classical* any pseudodifferential operator A in $\mathfrak{M}$ such that given any local chart (Ω, χ), the operator A^χ is a classical pseudodifferential operator in $\chi(\Omega)$ (Example 4.2). An important consequence of formula (4.35) is that *the transform under a diffeomorphism of a classical pseudodifferential operator is also classical.* We leave the proof of this statement as an exercise. Suppose that $A \in \Psi^m(\mathfrak{M})$ and let the formal series (4.1) define the symbol of A in the local chart $(\mathcal{O}, \chi) = (\mathcal{O}, x^1, \ldots, x^n)$. Each term $a_j(x, \xi)$ is positive-homogeneous of degree m_j and $m_j - m_{j+1}$ is a positive integer (see Example 4.1). Then the expression of $\sigma(A)$ in the local coordinates $x^1, \ldots, x^n$ is $a_o(x, \xi)$.

Appendix: Elliptic Pseudodifferential Operators on a Manifold

We have already remarked (on p. 42) that ellipticity is a property that is invariant under diffeomorphism. This enables us to define elliptic pseudodifferential operators on a C^∞ manifold $\mathfrak{M}$:

A pseudodifferential operator of order m in $\mathfrak{M}$, A, is said to be elliptic if given any local chart $(\mathcal{O}, \chi)$ in $\mathfrak{M}$, the symbol of A in that local chart is elliptic of degree m.

By availing ourselves of the notion of principal symbol it is easy to construct elliptic pseudodifferential operators on $\mathfrak{M}$ of any given degree. Indeed, let m be an arbitrary real number. Let $p(x\, \xi)$ be a C^∞ function in $T^*\mathfrak{M}\backslash 0$, positive-homogeneous of degree m with respect to the fiber variable ξ. Suppose that $p(x, \xi)$ is everywhere strictly positive in $T^*\mathfrak{M}\backslash 0$. (Such functions can easily be constructed: Equip $\mathfrak{M}$ with a Riemannian metric and define $p(x, \xi)$ for $|\xi| = 1$; then extend by positive-homogeneity of degree m.) Next, take a locally finite covering of $\mathfrak{M}$ by local charts $(\mathcal{O}_j, \chi_j)$ $(j = 1, 2, \ldots)$. Using the local coordinates in each $\mathcal{O}_j$ select a properly supported pseudodifferential operator in $\mathcal{O}_j$, A_j, whose symbol is equal to $p(x, \xi)$ in $T^*\mathcal{O}_j\backslash 0$. Finally use a C^∞ partition of unity in $\mathfrak{M}$, $\{g_j\}_{j=1,2,\ldots}$, subordinate to the covering $\{\mathcal{O}_j\}$ and consisting solely of nonnegative functions g_j, and define a pseudodifferential operator in $\mathfrak{M}$ by the formula

$$(5.15) \qquad Au = \sum_{j=1,2,\ldots} g_j A_j(u|_{\mathcal{O}_j}), \qquad u \in \mathcal{D}'(\mathfrak{M}).$$

It follows from Theorem 4.4 that A has a principal symbol, equal to $p(x, \xi)$. In particular it is elliptic of order m.

The construction of global parametrices for elliptic pseudodifferential operators can be effected along the same lines. Let A be a pseudodifferential operator on $\mathfrak{M}$, properly supported, elliptic of order m. Use the locally finite covering $\{\mathcal{O}_j\}$ and the partition of unity $\{g_j\}$ of the preceding argument. By means of the local coordinates in each $\mathcal{O}_j$ (which, we recall, is the domain of a local chart) we can construct a parametrix B_j of A in $\mathcal{O}_j$, properly supported in $\mathcal{O}_j$ (see the end of Section 4). Then set

$$Bu = \sum_{j=1,2,\dots} g_j B_j(u|_{\mathcal{O}_j}), \qquad u \in \mathcal{D}'(\mathfrak{M}). \tag{5.16}$$

We have $BA - I \in \Psi^{-\infty}(\mathfrak{M})$. On the other hand, in $\mathcal{O}_{j_o}$,

$$AB - I \sim \sum_j [A, g_j]B_j \tag{5.17}$$

is congruent to $\sum_j [A, g_j](B_j - B_{j_o}) \bmod \Psi^{-\infty}(\mathfrak{M})$, since $\sum_j g_j \equiv 1$.

But in a neighborhood of $\mathcal{O}_{j_o} \cap \operatorname{supp} g_j$ we have $B_j - B_{j_o} \in \Psi^{-\infty}(\Omega)$, by the uniqueness of the parametrix modulo regularizing operators. We reach the conclusion that (5.17) is regularizing. Thus B, defined in (5.16), is a left and right parametrix of A in $\mathfrak{M}$.

EXAMPLE 5.1. Let S^1 denote the unit circle, i.e., the set of complex numbers $e^{i\theta}$, $-\pi \le \theta \le \pi$. The Fourier series expansion

$$u \mapsto \{u_n\}_{n \in \mathbf{Z}}, \qquad u_n = (2\pi)^{-1} \int_{-\pi}^{\pi} e^{-in\theta} u(\theta)\, d\theta, \tag{5.18}$$

establishes an isomorphism of the space of distributions $\mathcal{D}'(S^1)$ onto s', the space of tempered complex sequences; this means that for some constants M, m, $|u_n| \le M(1 + n^2)^m$. It also establishes an isomorphism of $C^\infty(S^1)$ onto s, the space of rapidly decaying sequences; this means that, for each $k = 1, 2, \dots$, there is $C_k > 0$ such that $|u_n| \le C_k(1 + n^2)^{-k}$. The Sobolev space $H^s(S^1)$ is the space of distributions u such that

$$\|u\|_s^2 = \sum_{n \in \mathbf{Z}} (1 + n^2)^s |u_n|^2 < +\infty. \tag{5.19}$$

Consider the operator defined by

$$P^+u = \sum_{n \ge 0} u_n e^{in\theta}, \qquad u \in C^\infty(S^1). \tag{5.20}$$

By using the expression of u_n we obtain

$$P^+u(\theta) = (2\pi)^{-1} \int_{-\pi}^{\pi} [1 - e^{i(\theta - \theta')}]^{-1} u(\theta')\, d\theta'. \tag{5.21}$$

This is of course a generalized integral: the kernel $[1 - e^{i(\theta-\theta')}]^{-1}$ has a "pole" of order one at the diagonal $\theta = \theta'$. It is very regular, and thus P^+ is pseudolocal. We may apply Proposition 5.1; in order to check that P^+ is a pseudodifferential operator, it suffices to look at its restriction to the local chart $]-\pi/2, \pi/2[$ and, equivalently, to its rotates. Thus we take u a C^∞ function compactly supported in that arc of the circle and transfer everything to the real line, where the variable is now denoted by x (or by y). Let $g \in C_c^\infty(]-\pi, \pi[)$ be equal to one in $[-3\pi/4, 3\pi/4]$, and set

$$a(x, y) = (x - y)[1 - e^{i(x-y)}]^{-1} g(y).$$

We have

$$P^+u(x) = (2\pi)^{-1} \int \frac{a(x, y)}{x - y} u(y)\, dy. \tag{5.22}$$

The integral ought to be interpreted, not as a principal value, but as the limit:

$$(2\pi)^{-1} \lim_{\varepsilon \to +0} \int a(x, y)(x - y + i\varepsilon)^{-1} u(y)\, dy. \tag{5.23}$$

This is evident in formula (5.20), where one may replace θ by $\theta + i\varepsilon$ and then go to the limit as $\varepsilon \to +0$. Observe then that

$$(x - y + i\varepsilon)^{-1} = \int_0^{+\infty} e^{i\xi(x-y+i\varepsilon)}\, d\xi/i. \tag{5.24}$$

This shows that

$$P^+u(x) = \frac{1}{2i\pi} \int \int_{\xi>0} e^{i\xi(x-y)} a(x, y) u(y)\, dy\, d\xi. \tag{5.25}$$

Note that

$$a(x, x) = i. \tag{5.26}$$

Thus the principal symbol of the operator (5.25) is the *Heaviside function* $\mathcal{Y}(\xi)$ (which equals one for $\xi > 0$ and zero for $\xi < 0$). Note that the cotangent bundle T^*S^1 over S^1 is the product $S^1 \times \mathbb{R}_1$ with $\mathbb{R}_1$ oriented by the orientation on S^1. We may therefore talk of the part $\mathbb{R}_+$ in the cotangent spaces at points of the unit circle (they are defined by $\xi > 0$).

THEOREM 5.3. *The operator P^+, defined in* (5.20), *is a pseudodifferential of order zero on the unit circle S^1, classical, whose principal symbol is equal to one in the positive part of $T^*S^1\backslash 0$ and to zero on the negative part.*

In the local chart used before the *total* symbol of P^+ is equal to $\mathcal{Y}(\xi)$. Of course, we could also have studied

$$P^- u = \sum_{n \le 0} u_n e^{in\theta}. \tag{5.27}$$

It is a classical pseudodifferential operator of order zero in S^1 with principal symbol $\mathcal{Y}(-\xi)$. Of course we have $P^+ + P^- \sim I$, the identity. More precisely,

$$P^+ u + P^- u = u + u_0, \tag{5.28}$$

and the operator $u \mapsto u_0$ is obviously regularizing.

Let α, β be two C^∞ complex functions on S^1 that do not vanish at any point. The operator

$$\alpha(\theta)P^+ + \beta(\theta)P^- \tag{5.29}$$

is an *elliptic* pseudodifferential operator of order zero on S^1.

6. Microlocalization and Wave-Front Sets

We may refine the study of the singularities (or, equivalently, of the regularity) of a distribution, possibly the solution of a differential equation, by analyzing their location in the base space (for us, the open set $\Omega \subset \mathbb{R}^n$ or the manifold $\mathfrak{M}$) and the frequencies at which they occur—in other words, by lifting the analysis in the product set $\Omega \times \mathbb{R}_n$ or, respectively, in the cotangent bundle $T^*\mathfrak{M}$. Actually we shall deal mostly with distributions and operators in the open set Ω and briefly indicate how the concepts and results are extended to the case of a manifold $\mathfrak{M}$.

DEFINITION 6.1. *We say that a distribution u in Ω is C^∞ in the neighborhood of a point (x_o, ξ^o), $x_o \in \Omega$, $0 \neq \xi^o \in \mathbb{R}_n$, if there is a function $g \in C_c^\infty(\Omega)$ equal to one in a neighborhood U of x_o and an open cone Γ^o in $\mathbb{R}_n$, containing ξ^o, such that the following holds:*

(6.1) *To every number $M \ge 0$ there is a number $C_M \ge 0$ such that*

$$|(\widehat{gu})(\xi)| \le C_M(1 + |\xi|)^{-M}, \qquad \forall \xi \in \Gamma^o.$$

By a cone in $\mathbb{R}_n$ we always mean a subset of $\mathbb{R}_n$ stable under the dilations $\xi \mapsto \rho\xi$, $\rho > 0$. Because of the role of dilations in the present theory, it is convenient to adopt the following terminology:

Let $\mathfrak{M}$ be any C^∞ manifold. A subset Γ of $T^*\mathfrak{M}$ is said to be *conic* if it is stable under all dilations $(x, \xi) \mapsto (x, \rho\xi)$, $\rho > 0$.

Because of the importance of dilations in this context, some authors prefer to use the cosphere bundle over $\mathfrak{M}$, $S^*\mathfrak{M}$, instead of the cotangent bundle: $S^*\mathfrak{M}$ is the quotient of $T^*\mathfrak{M}\backslash 0$ modulo the equivalent relation

$$(6.2) \qquad (x, \xi) \sim (y, \eta) \quad \Leftrightarrow \quad x = y, \exists \rho > 0 \text{ such that } \eta = \rho\xi.$$

Let π be the canonical projection of $T^*\mathfrak{M}\backslash 0$ onto $S^*\mathfrak{M}$. A subset Γ of $T^*\mathfrak{M}\backslash 0$ is conic if and only if $\Gamma = \pi^{-1}(\pi(\Gamma))$. We shall say that Γ is *conically compact* if $\pi(\Gamma)$ is compact.

The simplest example of a conic open set Γ in $\Omega \times \mathbb{R}_n$ $(\cong T^*\Omega)$ is a product $U \times \Gamma^o$ with U and Γ^o as in Definition 6.1.

DEFINITION 6.2. *A distribution u in Ω is said to be C^∞ in a conic open subset Γ of $\Omega \times (\mathbb{R}_n\backslash\{0\})$ if it is C^∞ in the neighborhood of every point of Γ.*

The complement in $\Omega \times (\mathbb{R}_n\backslash\{0\})$ of the union of all conic open sets in which u is C^∞ is called the wave-front set of u and shall be denoted by WF(u).

For purposes of exposition it is convenient to immediately relate these concepts to analogous ones for pseudodifferential operators:

DEFINITION 6.3. *We say that a pseudodifferential operator A in Ω is regularizing in the neighborhood of (x_o, ξ^o), $x_o \in \Omega$, $0 \neq \xi^o \in \mathbb{R}_n$, if there is a function g and a cone Γ^o like those in Definition* 6.1 *such that the symbol $a(x, \xi)$ of A has the following property:*

(6.3) *To every $M \geq 0$ and every pair of n-tuples α, β, there is $C_{\alpha,\beta,M} \geq 0$ such that*

$$\sup_x |\partial_x^\alpha \partial_\xi^\beta[g(x)a(x, \xi)]| \leq C_{\alpha,\beta,M}(1 + |\xi|)^{-M}, \qquad \forall \xi \in \Gamma^o.$$

By the symbol of A we mean in Definition 6.3 any representative of the symbol class of A. Actually the definition continues to make sense if we replace A by its class mod $\Psi^{-\infty}(\Omega)$.

DEFINITION 6.4. *A pseudodifferential operator A in Ω is said to be regularizing in a conic open subset Γ of $\Omega \times (\mathbb{R}_n\backslash\{0\})$ if it is regularizing in the neighborhood of every point of Γ.*

The complement in $\Omega \times (\mathbb{R}_n\backslash\{0\})$ of the union of all conic open sets in which A is regularizing is called the microsupport of A and is denoted by μ supp A.

Both the wave-front set of a distribution and the microsupport of a pseudodifferential operator are *conic closed* subsets of $\Omega \times (\mathbb{R}_n \backslash \{0\})$.

In view of Definition 6.3 we may say that *a pseudodifferential operator in Ω in regularizing in the conic open set Γ if and only if its symbol class vanishes identically in Γ.*

EXAMPLE 6.1. Let Γ be an arbitrary conic open subset of $\Omega \times (\mathbb{R}_n \backslash \{0\})$, $g(x, \xi)$ a C^∞ function in $\Omega \times (\mathbb{R}_n \backslash \{0\})$, with support contained in Γ, positive-homogeneous of degree zero. Let us write, for any $u \in C_c^\infty(\Omega)$,

$$g(x, D)u(x) = (2\pi)^{-n} \int e^{ix\cdot\xi} g(x, \xi)\hat{u}(\xi)\, d\xi. \tag{6.4}$$

Any pseudodifferential operator A in Ω that differs from $g(x, D)$ by a regularizing operator has the property

$$\mu \operatorname{supp} A \subset \Gamma. \tag{6.5}$$

Suppose furthermore that $g(x, \xi)$ is identically one in a conic open set $\Gamma' \subset \Gamma$. Then $\Gamma' \subset \mu \operatorname{supp} A$; more precisely,

$$\Gamma' \cap \mu \operatorname{supp} (I - A) = \varnothing. \tag{6.6}$$

EXAMPLE 6.2. Let U and Γ^o be as in Definition 6.1. Let h be a C^∞ function in $\mathbb{R}_n \backslash \{0\}$ with support in Γ^o, positive-homogeneous of degree zero, equal to one in a conic open neighborhood of ξ^o, Γ'^o. Let U' denote the interior of the set of points in which $g(x) = 1$. Let us then denote by $h(D)g(x)$ the operator $v \mapsto h(D)(gv)$. The convolution operator $h(D)$ is a particular case of (6.4), and is defined likewise. Property (6.1) implies

$$h(D)(gu) \in C^\infty(\Omega) \qquad (\text{or } C^\infty(\mathbb{R}^n)). \tag{6.7}$$

Note that the symbol class of $h(D)g(x)$ is defined by the formal symbol

$$\sum_{\alpha \in \mathbb{Z}_+^n} \frac{1}{\alpha!} h^{(\alpha)}(\xi) D_x^\alpha g(x), \tag{6.8}$$

which is clearly supported (each term is supported) in $U \times \Gamma^o$ and is equal to one in $U' \times \Gamma'^o$. By constructing a true symbol $g(x, \xi)$ from the formal one (6.8) (by means of cutoff functions), we see that $h(D)g(x)$ is of the type of the operators A in Example 6.1, where $\Gamma = U \times \Gamma^o$, $\Gamma' = U' \times \Gamma'^o$.

Generally speaking, the functions h are easy to construct. One takes the intersection of the cone Γ^o with the unit sphere S_{n-1} in $\mathbb{R}_n$; this is an open set $\mathcal{O}$. One then takes a function $h_o \in C_c^\infty(\mathcal{O})$ equal to one in the intersection of Γ'^o with S_{n-1}, and sets $h(\xi) = h_o(|\xi|^{-1}\xi)$.

The case $n = 1$ is worth looking at a little closer. All functions h of the preceding kind are linear combinations of $\mathcal{Y}(\xi)$, where $\mathcal{Y}$ is the Heaviside function, and of $\mathcal{Y}(-\xi)$. If $\xi^o > 0$ (resp., <0), one can take $h(\xi) = \mathcal{Y}(\xi)$ (resp., $h(\xi) = \mathcal{Y}(-\xi)$). Observe that

$$(2\pi)^{-1} \int e^{ix\xi}\mathcal{Y}(\xi)\, d\xi = \lim_{\varepsilon \to +0} \int_0^{+\infty} e^{(ix-\varepsilon)\xi}\, d\xi$$

$$= \lim_{\varepsilon \to +0} \frac{-1}{2i\pi} \frac{1}{x + i\varepsilon}.$$

Let us set

$$pv^+\frac{1}{x} = \lim_{\varepsilon \to +0} (x + i\varepsilon)^{-1}, \qquad pv^-\frac{1}{x} = \lim_{\varepsilon \to +0} (x - i\varepsilon)^{-1}.$$

The inverse Fourier transform of $\mathcal{Y}(-\xi)$ is equal to $(2i\pi)^{-1}pv^-(1/x)$, and so

$$\frac{1}{2i\pi}\left(pv^-\frac{1}{x} - pv^+\frac{1}{x}\right) = \delta(x),$$

the Dirac distribution. On the other hand, $pv^+(1/x) + pv^-(1/x) = 2pv(1/x)$, the *Cauchy principal value*, and thus the "pseudodifferential" operator $\mathcal{Y}(D)$ is the convolution with

$$-\frac{1}{2i\pi} pv^+\frac{1}{x} = \frac{1}{2}\delta(x) - \frac{1}{2i\pi}pv\frac{1}{x}.$$ □

PROPOSITION 6.1. *Let U be a distribution in Ω, (x_o, ξ^o) a point of $\Omega \times (\mathbb{R}_n\backslash\{0\})$. Then the following properties are equivalent:*

(6.9) $(x_o, \xi^o) \notin \mathrm{WF}(u)$.

(6.10) *There is a C^∞ function $g(x, \xi)$ in $\Omega \times (\mathbb{R}_n\backslash\{0\})$, positive-homogeneous of degree zero, equal to one in some neighborhood of (x_o, ξ^o), such that if A is any properly supported pseudodifferential operator in Ω with symbol $g(x, \xi)$, we have $Au \in C^\infty(\Omega)$.*

(6.11) *There is a conic open neighborhood Γ of (x_o, ξ^o) in $\Omega \times (\mathbb{R}_n\backslash\{0\})$ such that $Bu \in C^\infty(\Omega)$ for any properly supported pseudodifferential operator B in Ω whose microsupport in contained in Γ.*

(6.11′) *Same statement as* (6.11) *except that the microsupport of the operator B is conically compact.*

PROOF. That (6.9) implies (6.10) follows at once from the considerations in Example 6.2. Let Γ be a conic open neighborhood of (x_o, ξ^o) in which the function $g(x, \xi)$ of (6.10) is equal to one, A is a pseudodifferential operator like the one in (6.10), B is one like that in (6.11). It is clear that the symbol of AB (resp., BA) is equal to that of B; hence $BA - B$ is regularizing. Therefore, if Au is C^∞ in Ω the same is true of Bu. Thus (6.10) implies (6.11), which implies trivially (6.11′). Suppose (6.11′) holds. For a suitable choice of U and Γ^o we may take $B = h(D)g(x)$ as in Example 6.2. But (6.7) implies (6.1) with Γ'^o substituted for Γ^o. □

Proposition 6.1 has numerous consequences; the first one is that the wave-front set is "invariant" under diffeomorphism:

PROPOSITION 6.2. *Let ϕ be a diffeomorphism of Ω onto an open set $\Omega' \subset \mathbb{R}^n$. Let ϕ_+ denote the associated diffeomorphism of $T^*\Omega$ onto $T^*\Omega'$, ϕ_* the associated isomorphism of $\mathscr{D}'(\Omega)$ onto $\mathscr{D}'(\Omega')$* (see Sections 3, 5). *Then, for any distribution u in Ω,*

$$\text{WF}(\phi_* u) = \phi_+ \text{WF}(u). \tag{6.12}$$

REMARK 6.1. The diffeomorphism ϕ_+ commutes with the positive dilations in the cotangent bundles.

Proposition 6.2 is an immediate consequence of the fact that (6.9) and (6.11) are equivalent and that the latter property is invariant under diffeomorphisms, by virtue of Theorem 3.3 and formula (4.35).

Proposition 6.2 enables us to extend Definitions 6.1 and 6.2 to a C^∞ manifold $\mathfrak{M}$.

PROPOSITION 6.3. *If $u \in \mathscr{D}'(\Omega)$, the base projection of* WF(u) *is exactly equal to* sing supp u.

PROOF. Suppose that u is C^∞ in an open neighborhood U of x_o in Ω. Then $gu \in C_c^\infty(U)$ whatever $g \in C_c^\infty(U)$ and (6.1) holds with $\Gamma^o = \mathbb{R}_n$.

Suppose that there is an open neighborhood U of x_o such that WF$(u) \cap T^*U = \varnothing$. According to Proposition 6.1, we can find a finite collection of C^∞ functions g_i in $\Omega \times (\mathbb{R}_n \backslash \{0\})$, $i = 1, \ldots, r$, with the following properties: (1) For eaeh i, $g_i(x, \xi)$ is positive-homogeneous of degree zero with respect to ξ. (2) $g_1(x, \xi) + \cdots + g_r(x, \xi) = 1$ for all x in some open neighborhood U' of x_o and all $\xi \in \mathbb{R}_n \backslash \{0\}$. (3) For every i and every properly supported pseudodifferential operator in Ω, A_i, with symbol $g_i(x, \xi)$, we have $A_i u \in C^\infty(\Omega)$. From property 2 we derive that $I - (A_1 + \cdots + A_r)$ is regularizing in U'; hence $u \in C^\infty(U')$. □

PROPOSITION 6.4. *A properly supported pseudodifferential operator A in Ω is regularizing in a conic open subset Γ of $\Omega \times (\mathbb{R}_n \backslash \{0\})$ if and only if Au is C^∞ in Γ for all distributions u in Ω.*

PROOF. Let (x_o, ξ^o) be an arbitrary point in Γ, $g(x, \xi)$ a C^∞ function in $\Omega \times (\mathbb{R}_n \backslash \{0\})$ positive-homogeneous of degree zero, with support conically compact and contained in Γ, equal to one in some neighborhood of (x_o, ξ^o). Let B be any properly supported pseudodifferential operator in Ω with symbol $g(x, \xi)$. If A is regularizing in Γ, the symbol of BA vanishes identically, and therefore this operator is regularizing. Thus BAu is C^∞ in Ω. By Proposition 6.1 we conclude that $(x_o, \xi^o) \notin \mathrm{WF}(Au)$.

Conversely, suppose that Au is C^∞ in Γ for all $u \in \mathscr{D}'(\Omega)$. Then $BAu \in C^\infty(\Omega)$, i.e., BA is regularizing in Ω, and thus (see Theorem 4.1) its symbol vanishes identically. There is a conic neighborhood of (x_o, ξ^o) in which that symbol coincides with the symbol of A. □

COROLLARY 6.1. *Let ϕ be as in Proposition 6.2, and let A^ϕ be the transfer via ϕ of a pseudodifferential operator A in Ω. We have*

$$\mu \operatorname{supp} A^\phi = \phi_+(\mu \operatorname{supp} A). \tag{6.13}$$

PROPOSITION 6.5. *Let A be a properly supported pseudodifferential operator in Ω and u a distribution in Ω. We have*

$$\mathrm{WF}(Au) \subset \mathrm{WF}(u) \cap \mu \operatorname{supp} A. \tag{6.14}$$

PROOF. If A is regularizing in some conic open subset Γ of $\Omega \times (\mathbb{R}_n \backslash \{0\})$, then we have $\mathrm{WF}(Au) \cap \Gamma = \varnothing$ by Proposition 6.4. Suppose that $\mathrm{WF}(u) \cap \Gamma = \varnothing$. Let Γ' be any conic open set whose closure is conically compact and contained in Γ. We may write $A = A_1 + B$ with A_1 regularizing in Γ' and B having its microsupport contained in Γ. By Proposition 6.1 we have $Bu \in C^\infty(\Omega)$, and by the first part, $\mathrm{WF}(A_1 u) \cap \Gamma' = \varnothing$; hence $\mathrm{WF}(Au) \cap \Gamma' = \varnothing$. By letting Γ' grow to Γ we obtain here also $\mathrm{WF}(Au) \cap \Gamma = \varnothing$. □

COROLLARY 6.2. *Pseudodifferential operators decrease the wave-front sets: if A is one and u is a distribution, both in Ω, and if either A is properly supported or u is compactly supported, then* $\mathrm{WF}(Au) \subset \mathrm{WF}(u)$.

PROPOSITION 6.6. *Let A, B be two properly supported pseudodifferential operators in Ω. We have*

(6.15) $\mu\ \mathrm{supp}(AB) \subset (\mu\ \mathrm{supp}\ A) \cap (\mu\ \mathrm{supp}\ B)$,

(6.16) $\mu\ \mathrm{supp}\ {}^t\!A = \{(x, \xi) \in \Omega \times \mathbb{R}_n;\ (x, -\xi) \in \mu\ \mathrm{supp}\ A\}$,

(6.17) $\mu\ \mathrm{supp}\ A^* = \mu\ \mathrm{supp}\ A$.

Follows at once from Definition 6.4 and Theorems 4.2 and 4.3.

As a distribution in $\Omega \times \Omega$, the kernel $A(x, y)$ associated to the pseudodifferential operator A has a wave-front set, WF $A(x, y)$. What is the relation between this wave-front set and the microsupport of the operator A? The answer is easily derived from the argument at the end of the proof of Theorem 4.1.

Let x_o be an arbitrary point in Ω and u a function belonging to $C_c^\infty(\Omega)$ and equal to one in some open neighborhood of x_o, U. In U we have

$$a(x, \xi) \sim \int A(x, y)u(y)\, e^{iy\cdot\xi}\, dy,$$

where $\sim$ stands for the natural asymptotic equivalence, for $|\zeta| \sim +\infty$. Let g be another function member of $C_c^\infty(\Omega)$ equal to one in a neighborhood of x_o, and Γ^o an open cone in $\mathbb{R}_n\backslash\{0\}$ containing ξ^o. Observe that (6.3) implies at once that the Fourier transform of $g(x)a(x, \xi)$ with respect to x, $\int e^{-ix\cdot\eta} g(x)a(x, \xi)\, dx$, decreases faster than any power of $(|\xi| + |\eta|)^{-1}$ when the latter tends to zero in $\Gamma^o \times \mathbb{R}_n$ (that is, ξ remains in Γ^o). This is of course equivalent to saying that

$$\iint \exp\{-i(x\cdot\eta - y\cdot\xi)\}g(x)A(x, y)u(y)\, dx\, dy$$

has the same property. Combining this with Definition 6.1 yields

PROPOSITION 6.6. *Let x_o belong to Ω, ξ^o to $\mathbb{R}_n\backslash\{0\}$. In order for (x_o, ξ^o) to belong to μ supp A, it is necessary and sufficient for there to be $\eta \in \mathbb{R}_n$ such that $(x_o, x_o, \eta, -\xi^o)$ belongs to the wave-front set of the distribution kernel $A(x, y)$.*

All preceding definitions and results extend to C^∞ manifolds, as we have already indicated. We now give some complements on the subject of wave-front sets and their transformations under C^∞ mappings. Here we deal with distributions on C^∞ manifolds.

Let $\mathfrak{M}$, $\mathfrak{M}'$ be two such manifolds and ϕ a C^∞ mapping of $\mathfrak{M}$ into $\mathfrak{M}'$. The pullback map $\phi^* : C^\infty(\mathfrak{M}') \to C^\infty(\mathfrak{M})$, that is, the map $f \mapsto f \circ \phi$, induces a linear map $C_c^\infty(\mathfrak{M}') \to C_c^\infty(\mathfrak{M})$ if and only if it is *proper*, i.e., if and only if preimages of compact subsets of $\mathfrak{M}'$ are compact sets in $\mathfrak{M}$. If this is the case, we may define the direct image map $\phi_* : \mathcal{D}'(\mathfrak{M}) \to \mathcal{D}'(\mathfrak{M}')$ as the transpose of ϕ^*.

PROPOSITION 6.7. *Suppose that the C^∞ mapping ϕ is proper, and let u be a distribution in $\mathfrak{M}$.*

A point (y, η) in $T^*\mathfrak{M}'\backslash 0$ belongs to $\mathrm{WF}(\phi_* u)$ if and only if there is x in $\operatorname{supp} u$ such that $\phi(x) = y$ and either $(x, {}^tJ_\phi(x)\eta) \in \mathrm{WF}(u)$ or else ${}^tJ_\phi(x)\eta = 0$.

As before we have denoted by ${}^tJ_\phi(x)$ the linear map $T^*_{\phi(x)}\mathfrak{M}' \to T^*_x\mathfrak{M}$ defined by ϕ.

We shall leave the proof of Proposition 6.7 as an exercise to the reader. Instead we look closer at the following particular case:

EXAMPLE 6.3. Suppose that $\mathfrak{M}$ is a C^∞ submanifold of $\mathfrak{M}'$ and ϕ is the natural injection of $\mathfrak{M}$ into $\mathfrak{M}'$. Then ϕ^* is simply the restriction to $\mathfrak{M}$ of C^∞ functions in $\mathfrak{M}'$. That ϕ is proper means exactly that $\mathfrak{M}$ is closed in $\mathfrak{M}'$. We assume this to be so.

For each x in $\mathfrak{M}$ the map ${}^tJ_\phi(x) : T^*_x\mathfrak{M}' \to T^*_x\mathfrak{M}$ is surjective. It is the transpose of the differential map at x, $\phi_* : T_x\mathfrak{M} \to T_x\mathfrak{M}'$. Its null space (or kernel) is the conormal space $N^*_x\mathfrak{M}$ to $\mathfrak{M}$ at x, that is, the orthogonal of $T_x\mathfrak{M}$ in $T_x\mathfrak{M}'$. As x ranges over $\mathfrak{M}$ the set-theoretical union of the linear spaces $N^*_x\mathfrak{M}$ is the *conormal bundle* over $\mathfrak{M}$, $N^*\mathfrak{M}$. This is a vector subbundle of the restriction of $T^*\mathfrak{M}'$ to $\mathfrak{M}$, $T^*\mathfrak{M}'|_{\mathfrak{M}}$.

Now, if u is a distribution in $\mathfrak{M}$, then $\phi_* u$ is the distribution

$$C_c^\infty(\mathfrak{M}') \ni g \mapsto \langle u, g|_{\mathfrak{M}} \rangle$$

on $\mathfrak{M}'$. Its wave-front set $\mathrm{WF}(\phi_* u)$ is the set of vectors (x, η) with $x \in \operatorname{supp} u \subset \mathfrak{M}$, and $\eta \in T^*_x\mathfrak{M}'\backslash 0$ such that either $\eta \in N^*_x\mathfrak{M}$ or else $(x, {}^tJ_\phi(x)\eta) \in \mathrm{WF}(u)$. In particular the piece of $N^*\mathfrak{M}\backslash 0$ that lies above the support of u (identified to a subset of $T^*\mathfrak{M}'$) is entirely contained in $\mathrm{WF}(\phi_* u)$.

EXAMPLE 6.4. Let $\mathfrak{M}$ be as in Example 6.3, a closed C^∞ submanifold of $\mathfrak{M}'$, and let $d\mu$ denote a C^∞ density nowhere vanishing on $\mathfrak{M}$. We regard it as a distribution on $\mathfrak{M}'$, $C_c^\infty(\mathfrak{M}') \ni g \mapsto \int_{\mathfrak{M}} g\, d\mu$. The wave-front set of this distribution is exactly equal to $N^*\mathfrak{M}$.

EXAMPLE 6.5. Let $\mathcal{Y}(t)$ denote the Heaviside function on the real line, i.e., the function that equals 1 for $t > 0$ and zero for $t < 0$. Let us take $\mathfrak{M} = \mathbb{R}^1$, $\mathfrak{M}' = \mathbb{R}^2$, and let ϕ be the injection $t \mapsto (t, 0)$. The direct image of $\mathcal{Y}(t)$ (as a distribution on $\mathbb{R}^1$) is the tensor product $\mathcal{Y}(x^1)\otimes\delta(x^2)$. We identify the cotangent bundle over $\mathbb{R}^2$ with $\mathbb{R}^2 \times \mathbb{R}_2$ where we denote the coordinates by (x^1, x^2, ξ_1, ξ_2). The conormal bundle to the image of $\mathbb{R}^1$ can then be identified to the set of points of the kind $(x^1, 0, 0, \xi_2)$. If we observe that the whole cotangent space at the origin to $\mathbb{R}^1$ belongs to the wave-front set of $\mathcal{Y}(t)$, we conclude that the wave-front set of $\mathcal{Y}(x^1)\otimes\delta(x^2)$ is equal to the subset of $\mathbb{R}^2 \times \mathbb{R}_2$ defined by the conditions

$$(6.18) \qquad x^1 > 0, x^2 = \xi_1 = 0, \xi_2 \neq 0 \quad \text{or} \quad x^1 = x^2 = 0, (\xi_1, \xi_2) \neq 0.$$

REMARK 6.2. When $\mathfrak{M}$ is a closed submanifold of $\mathfrak{M}'$, it is of course not true that every distribution in $\mathfrak{M}'$ whose support is contained in $\mathfrak{M}$ is of the form $\phi_* u$ with $u \in \mathcal{D}'(\mathfrak{M})$ and ϕ the natural injection $\mathfrak{M} \to \mathfrak{M}'$. For instance $\mathcal{Y}(x^1)\otimes\delta'(x^2)$ in $\mathfrak{M}' = \mathbb{R}^2$ is not of that form, when $\mathfrak{M}$ is the set of points $(t, 0)$ as in Example 6.5. Nevertheless, as one can easily check, the wave-front set of $Y(x^1)\otimes\delta'(x^2)$ is the set defined by (6.18).

REMARK 6.3. Suppose that $\mathfrak{M}$ is an open subset of $\mathfrak{M}'$ that is not closed, i.e., is not a union of connected components of $\mathfrak{M}'$. Then the natural injection map ϕ is obviously not proper. The direct image ϕ_* is defined only for compactly supported distributions in $\mathfrak{M}$. But we may consider the restriction to $\mathfrak{M}$ of distributions in $\mathfrak{M}'$. We have, with the obvious identifications,

$$(6.19) \qquad \mathrm{WF}(v|_{\mathfrak{M}}) = \mathrm{WF}(v)|_{\mathfrak{M}}, \qquad v \in \mathcal{D}'(\mathfrak{M}').$$

The most precise way of formulating the preceding microlocal concepts and results is by way of sheaves, the sheaf of microdistributions and the sheaf of pseudodifferential operators on a C^∞ manifold $\mathfrak{M}$. The reader who is not familiar with the basic definitions of sheaf theory (to be found, for instance, in Godement [1]) may skip the next two paragraphs.

For any open subset $\mathcal{U}$ of $T^*\mathfrak{M}\backslash 0$ we denote by $C^\infty((\mathcal{U}))$ the space of distributions u in $\mathfrak{M}$ which are C^∞ in the conic span of $\mathcal{U}$, that is, in the smallest conic set containing $\mathcal{U}$. We form the quotient space

$$(6.20) \qquad \mathcal{D}'((\mathcal{U})) = \mathcal{D}'(\mathfrak{M})/C^\infty((\mathcal{U})).$$

If $\mathscr{V} \subset \mathscr{U}$ is also open, we have $C^\infty((\mathscr{U})) \subset C^\infty((\mathscr{V}))$ and thus there is a natural "restriction map"

$$r^{\mathscr{U}}_{\mathscr{V}}: \mathscr{D}'((\mathscr{U})) \to \mathscr{D}'((\mathscr{V})). \tag{6.21}$$

The presheaf $(\mathscr{D}'((\mathscr{U})), r^{\mathscr{U}}_{\mathscr{V}})$ defines a sheaf over $T^*\mathfrak{M}\backslash 0$ called the *sheaf of microdistributions* in $\mathfrak{M}$, which we denote by $\mathscr{D}'(\mathfrak{M})$. Any distribution u in $\mathfrak{M}$ defines a section $u^\#$ of this sheaf. The support of $u^\#$, i.e., the set of points (x, ξ), $\xi \neq 0$, such that $u^\#(x, \xi) \neq 0$, is equal to $\mathrm{WF}(u)$.

Let us now denote by $\Psi^{-\infty}((\mathscr{U}))$ the space of pseudodifferential operators in $\mathfrak{M}$ which are regularizing in the conic span of $\mathscr{U}$, and define the quotient space

$$\Psi^m((\mathscr{U})) = \Psi^m(\mathfrak{M})/[\Psi^m(\mathfrak{M}) \cap \Psi^{-\infty}((\mathscr{U}))]. \tag{6.22}$$

If $\mathscr{V} \subset \mathscr{U}$ we have $\Psi^{-\infty}((\mathscr{U})) \subset \Psi^{-\infty}((\mathscr{V}))$ whence the "restriction map"

$$R^{\mathscr{U}}_{\mathscr{V}}: \Psi^m((\mathscr{U})) \to \Psi^m((\mathscr{V})). \tag{6.23}$$

The presheaf $(\Psi^m((\mathscr{U})), R^{\mathscr{U}}_{\mathscr{V}})$ defines a sheaf over $T^*\mathfrak{M}\backslash 0$, called the *sheaf of pseudodifferential operators of order m* in $\mathfrak{M}$, which we denote by $\boldsymbol{\Psi}^m(\mathfrak{M})$. The union of the sheaves $\boldsymbol{\Psi}^m(\mathfrak{M})$ as m ranges over $\mathbb{R}$ makes up *the sheaf* $\boldsymbol{\Psi}(\mathfrak{M})$ *of pseudodifferential operators in* $\mathfrak{M}$. A section $A^\#$ of $\boldsymbol{\Psi}^m(\mathfrak{M})$ can be made to act upon a section $u^\#$ of $\mathscr{D}'(\mathfrak{M})$, assuming that both are sections over the same subset of $T^*\mathfrak{M}\backslash 0$, Σ. Indeed for each point (x, ξ) of Σ we can find an open neighborhood of that point, which we may assume to be conic, Γ, such that $A^\#(x, \xi)$ and $u^\#(x, \xi)$ both have representatives in $\Psi^m((\Gamma))$ and $\mathscr{D}'((\Gamma))$ respectively, denoted by $\tilde{A}$ and $\tilde{u}$; in turn these have representatives, denoted by A and u, in $\Psi^m(\mathfrak{M})$ and $\mathscr{D}'(\mathfrak{M})$ respectively. Nothing is changed if we take A to be properly supported. We may then form Au and take its canonical image first in $\mathscr{D}'((\Gamma))$, then in the stalk of $\mathscr{D}'(\mathfrak{M})$ at (x, ξ). By definition this is the value of $A^\# u^\#$ at that point. In this notation the relation (6.14) simply reads

$$\mathrm{supp}(A^\# u^\#) \subset (\mathrm{supp}\, A^\#) \cap (\mathrm{supp}\, u^\#). \tag{6.24}$$

One can add precision to the evaluation of regularity in the cotangent bundle by introducing intermediate spaces between $C^\infty((\Gamma))$ and $\mathscr{D}'((\Gamma))$ (for any conic open subset Γ of $T^*\mathfrak{M}\backslash 0$). For instance, it is easy to define the Sobolev spaces $H^s_{\mathrm{loc}}((\Gamma))$ for any real number s:

Definition 6.5. *Let s be any real number, Γ any conic open subset of $T^*\mathfrak{M}\backslash 0$. We say that a distribution u in $\mathfrak{M}$ belongs to $H^s_{\mathrm{loc}}((\Gamma))$ if $Bu \in H^s_{\mathrm{loc}}((\mathfrak{M}))$ for any properly supported pseudodifferential operator of order zero in $\mathfrak{M}$ whose microsupport is conically compact and contained in Γ.*

By virtue of the equivalence of (6.9) and (6.11') (Proposition 6.1), we see that the intersection of all the spaces $H^s_{\mathrm{loc}}((\Gamma))$ as s ranges over $\mathbb{R}$ is equal to $C^\infty((\Gamma))$.

This definition is invariant under diffeomorphism, by virtue of Theorem 3.3 and Corollary 6.1.

The next statement "microlocalizes" Theorem 2.1:

PROPOSITION 6.8. *Let s and Γ be as in Definition 6.5. If A is any properly supported pseudodifferential operator of order m, $AH^s_{\mathrm{loc}}((\Gamma)) \subset H^{s-m}_{\mathrm{loc}}((\Gamma))$.*

PROOF. If Γ' is any conically compact conic open subset of Γ, we may write $A = A_o + A_1$ with A_o and A_1 two properly supported pseudodifferential operators of order m in $\mathfrak{M}$, such that the microsupport of A_o is conically compact and contained in Γ, and A_1 is regularizing in Γ'. Let E and F be two properly supported elliptic operators in $\mathfrak{M}$, of order m and $-m$ respectively, such that $EF - I$ and $FE - I$ are regularizing in the whole of Γ (see end of Section 5). Let us set $B_o = FA_o$; A is congruent modulo $\Psi^{-\infty}(\mathfrak{M})$ to $EB_o + A_1$. If $u \in H^s_{\mathrm{loc}}((\Gamma))$, we know that $B_o u \in H^s_{\mathrm{loc}}(\mathfrak{M})$; hence $EB_o u \in H^{s-m}_{\mathrm{loc}}(\mathfrak{M})$. On the other hand, $A_1 u \in C^\infty((\Gamma))$, and hence $BA_1 u \in C^\infty(\mathfrak{M})$ for any pseudodifferential operator B whose microsupport is conically compact and contained in Γ'. We reach the conclusion that if the order of B is equal to zero, $BAu \in H^{s-m}_{\mathrm{loc}}(\mathfrak{M})$; hence $Au \in H^{s-m}_{\mathrm{loc}}((\Gamma'))$. By letting Γ' grow to Γ we obtain the desired result. □

REMARK 6.4. If the conic set Γ is of the form $T^*\mathcal{O}\backslash 0$, with $\mathcal{O}$ any open subset of $\mathfrak{M}$, then $C^\infty((\Gamma))$ (resp., $H^s((\Gamma))$ is the space of distributions in $\mathfrak{M}$ whose restriction to $\mathcal{O}$ belongs to $C^\infty(\mathcal{O})$ (resp., to $H^s_{\mathrm{loc}}(\mathcal{O})$). One should be careful not to identify this space with $C^\infty(\mathcal{O})$ (resp., with $H^s_{\mathrm{loc}}(\mathcal{O})$), except in a few trivial situations.

One can also microlocalize the notion of *ellipticity*. We shall do it in the cotangent bundle over the open subset Ω of $\mathbb{R}^n$. The generalization to an arbitrary C^∞ manifold is done by using local charts and the symbols in those charts. It is straightforward and we leave it to the reader.

DEFINITION 6.6. *We say that $a(x, \xi) \in S^m(\Omega)$ is elliptic in a conic open subset Γ of $\Omega \times (\mathbb{R}_n\backslash\{0\})$ if, given any conic subset Γ' of Γ whose closure in Γ is conically compact, there are numbers $\rho, c > 0$ such that*

$$|a(x, \xi)| \geq c|\xi|^m, \qquad \forall (x, \xi) \in \Gamma', |\xi| > \rho. \tag{6.25}$$

The same description "elliptic in the conic open set Γ" applies to any symbol class belonging to $\dot{S}^m(\Omega)$ having a representative such as $a(x, \xi)$, while we shall call "elliptic of order m in Γ" any pseudodifferential operator or class of pseudodifferential operators mod $\Psi^{-\infty}(\Omega)$ whose symbol class has such a representative.

Proposition 4.1 and its corollaries have microlocal generalizations, with a slight constraint: pseudodifferential operators that are elliptic in the conic open set Γ can be inverted but only in conic sets whose closure is conically compact and contained in Γ.

PROPOSITION 6.9. *The following properties of* $A \in \Psi^m(\Omega)$ *are equivalent*:

(i) *A is elliptic of order m in the conic open set Γ.*

(ii) *Given any conic open subset Γ' of Γ whose closure in Γ is conically compact, there is $B \in \Psi^{-m}(\Omega)$ such that $AB - I$ and $BA - I$ are regularizing in Γ'.*

PROOF. Suppose that the symbol $a(x, \xi)$ of A is elliptic in Γ (see Definition 6.6). We determine recursively the terms b_j in the formal symbol $\sum_{j=0}^{+\infty} b_j(x, \xi)$ by formula (4.39) and either (4.40) or (4.41). We can do this only in Γ. Then let $g \in C^\infty(\Omega \times (\mathbb{R}_n \backslash \{0\}))$ be positive-homogeneous of degree zero, have its support conically compact and contained in Γ, and be equal to one in Γ'. Then $\sum_{j=0}^{+\infty} g(x, \xi) b_j(x, \xi)$ is a bona fide formal symbol over Ω, of degree $-m$, which defines a symbol class $\dot{b}$, which in turn defines a class of pseudodifferential operators $\dot{B} \in \Psi^m(\Omega)$. Any representative B of $\dot{B}$ has the property required in (ii).

The proof of the converse, that (ii) implies (i), is a routine generalization of the proofs of the implications (ii) $\Rightarrow$ (i), (iii) $\Rightarrow$ (i) in Proposition 4.1. $\square$

REMARK 6.5. If, instead of dealing with a pseudodifferential operator A in Ω, we had dealt with a section $A^\#$ of the sheaf $\boldsymbol{\Psi}^m(\Omega)$ over the conic open set Γ, then condition (ii) could have been replaced by the following, stronger condition:

(ii′) *There exists a section $B^\#$ over Γ of the sheaf $\boldsymbol{\Psi}^{-m}(\Omega)$ such that* $A^\# B^\# = B^\# A^\# = I^\#$ (over Γ; $I^\#$ is the section "identity").

REMARK 6.6. If A is elliptic of order m in Γ, the operator B in (ii) is elliptic of order $-m$ in Γ'.

Proposition 6.10. *Let A be a properly supported pseudodifferential operator in Ω, elliptic in the conic open set $\Gamma \subset \Omega \times (\mathbb{R}_n \backslash \{0\})$. Then, whatever the distribution u in Ω,*

$$\mathrm{WF}(Au) \cap \Gamma = \mathrm{WF}(u) \cap \Gamma. \tag{6.26}$$

Proof. We have $\mathrm{WF}(Au) \cap \Gamma \subset \mathrm{WF}(u) \cap \Gamma$ by Corollary 6.2. Let B be as in (ii) of Proposition 6.9. We know that $u - BAu$ is C^∞ in Γ'; hence $\mathrm{WF}(u) \cap \Gamma' = \mathrm{WF}(BAu) \cap \Gamma' \subset \mathrm{WF}(Au) \cap \Gamma'$. By letting Γ' grow to Γ we see that $\mathrm{WF}(u) \cap \Gamma \subset \mathrm{WF}(Au) \cap \Gamma$. □

Remark 6.7. Proposition 6.10 confirms the fact, already obvious by Definition 6.6, that if A is elliptic in the conic open set Γ, then $\Gamma \subset \mu$ supp A (cf. Proposition 6.4).

Example 6.6. Let $g(x, \xi)$ be as in Example 6.1: C^∞ in $\Omega \times (\mathbb{R}_n \backslash \{0\})$, positive-homogeneous of degree zero, supp $g \subset \Gamma$. Suppose furthermore that $g \equiv 1$ in $\Gamma' \subset \Gamma$. Then every pseudodifferential operator congruent to $g(x, D)$ (see (6.6)) mod $\Psi^{-\infty}(\Omega)$ is elliptic of order zero in Γ'. In particular the operator $h(D)g(x)$ of Example 6.2 is elliptic of order zero in $U' \times \Gamma'^o$.

Consider now a classical pseudodifferential operator P of degree m (see Examples 4.2 and 5.1) on a C^∞ manifold $\mathfrak{M}$, or at any rate a pseudodifferential operator having a principal symbol $p(x, \xi)$. We recall that $p(x, \xi)$ is a C^∞ function in $T^*\mathfrak{M}\backslash 0$, positive-homogeneous of degree m and that if A is any pseudodifferential operator in any open subset U of $\mathfrak{M}$ with principal symbol equal to $p(x, \xi)$ there, then $P - A$ is a pseudodifferential operator of order $<m$ in U.

Definition 6.7. *The subset of $T^*\mathfrak{M}\backslash 0$ defined by $p(x, \xi) = 0$ is called the characteristic set of the operator P. We shall denote it by* Char P.

Definition 6.7 generalizes the traditional terminology for differential operators. Char P is a closed conic set; in its complement P is elliptic of order m. Then Proposition 6.10 implies the following Corollary.

Corollary 6.7. *Let $u \in \mathcal{D}'(\mathfrak{M})$ such that $Pu \in C^\infty(\mathfrak{M})$. Then*

$$\mathrm{WF}(u) \subset \mathrm{Char}\, P. \tag{6.27}$$

Appendix: Traces and Multiplication of Distributions Whose Wave-Front Sets Are in Favorable Positions

Let us separate the coordinates in $\mathbb{R}^n$ into two sets, $x' = (x^1, \ldots, x^p)$, $x'' = (x^{p+1}, \ldots, x^n)$, with $1 \le p \le n-1$. Consider a compactly supported distribution u in $\mathbb{R}^n$ that has the following property:

(6.28) *There is a number $\varepsilon > 0$ such that to every $M > 0$ there is a constant $C_M > 0$ such that*

$$|\hat{u}(\xi)| \le C_M(1+|\xi|)^{-M} \qquad \textit{if } |\xi'| \le \varepsilon |\xi''|. \tag{6.29}$$

Of course, $\xi' = (\xi_1, \ldots, \xi_p)$, $\xi'' = (\xi_{p+1}, \ldots, \xi_n)$. Since the Fourier transform $\hat{u}$ of u is tempered, for a suitable choice of C, m, we have

$$|\hat{u}(\xi)| \le C(1+|\xi|)^m, \qquad \xi \in \mathbb{R}_n. \tag{6.30}$$

This implies that if $|\xi''| < \varepsilon^{-1}|\xi'|$, then for a suitable choice of C,

$$|\hat{u}(\xi)| \le C(1+|\xi'|)^{m+n+1}(1+|\xi''|)^{-n-1}. \tag{6.31}$$

By (6.28) this is certainly also valid if $|\xi''| \ge \varepsilon^{-1}|\xi'|$. It follows from this that

$$\hat{u}_o(\xi') = (2\pi)^{-n+p} \int \hat{u}(\xi', \xi'')\, d\xi'' \tag{6.32}$$

is tempered. The inverse Fourier transform of $\hat{u}_o(\xi')$,

$$u_o(x') = (2\pi)^{-p} \int e^{ix'\cdot\xi'} \hat{u}_o(\xi')\, d\xi', \tag{6.33}$$

can be regarded as the *trace* of the distribution u on the vector subspace $x'' = 0$. If $u \in C_c^\infty(\mathbb{R}^n)$, we indeed have $u_o(x') = u(x', 0)$; if u is a distribution in $\mathscr{E}'(\mathbb{R}^n)$ satisfying (6.28), with u the limit of a sequence of elements $u_{(j)}$ of $C_c^\infty(\mathbb{R}^n)$, then the $u_{(j)}(x', 0)$ converge to $u_o(x')$ in $\mathscr{E}'(\mathbb{R}^p)$. One can also observe that the inequality (6.31) implies that u is a continuous function of x'' with values in the space of tempered distributions $v(x')$ such that $(1+|\xi'|)^{-m-n-1}\hat{v}(\xi') \in L^\infty(\mathbb{R}^p)$ and, as such, has a trace when $x'' = 0$.

In order to microlocalize the preceding remark we note that (6.28) implies that the wave-front set of u does not intersect the subspace $\xi' = 0$ of $\mathbb{R}_n$. This subspace can be identified to the conormal bundle of the subspace $x'' = 0$. Now let v be an arbitrary distribution in $\mathbb{R}^n$. Suppose that given any point x_o such that $x_o'' = 0$, there is a function $g \in C_c^\infty(\mathbb{R}^n)$ equal to one in a neighborhood of x_o such that gv has property (6.28). Then we may define the trace of gv on the subspace $x'' = 0$, hence that of v itself in the neighborhood of x_o and finally that of v on the whole subspace, by letting x_o

vary in it. In other words, if the wave-front set of a distribution v does not intersect the conormal bundle of a vector subspace, the trace of v on this subspace is defined. But since all the notions we are using are invariant under diffeomorphism, we may state

PROPOSITION 6.11. *Let $\mathfrak{M}'$ be a C^∞ submanifold of the C^∞ manifold $\mathfrak{M}$, v a distribution in $\mathfrak{M}$ whose wave-front set does not intersect the conormal bundle of $\mathfrak{M}'$. Then the trace of v on $\mathfrak{M}'$ is well defined.*

Inspection of this argument shows that v can be regarded as a continuous function of variables transversal to $\mathfrak{M}'$ with values in the space of distributions on $\mathfrak{M}'$, this in a suitable neighborhood of $\mathfrak{M}'$. The trace of v is then the value of that continuous function when the transversal variables take the values that define $\mathfrak{M}'$ within the neighborhood in question.

We shall now give an important application of Proposition 6.11. Let u, v be two distributions in $\mathfrak{M}$. Let us regard u as a distribution on $\mathfrak{M}$, v as a distribution on a copy of $\mathfrak{M}$; in order to distinguish them, we call $u(x)$ the first one and $v(y)$ the second one. We may then form the tensor product $u(x) \otimes v(y)$, which is a distribution (kernel) on $\mathfrak{M} \times \mathfrak{M}$. We leave the proof of the following statement as an exercise:

PROPOSITION 6.12. *Let u be a distribution on a C^∞ manifold X, v a distribution on a C^∞ manifold Y. Then*

$$\text{(6.34)} \quad \mathrm{WF}(u \otimes v) = [\mathrm{WF}(u) \times \mathrm{WF}(v)] \cup [\mathrm{WF}(u) \times \{0\}] \cup [\{0\} \times \mathrm{WF}(v)]$$

({0}: zero section).

Then taking $X = \mathfrak{M}$, $Y = \mathfrak{M}$ as before, we introduce the following condition:

(6.35) $\mathrm{WF}(u) \times \mathrm{WF}(v)$ *does not intersect the conormal bundle of the diagonal $\Delta_{\mathfrak{M}}$ in $\mathfrak{M} \times \mathfrak{M}$.*

The vectors tangent to the diagonal $\Delta_{\mathfrak{M}}$ are of the form (in local coordinates)

$$\sum_{j=1}^{n} \alpha^j \left(\frac{\partial}{\partial x^j} + \frac{\partial}{\partial y^j} \right);$$

the covectors at a point (x, x) orthogonal to these vectors are of the form $(\xi, -\xi)$. In other words, the conormal bundle of $\Delta_{\mathfrak{M}}$ is the subset of $T^*(\mathfrak{M} \times \mathfrak{M})$ consisiting of the points of the form $(x, x, \xi, -\xi)$. Let us then call $\mathrm{WF}'(v)$ the image of $\mathrm{WF}(v)$ under the symmetry $(x, \xi) \mapsto (x, -\xi)$. By virtue of Proposition (6.11) we may state the following proposition:

PROPOSITION 6.13. *If u, v are two distributions in $\mathfrak{M}$ such that*

$$\mathrm{WF}(u) \cap \mathrm{WF}'(v) = \varnothing, \tag{6.36}$$

the product uv is well defined.

Indeed $u(x)v(x)$ is the trace of $u(x) \otimes v(y)$ on the diagonal of $\mathfrak{M} \times \mathfrak{M}$.

Proposition 6.13 generalizes the known property that one can multiply two distributions whose singular supports are disjoint.

EXAMPLE 6.7. Consider the distributions on the real line considered on p. 61:

$$pv^{+}\frac{1}{x} = \lim_{\varepsilon \to +0} (x + i\varepsilon)^{-1}, \qquad pv^{-}\frac{1}{x} = \lim_{\varepsilon \to +0} (x - i\varepsilon)^{-1}. \tag{6.37}$$

The wave-front set of $pv^{+}(1/x)$ is the half-line $\{(x, \xi); x = 0, \xi > 0\}$, that of $pv^{-}(1/x)$ is $\{(x, \xi); x = 0, \xi < 0\}$. Thus we have the right to multiply $pv^{+}(1/x)$ with itself, $pv^{-}(1/x)$ with itself, but not $pv^{+}(1/x)$ with $pv^{-}(1/x)$.

7. Standard Pseudodifferential Operators Acting on Vector-Valued Distributions and on Sections of Vector Bundles

We continue to denote by $\mathfrak{M}$ a C^∞ manifold of dimension n, countable at infinity. Let E denote a complex vector space, of dimension $d < +\infty$. The E-valued distributions in $\mathfrak{M}$ are the continuous linear mappings of the space of test functions in $\mathfrak{M}$, $C_c^\infty(\mathfrak{M})$, into E. They can be identified to the (classes) of finite sums† $S_1 v_1 + \cdots + S_r v_r$ where the v_j's are vectors in E and the S_j's are scalar distributions in $\mathfrak{M}$. Of course, it is convenient to take vectors v_j that make up a basis of E (which insure the uniqueness of the preceding finite-sum representations; then $r = d$) and regard E-valued distributions as systems of d scalar ones. However, in many important applications, one needs to be free to change basis, and thus the invariant definition is preferable. We shall denote by $\mathscr{D}'(\mathfrak{M}; E)$ the space of E-valued distributions in $\mathfrak{M}$, by $\mathscr{E}'(\mathfrak{M}; E)$ the subspace of the compactly supported ones. The topologies in these spaces are obvious generalizations of those in the scalar case.

Let F be another complex vector space, of dimension d', and let $L(E; F)$ denote the space of linear mappings $E \to F$. The continuous linear mappings $\mathscr{E}'(\mathfrak{M}; E) \to \mathscr{D}'(\mathfrak{M}; F)$ can be identified with the finite sums

† The right way to define vector-valued functions and distributions is by using tensor products (see Treves [2], Part III).

$A_1f_1 + \cdots + A_sf_s$ where the f_j's are linear mappings $E \to F$ and A_j, continuous linear mappings $\mathscr{E}'(\mathfrak{M}) \to \mathscr{D}'(\mathfrak{M})$. If we are willing to use a basis in E and one in F, the linear mappings $E \to F$ can be represented by $d \times d'$ matrices, and any continuous linear map $A : \mathscr{E}'(\mathfrak{M}; E) \to \mathscr{D}'(\mathfrak{M}; F)$ can be represented by a $d \times d'$ matrix whose entries are continuous linear mappings $\mathscr{E}'(\mathfrak{M}) \to \mathscr{D}'(\mathfrak{M})$, A^j_k. We say that A is a *pseudodifferential operator valued in* $L(E; F)$ if each A^j_k is a (scalar) pseudodifferential operator. If all the A^j_k have order m, we say that the order of A is m.

We shall denote by $\Psi^m(\mathfrak{M}; L(E; F))$ the space of pseudodifferential operators in the manifold $\mathfrak{M}$, of order m, valued in $L(E; F)$. Of course,

$$\Psi^m(\mathfrak{M}; L(E; F)) = \Psi^m(\mathfrak{M}) \otimes L(E; F). \tag{7.1}$$

The theory developed so far for scalar pseudodifferential operators extends practically without change to operators in (7.1). As we indicate later some care should be taken in dealing with matrices that do not commute.

Thus a pseudodifferential operator in an open subset Ω of $\mathbb{R}^n$, A, valued in $L(E; F)$, is an operator Op a, where a is an amplitude valued in $L(E; F)$. The definition of such amplitudes is obvious: insert in Definition 2.1 that the C^∞ function $a(x, y, \xi)$ in $\Omega \times \Omega \times \mathbb{R}_n$ should now be valued in $L(E; F)$, and replace the absolute values on the left-hand side of (2.5) by the operator (or matrix) norm.

Let $H^s(\mathbb{R}^n; E)$ denote the sth Sobolev space of E-valued distributions; they are finite sums $S_1v_1 + \cdots + S_rv_r$ with $S_j \in H^s(\mathbb{R}^n)$. We define exactly as in the scalar case the "derived" spaces $H^s_c(\Omega; E)$, $H^s_{\text{loc}}(\Omega; E)$, etc. Theorems 2.1 and 2.2 extend routinely: thus if $A \in \Psi^m(\Omega; L(E; F))$, it defines a continuous linear map $H^s_c(\Omega; E) \to H^{s-m}_{\text{loc}}(\Omega; F)$; and if $u \in \mathscr{E}'(\Omega)$ is a C^∞ function in an open subset U of Ω valued in E, then Au is a C^∞ function in U valued in F.

Likewise the results of Section 3 extend routinely, with the obvious modification: if $A \in \Psi^m(L(E; F))$, its transpose ${}^tA$ and its adjoint A^* belong to $\Psi^m(L(F^*; E^*))$, where the upper asterisk * indicates the *dual space*. If G is a third complex vector space (also of finite dimension) and $B \in \Psi^{m'}(L(F; G))$, and if we assume that either A or B is properly supported, then $B \circ A \in \Psi^{m+m'}(L(E; G))$.

The symbolic calculus for matrix-valued pseudodifferential operators is a routine extension of that for scalar ones, as described in Section 4. The symbols, and the various terms in the formal symbols are now matrices or (in base-free definition) linear mappings. The only difference between the vector and the scalar situations arise from the possible noncommutation of matrices.

For instance, let us take $F = E$ and denote by $L(E)$ the space of automorphisms of the vector space E. Suppose that A and B are pseudodifferential operators in Ω (of degrees m, m' respectively) with values in $L(E)$. If $\dim E > 1$ and if one does not make the explicit assumption that the symbols (which are $d \times d$ matrices) $a(x, \xi)$, $b(x, \xi)$ of A and B commute for all x, ξ, then Corollary 4.2 fails to hold. The commutator $[A, B]$ is not of order $m + m' - 1$; in general it will be of order $m + m'$. All that can be said about its symbol is that it is congruent modulo $\Psi^{m+m'-2}(\Omega)$ to

$$[a(x, \xi), b(x, \xi)] + \frac{1}{\sqrt{-1}}\{a(x, \xi), b(x, \xi)\}, \tag{7.2}$$

where the first term is the commutation bracket between matrices and the second one is the Poisson bracket between matrix-valued functions.

Certain applications require one to deal with pseudodifferential operators valued in spaces of bounded linear operators on infinite-dimensional Banach spaces (see, e.g., Treves [5]). We shall not dwell on this topic. Suffice it to say that the extension to this case presents no major difficulties. On the subject of distributions valued in Banach spaces refer to Treves [3], Section 39. The tensor product representation (7.1) is no longer valid; one must take a suitable completion of the right-hand side, but this can be done if needed.

Another generalization needed in the applications is that to vector bundles, specifically, to real or complex vector bundles over one and the same base manifold, for us $\mathfrak{M}$. Let E be a vector space, of dimension d, either over the real field $\mathbb{R}$ or over the complex field $\mathbb{C}$ (we shall refer to this field as the *scalar* field). Let then $\mathbb{E}$ be a *real* C^∞ manifold. We suppose that there is a C^∞ mapping π of $\mathbb{E}$ *onto* $\mathfrak{M}$, called the base projection, and an open covering $\{\mathcal{O}_j\}_{j\in J}$ of $\mathfrak{M}$, and, for each index $j \in J$, a *diffeomorphism* g_j *of* $\overset{-1}{\pi}(\mathcal{O}_j)$ *onto* $\mathcal{O}_j \times E$, with the following properties (valid for all indices, j, j' in J):

(7.3) *Let* $\omega \in \mathbb{E}$, $\pi(\omega) = x \in \mathcal{O}_j$; *then* $g_j(\omega) = (x, e)$ *for some* $e \in E$ (in other words, g_j commutes with the base projections).

(7.4) *If x belongs to* $\mathcal{O}_j \cap \mathcal{O}_{j'}$ *and e is any vector in E, define* $g_{jj'}(x)e$ *as the unique vector* $\tilde{e}$ *of E such that* $(x, \tilde{e}) = (g_j \circ \overset{-1}{g}_{j'})(x, e)$; *then* $g_{jj'}$ *is a* C^∞ *function of x in* $\mathcal{O}_j \cap \mathcal{O}_{j'}$ *valued in the group of automorphisms of E.*

Under these conditions we refer to $\mathbb{E}$ as a (complex or real) vector bundle over $\mathfrak{M}$. If E is a real vector space, then the dimension of the manifold $\mathbb{E}$ is $m + d$; if E is complex, the dimension is $m + 2d$. For each $x \in \mathfrak{M}$, $\overset{-1}{\pi}(x) = \mathbb{E}_x$

is the *fiber* at x; the map $\omega \mapsto g_j(\omega) = (x, e) \mapsto e$ allows us to equip $\mathbb{E}_x$ with a linear space structure; thus $\mathbb{E}_x$ is a copy of E. We shall call a *local chart* of the bundle $\mathbb{E}$ any pair $(\mathcal{O}, g)$ made up of an open subset $\mathcal{O}$ of $\mathfrak{M}$ and of a diffeomorphism g of $\overset{-1}{\pi}(\mathcal{O})$ onto $\mathcal{O} \times E$ such that (1) g commutes with the base projection into $\mathcal{O}$; (2) given any pair $(\mathcal{O}_j, g_j)$, any point x of $\mathcal{O}_j \cap \mathcal{O}$, the map that assigns to $e \in E$ the unique $\tilde{e} \in E$ such that $g_j(x, e) = g(x, \tilde{e})$ is an automorphism of E (which depends smoothly on x).

This definition suggests a method for *constructing* fiber bundles over the manifold $\mathfrak{M}$ with typical fiber E: again let $\{\mathcal{O}_j\}_{j \in J}$ be an open covering of $\mathfrak{M}$ and suppose that for each pair j, j' of indices in J, we are given a C^∞ function $g_{jj'}(x)$ of $\mathcal{O}_j \cap \mathcal{O}_{j'}$ into the group of automorphisms of E, submitted to the following "coherence" conditions:

$$(7.5) \qquad g_{jj} = \textit{Identity of } E; \qquad g_{j'j} = \overset{-1}{g}_{jj'}; \qquad g_{jj''} = g_{jj'} \circ g_{j'j''}$$

if j'' is also an index in J (the last equality must hold for x in $\mathcal{O}_j \cap \mathcal{O}_{j'} \cap \mathcal{O}_{j''}$). The $g_{jj'}$ are sometimes called "transition functions." Let $\tilde{\mathbb{E}}$ denote the set-theoretical union of all the product sets $\mathcal{O}_j \times E, j \in J$. Let us say that $(x, e) \in \mathcal{O}_j \times E$ and $(x', e') \in \mathcal{O}_{j'} \times E$ are equivalent if $x = x'$ and if $g_{jj'}(x)e' = e$. One can easily ascertain that the quotient set $\mathbb{E}$ of $\tilde{\mathbb{E}}$ modulo this equivalence relation can be equipped with a structure of vector bundle over $\mathfrak{M}$. Application: take $E = \mathbb{R}^n$ (with its canonical basis) and assume that each open set $\mathcal{O}_j$ is the domain of local coordinates $x_j^1, \ldots, x_j^n$; then let $g_{jj'}$ be the transpose of the Jacobian matrix defined by the change of coordinates $x_j \to x_{j'}$. The vector bundle thus obtained is isomorphic to the cotangent bundle $T^*\mathfrak{M}$.

The vector bundle $\mathbb{E}$ is said to be *trivial* if there is a diffeomorphism of $\mathbb{E}$ onto $\mathfrak{M} \times E$ which commutes with the base projections.

If one uses a linear basis in E (over its scalar field, be it $\mathbb{R}$ or $\mathbb{C}$), then the automorphisms $g_{jj'}$ of (7.4) are represented by invertible $d \times d$ matrices. The entries of these matrices are smooth functions in $\mathcal{O}_j \cap \mathcal{O}_{j'}$.

Sections of the bundle $\mathbb{E}$, say over an open subset U of $\mathfrak{M}$, are easy to define: they are mappings $f: U \mapsto \mathbb{E}$ such that $\pi \circ f$ is the identity of U. Let us denote by $C^\infty(\mathfrak{M}; E)$ the space of C^∞ sections of $\mathbb{E}$ over $\mathfrak{M}$, by $C_c^\infty(\mathfrak{M}; E)$ the subspace of the compactly supported ones; both spaces carry natural topologies. *Distribution sections* of the bundle $\mathbb{E}$ have a slightly more complicated definition. Let $\{\mathcal{O}_j\}_{j \in J}$ be the (or any) open covering of $\mathfrak{M}$ and $\{g_j\}$ the collection of associated diffeomorphisms, used to define the bundle structure on $\mathbb{E}$. Let there be given for each index $j \in J$ an E-valued distribution T_j in $\mathcal{O}_j$; the collection $\{T_j\}$ can be regarded as a distribution section of $\mathbb{E}$ if given any pair of indices j, j' in J, we have, with the notation of (7.4),

$$g_{jj'}(x)T_{j'} = T_j \qquad \textit{in } \mathcal{O}_j \cap \mathcal{O}_{j'}. \tag{7.6}$$

It is clear that the coupling of a distribution in $\mathcal{O}_j \cap \mathcal{O}_{j'}$ valued in E with a linear operator in E that depends in C^∞ fashion on x in $\mathcal{O}_j \cap \mathcal{O}_{j'}$ is well defined. (We could have defined a distribution section of $\mathbb{E}$ as a continuous linear functional on the space of compactly supported C^∞ sections of $\mathbb{E}^*$, the dual bundle—that is, the bundle with typical fiber E^*, dual space of E. But the latter definition does not generalize to fiber bundles with typical fiber a nonreflexive Banach space.) We denote by $\mathcal{D}'(\mathfrak{M}; \mathbb{E})$ the space of distribution sections of $\mathbb{E}$, by $\mathcal{E}'(\mathfrak{M}, \mathbb{E})$ the subspace of the compactly supported ones. Both spaces can be equipped with natural topologies, mimicking those in the scalar case.

Let $\mathbb{E}$, $\mathbb{F}$ be two complex vector bundles on the same manifold $\mathfrak{M}$; let E and F be their respective typical fibers. A *bundle homomorphism* of $\mathbb{E}$ into $\mathbb{F}$ is a C^∞ mapping f of the underlying C^∞ manifolds that commutes with the base projections onto $\mathfrak{M}$. Furthermore, if $(\mathcal{O}_j, g_j)$ and $(\mathcal{O}'_k, g'_k)$ are two local charts for the bundle structure on $\mathbb{E}$ and $\mathbb{F}$, respectively, then the restrictions to $\{x\} \times E$, $x \in \mathcal{O}_j \cap \mathcal{O}'_k$, of the C^∞ mapping

$$g'_k \circ f \circ \overset{-1}{g_j} : (\mathcal{O}_j \cap \mathcal{O}'_k) \times E \to (\mathcal{O}_j \cap \mathcal{O}'_k) \times F$$

induce a linear map $E \to F$ (which of course is a C^∞ function of x in $\mathcal{O}_j \cap \mathcal{O}'_k$). We see that any bundle homomorphism of $\mathbb{E}$ into $\mathbb{F}$ can be regarded as a C^∞ section of the bundle $\mathrm{Hom}_{\mathfrak{M}}(\mathbb{E}; \mathbb{F})$ over $\mathfrak{M}$; this is the bundle with typical fiber $L(E; F)$, defined in the obvious manner.

We may also deal with the bundle $\mathrm{Hom}_{\mathfrak{M}\times\mathfrak{M}}(\mathbb{E}; \mathbb{F})$ over $\mathfrak{M} \times \mathfrak{M}$: its typical fiber is also $L(E; F)$ but the base manifold is $\mathfrak{M} \times \mathfrak{M}$ instead of $\mathfrak{M}$. Let us define the transition functions: let $(\mathcal{O}_j, g_j)$, $(\mathcal{O}_{j'}, g_{j'})$ be two local charts in $\mathbb{E}$; $(\mathcal{O}'_k, g'_k)$, $(\mathcal{O}'_{k'}, g'_{k'})$ two local charts in $\mathbb{F}$. Suppose that (x, y) belongs to $(\mathcal{O}_j \times \mathcal{O}'_k) \cap (\mathcal{O}_{j'} \times \mathcal{O}'_{k'})$; we define an automorphism $G_{jk,j'k'}(x, y)$ of $L(E; F)$ by the formula

$$f \mapsto g_k(y) \circ \overset{-1}{g}_{k'}(y) \circ f \circ \overset{-1}{g}_{j'}(x) \circ g_j(x). \tag{7.7}$$

The coherence conditions (7.5) are obviously satisfied. This is the vector bundle one needs in order to generalize the Schwartz kernels theorem: every continuous linear map $K : C_c^\infty(\mathfrak{M}; \mathbb{E}) \mapsto \mathcal{D}'(\mathfrak{M}; \mathbb{F})$ is associated with a unique kernel $K(x, y)$, which is a distribution section of $\mathrm{Hom}_{\mathfrak{M}\times\mathfrak{M}}(\mathbb{E}; \mathbb{F})$, in such a way that $Ku(x) = \int K(x, y)u(y)\,dy$.†

† This is a generalized integral. Here more than ever one should use currents (or else distribution densities, see Section 3, Chapter VII) rather than plain distributions.

We may then transfer K by means of local charts $(\mathcal{O}_j, g_j)$, $(\mathcal{O}'_k, g'_k)$ like the preceding ones to a continuous linear operator $C_c^\infty(\mathcal{O}_j; E) \to \mathcal{D}'(\mathcal{O}'_k; F)$, K_{jk}. Let us say that K_{jk} is a *pseudodifferential operator of order m* if there is a pseudodifferential operator $K^{\#}_{jk}$ in $\mathfrak{M}$, with values in $L(E; F)$ such that given any $u \in C_c^\infty(\mathcal{O}_j; E)$, the restriction to $\mathcal{O}'_k$ of $K^{\#}_{jk}u$ is equal to $K_{jk}u$. Finally, we say that K itself is a *pseudodifferential operator of order m from* $\mathbb{E}$ *to* $\mathbb{F}$ if for all possible choices of pairs of local charts $(\mathcal{O}_j, g_j)$ and $(\mathcal{O}'_k, g'_k)$, the operator K_{jk} defined as we have just done is a pseudodifferential operator of order m.

Given a vector bundle $\mathbb{E}$ over $\mathfrak{M}$, we can associate with it a vector bundle $\mathbb{E}^{(*)}$ over the cotangent bundle $T^*\mathfrak{M}$, with the same typical fiber E. It suffices to specify that the fiber at any point (x, ξ) of $T^*\mathfrak{M}$ is the same as the one at x. Thus $\mathbb{E}^{(*)}$ can be regarded as the manifold of points (ω, ξ) with ξ a cotangent vector at $x = \pi(\omega)$; the projection into the base, which is here $T^*\mathfrak{M}$, is the map $(\omega, \xi) \mapsto (x, \xi)$. Such a vector bundle, actually with typical fiber $L(E; F)$, is needed to deal with the *principal symbol* of a pseudodifferential operator from $\mathbb{E}$ to $\mathbb{F}$. Observe that it makes sense to say that such a pseudodifferential operator P (say, of order m) is *classical*. Indeed, it suffices to give it a meaning when $\mathbb{E} = \mathfrak{M} \times E$ and $\mathbb{F} = \mathfrak{M} \times F$; for by using local charts in $\mathbb{E}$ and $\mathbb{F}$, we may always reduce the analysis to such a situation. Then it suffices to use linear bases in E and F and to require that the matrix that represents the operator has entries that are classical scalar pseudodifferential operators of order m in $\mathfrak{M}$. This reasoning also reveals the meaning of the principal symbol of P. Locally, and if we use bases in E and F, it will be a matrix whose entries are C^∞ functions in $T^*\mathfrak{M}\backslash 0$, positive-homogeneous of degree m with respect to ξ; globally it will be a C^∞ section over $T^*\mathfrak{M}\backslash 0$, again positive-homogeneous of degree m with respect to ξ, of the vector bundle over $T^*\mathfrak{M}$ with typical fiber $L(E; F)$, obtained by "extending" $\mathrm{Hom}_{\mathfrak{M}}(\mathbb{E}, \mathbb{F})$.

We shall now describe briefly some of the simplest examples; they all happen to be differential operators. In applications discussed in Section 2.2 of Chapter II we shall encounter truly pseudodifferential operators from one vector bundle to another.

EXAMPLE 7.1. We may regard the differential d as an operator from the trivial line bundle $\mathfrak{M} \times \mathbb{C}$ to the complexified cotangent bundle over $\mathfrak{M}$, $T^{*\mathbb{C}}\mathfrak{M}$. The latter is regarded as a complex vector bundle over $\mathfrak{M}$, with typical fiber $\mathbb{C}^n$. If $\mathcal{O}$ is the domain of local coordinates $x^1, \ldots, x^n$, then a diffeomorphism of $\overset{-1}{\pi}(\mathcal{O})$ onto $\mathcal{O} \times \mathbb{C}^n$ is given by $(x, \zeta) \mapsto (x, \zeta_1, \ldots, \zeta_n)$ where the ζ_j are the coordinates of ζ in the base $dx^1, \ldots, dx^n$ of the cotangent space at x (duly complexified). We note that $L(\mathbb{C}; \mathbb{C}^n)$ can be identified with $\mathbb{C}^n$. Thus the principal symbol of d is a section over $T^*\mathfrak{M}\backslash 0$ (actually over $T^*\mathfrak{M}$ because we are dealing with a differential operator) of the "extension" of the bundle $T^{*\mathbb{C}}\mathfrak{M}$; the latter is the preceding bundle $\mathbb{E}$, and its extension is what we have denoted by $\mathbb{E}^{(*)}$. Now in the preceding local chart, the operator d is represented by the map

$$(x, f(x)) \mapsto \left(x, \frac{\partial f}{\partial x^1}(x), \ldots, \frac{\partial f}{\partial x^n}(x)\right).$$

We thus see that the principal symbol of d at the point (x, ξ) of $\overset{-1}{\pi}(\mathcal{O})$ is the map $z \mapsto \sqrt{-1}z(\xi_1, \ldots, \xi_n)$ from $\mathbb{C}$ to $\mathbb{C}^n$.

EXAMPLE 7.2. We refer the reader who is not familiar with the bundles discussed in this example to Chapter VII, Section 2.1. We denote by $\Lambda^p T^*\mathfrak{M}$ the pth exterior power of the cotangent bundle over $\mathfrak{M}$; its sections are the p-forms over $\mathfrak{M}$. In a local chart $(\mathcal{O}, x^1, \ldots, x^n)$ they are represented by sums $\alpha = \sum_{|J|=p} \alpha_J\, dx^J$, with J an *ordered* set of indices $j_1 < \cdots < j_p$ (each j is one of the integers $1, \ldots, n$); $dx^J = dx^{j_1} \wedge \cdots \wedge dx^{j_p}$. Our operator will then be the *exterior derivative* acting from $\Lambda^p T^*\mathfrak{M}$ to $\Lambda^{p+1} T^*\mathfrak{M}$. In the preceding local coordinates,

$$\text{(7.8)} \qquad d\alpha = \sum_{j=1}^{n} \sum_{|J|=p, j \notin J} \frac{\partial \alpha_J}{\partial x^j} dx^j \wedge dx^J = \sum_{|J'|=p+1} \varepsilon^{jJ}_{J'} \frac{\partial \alpha_J}{\partial x^j} dx^{J'},$$

where we have used the following notation: as a set $J' = \{j\} \cup J$, but it is ordered (we are assuming that $j \notin J$); $\varepsilon^{jJ}_{J'}$ is equal to $+1$ or to -1 according to the parity of the permutation that transforms $\{j\} \cup J$ into J'. Let (x, ξ) be a point in $T^*\mathfrak{M}$ with x in $\mathcal{O}$, as in Example 7.1. The principal symbol of d at the point (x, ξ) is a linear map $\Lambda^p\mathbb{C}^n \to \Lambda^{p+1}\mathbb{C}^n$. As before we use the basis $dx^1, \ldots, dx^n$ to identify ξ to a vector $(\xi_1, \ldots, \xi_n)$ of $\mathbb{R}^n$. If $\zeta = (\zeta_J)_{|J|=p}$ denotes an arbitrary vector in $\Lambda^p\mathbb{C}^n$, its image under the map equal to the principal symbol of d is the vector $\tilde{\zeta} = (\tilde{\zeta}_{J'})_{|J'|=p+1}$ with

$$\tilde{\zeta}_{J'} = \sqrt{-1} \sum_{j,J} \varepsilon^{jJ}_{J'} \xi_j \zeta_J,$$

where the summation ranges over all pairs (j, J) with J an ordered multi-index of length p, j one of the integers $1, \ldots, n$ such that $j \notin J$, and J' and $\{j\} \cup J$ are equal as sets. This means that $\tilde{\zeta}$ is the exterior product $\sqrt{-1}\, \xi \wedge \zeta$, when we identify ξ to an element of $\Lambda^1\mathbb{C}^n$. Thus *the value at* (x, ξ) *of the principal symbol of the exterior derivative is the exterior multiplication from the left with* $\sqrt{-1}\xi$. Notice that this formulation does not require coordinates at all; this is due to the peculiar feature of the bundles under consideration, in which the fiber at one point (x, ξ) is an exterior power of the cotangent space $T_x^* \mathfrak{M}$.

In the remainder of this section we deal with classical pseudodifferential operators on the manifold $\mathfrak{M}$. Suppose that P is such an operator, valued in $L(E; F)$, and let $p(x, \xi)$ denote its principal symbol: $p(x, \xi)$ is a C^∞ function on $T^*\mathfrak{M}\backslash 0$ valued in $L(E; F)$.

The *characteristic set* of P is the subset of $T^*\mathfrak{M}\backslash 0$, which we shall denote by Char P (as in the scalar case), defined by the property

$$p(x, \xi): E \to F \textit{ is not injective.} \tag{7.9}$$

When $E = F$, this is equivalent to saying

$$\det p(x, \xi) = 0. \tag{7.10}$$

DEFINITION 7.1. *The operator P, defined in $\mathfrak{M}$ and valued in $L(E; F)$, is said to be elliptic in $\mathfrak{M}$ (resp., in a conic open subset Γ of $T^*\mathfrak{M}\backslash 0$) if for every (x, ξ) in $T^*\mathfrak{M}\backslash 0$ (resp., in Γ) the linear map $p(x, \xi): E \to F$ is injective.*

By using local trivializations, one can extend in an obvious manner this definition to classical pseudodifferential operators from one complex vector bundle $\mathbb{E}$ over $\mathfrak{M}$ to another one, $\mathbb{F}$.

EXAMPLE 7.3. The exterior derivative, operating from $\mathcal{D}'(\mathfrak{M}; \Lambda^p)$ to $\mathcal{D}'(\mathfrak{M}; \Lambda^{p+1})$ (see Examples 7.1, 7.2), defines an elliptic operator in $\mathfrak{M}$ if $p = 0$, a nonelliptic one if $p > 0$. When $p > 0$, the null space of the map corresponding to (7.9), which here is exterior multiplication from the left by $i\xi$, is the set of p-vectors of the form $\xi \wedge v$ with v a $(p-1)$-vector; in other words, it is the image of the same mapping but acting from $\Lambda^{p-1} T_x^*$ to $\Lambda^p T_x^*$.

Example 7.3 suggests that we consider sequences of classical pseudodifferential operators of the following kind:

$$\mathbb{E}_j \xrightarrow{P^j} \mathbb{E}_{j+1}, \qquad j = 0, 1, \dots, N, \tag{7.11}$$

where each $\mathbb{E}_j$ is a vector bundle over $\mathfrak{M}$. The sequence (7.11) is said to be a *complex* if

$$\operatorname{Im} P^j \subset \operatorname{Ker} P^{j+1}, \qquad j = 0, 1, \dots, N-1. \tag{7.12}$$

The complex (7.11) is said to be an *elliptic complex* if for every (x, ξ) in $T^*\mathfrak{M}\backslash 0$ we have

$$\operatorname{Im} p^j(x, \xi) = \operatorname{Ker} p^{j+1}(x, \xi), \qquad j = 0, 1, \dots, N-1. \tag{7.13}$$

One also expresses this property by saying that the principal symbol sequence associated with (7.11) is *exact.*

The points $(x, \xi) \in T^*\mathfrak{M}\backslash 0$ such that (7.13) is not valid make up the *characteristic set* of the complex (7.11).

EXAMPLE 7.4. Take $\mathbb{E}_j = \Lambda^j T^*\mathfrak{M}$, $j = 0, \ldots, n$, and $P^j = d$, the exterior derivative. This yields an elliptic complex, called the *De Rham complex.*

REMARK 7.1. We have assumed so far that the principal symbol $p(x, \xi)$ of a pseudodifferential operator P valued in $L(E; F)$ had a well-defined degree. But this is not always so; one encounters situations in which, if one uses a basis in E and one in F, $p(x, \xi)$ is represented by a matrix whose entries $p_{jk}(x, \xi)$ are positive-homogeneous of possibly different degrees m_{jk} ($j = 1, \ldots, d = \dim E$, $k = 1, \ldots, d' = \deg F$). In the case where $d = d'$ one then requires that the determinant of that matrix be homogeneous. In order for this to be the case, it is necessary and sufficient that there be $2d$ real numbers $m_1, \ldots, m_d, m'_1, \ldots, m'_d$, such that

$$m_{jk} = m_j + m'_k. \tag{7.14}$$

In the case where $d \neq d'$ one may also require that in suitable bases of E and F, the homogeneity degrees of the $p_{jk}(x, \xi)$, m_{jk}, satisfy (7.14), where now j varies from 1 to d and k from 1 to d'. If this is so we may replace $p(x, \xi)$ by $p_o(x, \xi) = \delta(x, \xi)p(x, \xi)\delta'(x, \xi)$, where $\delta(x, \xi)$ is the diagonal $d \times d$ matrix with diagonal entries $|\xi|^{-m_j}$ and $\delta'(x, \xi)$ is the diagonal $d' \times d'$ matrix with entries $|\xi|^{-m'_k}$; $p_o(x, \xi)$ then has homogeneity degree zero, and we are back in the standard situation.

II

Special Topics and Applications

Pseudodifferential operators are tools. How efficient such tools can be was first evidenced by their application to uniqueness in the Cauchy problem by A. Calderon (*ca.* 1957, see Calderon [1]) and their role in the proof of the Atiyah–Singer index formula (*ca.* 1963, see Atiyah and Singer [1]). As often happens in mathematics, they had already been used earlier in disguised form. But it is those two spectacular successes that convinced several analysts (J. J. Kohn and L. Nirenberg [1], L. Schwartz, A. Unterberger, and J. Bokobza [1, 2], Calderon himself, R. Seeley [1], L. Hörmander [4]) of the timeliness of a comprehensive and more precise theory.

The first two sections in this chapter state and prove all the basic facts about compact and Fredholm operators, at first general, then pseudodifferential, that are needed in the proof of the index theorem, before the topology takes off. Section 3 gives a complete proof of the simplest version of the Calderon theorem on uniqueness in the Cauchy problem. Section 4 is devoted to the statement and proof of Friedrich's lemma, a kind of strong continuity result, with improvement by one degree of differentiability, about the commutator of a pseudodifferential operator with a Friedrich's mollifier. Friedrich's lemma is used in Section 5, which presents the most recent, and very likely the simplest to date, proof of Hörmander's sum-of-squares theorem (first stated and proved in Hörmander [6]). The proof in this book is taken from Kohn [3]. Generalizations of this result to equations of any order, also based on the use of pseudodifferential operators, can be found in Oleinik and Radkevitch [1].

1. Compact Pseudodifferential Operators

We begin by characterizing the pseudodifferential operators of order *zero* in an open subset Ω of $\mathbb{R}^n$ which define compact linear maps $L_c^2(\Omega) \to L_{\text{loc}}^2(\Omega)$.

We recall the definition of such maps. If E, F are two Banach spaces, a linear map $u: E \to F$ is said to be compact if the image under u of any ball in E (with finite radius) has a compact closure in F. Compact linear maps are automatically continuous, and in fact they form a *closed* vector subspace of the Banach space $L(E; F)$ of all continuous linear maps $E \to F$. In the old terminology they are often called *completely continuous.* The definition extends in an evident manner to the case where F is any locally convex Hausdorff topological vector space, such as a Fréchet space.

Now, for any compact subset K of Ω, set $L_c^2(K) = \{f \in L^2(\Omega); \operatorname{supp} f \subset K\}$. A linear map $u: L_c^2(\Omega) \to L_{\text{loc}}^2(\Omega)$ is compact, by definition, if for any choice of the compact set K, the restriction of u to $L_c^2(K)$ maps the latter compactly into $L_{\text{loc}}^2(\Omega)$. It is a classical result that the latter can also be formulated as follows: Given any sequence of functions $f_j \in L_c^2(K)$ $(j = 1, 2, \ldots)$ converging weakly to zero, their images $u(f_j)$ converge strongly to zero in $L_{\text{loc}}^2(\Omega)$, which in turn means that given any compact subset K' of Ω,

$$\int_{K'} |u(f_j)|^2 \, dx \to 0 \qquad \text{as } j \to +\infty.$$

Proposition 1.1. *Every regularizing operator in Ω induces a compact linear map $L_c^2(\Omega) \to L_{\text{loc}}^2(\Omega)$.*

Proof. Let $A(x, y) \in C^\infty(\Omega \times \Omega)$, $\{f_j\}$ be a sequence of L^2 functions in Ω, converging weakly to zero there and having their supports contained in one and the same compact set $K \subset \Omega$. The L^2 norms of the f_j are bounded. Consequently, for each x in Ω, $|\int A(x, y) f_j(y)\, dy|^2$ converges to zero as $j \to +\infty$. By the Cauchy–Schwarz inequality, it is bounded by $C \int_K |A(x, y)|^2 \, dy$, which is an integrable function of x on any compact subset of Ω. The Lebesgue dominated convergence theorem applies and implies what we want. ∎

Remark 1.1. We have of course proved that if $A(x, y) \in L_{\text{loc}}^2(\Omega \times \Omega)$, the operator $L_c^2(\Omega) \to L_{\text{loc}}^2(\Omega)$ defined by $A(x, y)$ is compact. Such operators can be said to be *Hilbert–Schmidt.*

REMARK 1.2. Let A be a continuous linear map $\mathscr{E}'(\Omega) \to C^\infty(\Omega)$ and $\{u_j\}_{j=1,2,\dots}$ be a sequence of distributions in Ω whose support is contained in one and the same compact set $K \subset \Omega$, weakly converging to zero (i.e., such that $\langle u_j, f\rangle \to 0$ as $j \to +\infty$ for each $f \in C^\infty$). Then Au_j converges to zero in $C^\infty(\Omega)$, simply because, for sequences of distributions, strong and weak convergence are the same. This is a much stronger result than Proposition 1.1, and we shall use it later.

According to these results, all pseudodifferential operators in one class mod $\Psi^{-\infty}(\Omega)$ are compact from $H^s_c(\Omega)$ to $H^{s'}_{\text{loc}}(\Omega)$ if this is true of one of them. We therefore limit ourselves to operators of the kind

$$p(x, D)u(x) = (2\pi)^{-n} \int e^{ix\cdot\xi} p(x, \xi)\hat{u}(\xi)\, d\xi, \tag{1.1}$$

first with $p(x, \xi) \in S^0(\Omega)$.

THEOREM 1.1. *The following two properties are equivalent*:

(1.2) $p(x, D)$ *induces a compact linear map* $L^2_c(\Omega) \to L^2_{\text{loc}}(\Omega)$.

(1.3) *Given any compact subset* K *of* Ω,

$$\lim_{|\xi|\to+\infty} \sup_{x\in K} |p(x, \xi)| = 0.$$

PROOF. We first assume that (1.3) holds and show that this implies (1.2). Let K, K' be any two compact subsets of $\Omega, \{u_j\}$ $(j = 1, 2, \dots)$ a sequence in $L^2(\Omega)$, weakly converging to 0, with supp $u_j \subset K$ for all j. We must show then that

$$\lim_{j\to+\infty} \int_{K'} |p(x, D)u_j|^2\, dx = 0. \tag{1.4}$$

Clearly, in proving this, we may assume that the symbol $p(x, \xi)$ vanishes identically when x is outside a compact neighborhood of K' in Ω.

Let us call $\hat{p}(\eta, \xi)$ the Fourier transform of $p(x, \xi)$ with respect to x. If $v_j(x) = p(x, D)u_j(x)$, then we have

$$|\hat{v}_j(\eta)| \le (2\pi)^{-n} \int |\hat{p}(\eta - \xi, \xi)\hat{u}_j(\xi)|\, d\xi, \tag{1.5}$$

and therefore

$$(2\pi)^{2n} \int |v_j(\eta)|^2\, d\eta \le \iint a(\xi, \xi')|\hat{u}_j(\xi)\hat{u}_j(\xi')|\, d\xi\, d\xi', \tag{1.6}$$

where we have set

$$a(\xi, \xi') = \int |\hat{p}(\eta - \xi, \xi)\hat{p}(\eta - \xi', \xi')|\, d\eta. \tag{1.7}$$

We apply the Cauchy–Schwarz inequality to (1.6), where we distinguish the factors $a(\xi, \xi')^{1/2}|\hat{u}_j(\xi)|$, $a(\xi, \xi')^{1/2}|\hat{u}_j(\xi')|$ and exploit the symmetry of a. Thus we get

$$(2\pi)^{2n} \int |\hat{v}_j(\eta)|^2\, d\eta \leq \iint a(\xi, \xi')|\hat{u}_j(\xi)|^2\, d\xi\, d\xi'. \tag{1.8}$$

Since the u_j converge weakly to zero in L^2, their Fourier transforms $\hat{u}_j(\zeta)$ converge pointwise to zero at each $\zeta \in \mathbb{C}_n$. On the other hand,

$$|\hat{u}_j(\zeta)| \leq \int e^{|x||\mathrm{Im}\,\zeta|}|u_j(x)|\, dx \leq \left\{\int_K e^{2|x||\mathrm{Im}\,\zeta|}\, dx\right\}^{1/2} \left\{\int |u_j|^2\, dx\right\}^{1/2}.$$

Since the L^2 norms of the u_j are bounded, it follows that the entire functions $\hat{u}_j(\zeta)$ are bounded in any slab $|\mathrm{Im}\,\zeta| \leq M < +\infty$. By the Montel theorem on normal families, this means that the closure of the set of those functions, $\{\hat{u}_j\}$, is compact for the uniform convergence on each compact subset of such a slab. But any subsequence that converges must converge to zero (since it does so pointwise), and therefore the whole sequence $\{\hat{u}_j\}$ actually converges to zero, in particular uniformly on each compact subset of the real space $\mathbb{R}_n$. Thus:

(1.9) *Given any numbers R, $\varepsilon > 0$, there is $j_o = j_o(R, \varepsilon)$ such that*

$$\sup_{|\xi| \leq R} |\hat{u}_j(\xi)| \leq \varepsilon, \qquad \forall j \geq j_o.$$

We take now a look at $a(\xi, \xi')$. We contend that under the hypothesis (1.3), we have the following:

(1.10) *To every $\varepsilon > 0$ there is $R > 0$ such that*

$$\sup_{|\xi| \geq R} \int a(\xi, \xi')\, d\xi' \leq \varepsilon.$$

PROOF OF (1.10). First we select a continuous function of $\varepsilon > 0$, $r(\varepsilon) > 0$, such that if $|\xi| \geq r(\varepsilon)$, then for all $\theta \in \mathbb{R}^n$,

$$|\hat{p}(\theta, \xi)| \leq \int |p(x, \xi)|\, dx \leq \varepsilon. \tag{1.11}$$

Note, on the other hand, that

$$(1.12)\qquad (1+|\theta|^2)^k|\hat{p}(\theta,\xi)| \le \int |(1-\Delta_x)^k p(x,\xi)|\,dx \le C_k < +\infty.$$

For later reference note that this implies

$$\int a(\xi,\xi')\,d\xi' \le C_k^2 \iint (1+|\eta-\xi|^2)^{-k}(1+|\eta-\xi'|^2)^{-k}\,d\xi'\,d\eta;$$

hence by taking k large in comparison with $n/2$,

$$(1.13)\qquad \int a(\xi,\xi')\,d\xi' \le C_o < +\infty.$$

And if we write

$$I_1(\xi) = \iint_{|\eta-\xi|\ge\delta|\xi|} |\hat{p}(\eta-\xi,\xi)\hat{p}(\eta-\xi',\xi')|\,d\xi'\,d\eta,$$

then a similar argument yields

$$I_1(\xi) \le (1+\delta|\xi|)^{-1}C_k^2 \iint (1+|\eta-\xi|^2)^{-k+1}(1+|\eta-\xi'|^2)^{-k}\,d\xi'\,d\eta$$
$$\le C'(1+\delta|\xi|)^{-1}.$$

We henceforth choose $\delta = \varepsilon^{-1/2n}/r(\varepsilon)$ and conclude:

(1.14) *If* $|\xi| = r(\varepsilon)$, *then* $I_1(\xi) \le C'(1+\varepsilon^{-1/2n})^{-1}$.

Call $I_2(\xi)$ the same integral as $I_1(\xi)$ except that integration with respect to η is performed over the region $|\eta-\xi| \le \delta|\xi|$. In other words,

$$\int a(\xi,\xi')\,d\xi' = I_1(\xi) + I_2(\xi).$$

When $|\xi| = r(\varepsilon)$, we have, by (1.11), (1.12),

$$I_2(\xi) \le (\delta|\xi|)^n \sup_\theta |\hat{p}(\theta,\xi)| \sup_\eta \int |\hat{p}(\eta-\xi',\xi')|\,d\xi'$$
$$\le \varepsilon(\delta|\xi|)^n C_k \int (1+|\xi'|^2)^{-k}\,d\xi'.$$

Of course we take $k > n/2$. Our choice of δ then yields

(1.15) *If* $|\xi| = r(\varepsilon)$, *then* $I_2(\xi) \le C'\varepsilon^{1/2}$.

Combining (1.14) and (1.15) yields (1.10).

We can now complete the proof of the implication (1.3) $\Rightarrow$ (1.2) in Theorem 1.1.

Select $\varepsilon > 0$ arbitrary and then R according to (1.10). Take $j \geq j_o$, the number in (1.9). We have, according to (1.8),

$$(1.16)\qquad (2\pi)^{2n}\int |\hat{v}_j(\eta)|^2\, d\eta \leq$$

$$\varepsilon \int_{|\xi|\leq R} a(\xi, \xi')\, d\xi\, d\xi' + \sup_{|\xi|\geq R}\int a(\xi, \xi')\, d\xi' \int |u_j|^2\, dx$$

$$\leq \varepsilon(C_o + C''),$$

where we have applied (1.13) and written $C'' = \sup_j \int |u_j|^2\, dx$.

Next we show that (1.2) $\Rightarrow$ (1.3). We reason by contradiction and assume that there is a sequence of points x_j ($j = 1, 2, \ldots$) converging to some point x_o in Ω and a sequence of vectors ξ^j in $\mathbb{R}_n$, $\rho_j = |\xi^j| \to +\infty$, such that $p(x_j, \xi^j) \to z_o \neq 0$. We show that such an assumption is incompatible with (1.2). Let $\mathscr{B}_o$ be an open ball centered at x_o, with closure contained in Ω. Let $g \in C_c^\infty(\Omega)$ be equal to one in a neighborhood of the closure of $\mathscr{B}_o$. If $\{u_j\}$ is any sequence in $L_c^2(\mathscr{B}_o)$ that converges weakly to zero in L^2, it follows from Theorem 2.2 of Chapter I and from Proposition 1.1 that $[1 - g(x)]p(x, D)u_j$ converges strongly to zero in $L^2_{\text{loc}}(\Omega)$. Therefore (1.2) implies that $g(x)p(x, D)u_j$ converges strongly in L^2. We may as well assume that $p(x, \xi)$ vanishes identically outside supp g. We select a function $u \in C_c^\infty(\mathbb{R}^n)$, not identically zero. We note that if j is large enough, the support of

$$u_j(x) = \rho_j^{n/4} u(\rho_j^{1/2}(x - x_j))\, e^{ix\cdot\xi^j}$$

will be contained in $\mathscr{B}_o$. We have $\int |u_j|^2\, dx = \int |u|^2\, dx$ whatever j, and given any $w \in C_c^\infty$,

$$\int u_j(x)w(x)\, dx = \rho_j^{-n/4}\int u(y)w(x_j + \rho_j^{-1/2}y)\, dy \to 0,$$

which shows that the u_j converge weakly to zero in L^2. Therefore, under our hypothesis of compactness of $p(x, D)$, we must have

$$(1.17)\qquad \lim_j \|p(x, D)u_j\|_{L^2} = 0.$$

We have

$$\hat{u}_j(\xi) = \rho_j^{-n/4} \exp\{-ix_j\cdot(\xi - \xi^j)\}\hat{u}(\rho_j^{-1/2}(\xi - \xi^j)),$$

whence

$$p(x, D)u_j(x) = \rho_j^{n/4}\, e^{ix\cdot\xi^j} v_j(\rho_j^{1/2}(x - x_j)),$$

where

$$v_j(x) = (2\pi)^{-n} \int e^{ix\cdot\eta} p(x_j + \rho_j^{-1/2}x, \xi^j + \rho_j^{1/2}\eta)\hat{u}(\eta)\, d\eta.$$

Given an arbitrary compact subset $\mathcal{K}$ of $\mathbb{R}^n \times \mathbb{R}_n$, there are positive constants C'', J such that for all $j \geq J$ and all $(x, \eta) \in \mathcal{K}$,

$$|p(x_j + \rho_j^{-1/2}x, \xi^j + \rho_j^{1/2}\eta) - p(x_j, \xi^j)| \leq C''\rho_j^{-1/2}.$$

This follows from the mean value theorem and from the inequalities defining symbols of degree zero. Since the integrand in the integral representing v_j is bounded by a constant times $|\hat{u}(\eta)|$, it follows that when $j \to +\infty$, the integrand converges to $z_o\hat{u}(\eta)$ uniformly with respect to (x, η) in $\mathcal{K}$, and therefore $v_j(x) \to z_o u(x)$ uniformly on compact subsets of $\mathbb{R}^n$. From (1.17) and from the fact the L^2 norm of $p(x, D)u_j$ is equal to that of v_j, we deduce

$$|z_o|^2 \int |u|^2\, dx \leq \lim_{j\to+\infty} \int |v_j|^2\, dx = 0,$$

whence $z_0 = 0$, a contradiction.

The proof of Theorem 1.1 is complete. □

COROLLARY 1.1. *Suppose that the symbol $p(x, \xi)$ of $P \in \Psi^m(\Omega)$ satisfies the following condition*:

(1.18) *Given any compact subset K of Ω,*

$$\lim_{|\xi|\to+\infty} \sup_{x\in K} \frac{|p(x, \xi)|}{(1 + |\xi|)^m} = 0.$$

Then for all real s, P induces a compact linear map $H_c^s(\Omega) \to H_{\text{loc}}^{s-m}(\Omega)$.

PROOF. For each real t, let U_t be a properly supported pseudodifferential operator in Ω, elliptic of order t (Chapter I, Definitions 3.1, 4.4), U_t^{-1} a properly supported parametrix of U_t (Chapter I, Corollary 4.3). It follows from (4.24) of Chapter I that the symbol of $Q = U_{s-m}PU_{-s}$ satisfies the analogue of (1.3); consequently $Q: L_c^2(\Omega) \to L_{\text{loc}}^2(\Omega)$ is compact, and so is $U_{s-m}^{-1}QU_{-s}^{-1} : H_c^s(\Omega) \to H_{\text{loc}}^{s-m}(\Omega)$, which is congruent to P mod $\Psi^{-\infty}(\Omega)$. It suffices then to use Remark 1.2. □

COROLLARY 1.2. *The natural embedding* $H_c^s(\Omega) \to H_c^{s-\delta}(\Omega)$ $(\delta > 0)$ *is compact.*

PROOF. The statement is that given any compact subset K of Ω, if a sequence u_j in $H_c^s(K)$ converges weakly, then it converges strongly in $H_c^{s-\delta}(\Omega)$. Using the same notation as in the proof of Corollary 1.1, we see that by that same corollary $U_{-\delta}u_j$ converges strongly in $H_{\text{loc}}^s(\Omega)$; therefore $U_{-\delta}^{-1}U_{-\delta}u_j$ does the same in $H_{\text{loc}}^{s-\delta}(\Omega)$. But $u_j = U_{-\delta}^{-1}U_{-\delta}u_j + Ru_j$, where R is regularizing. Once again we apply Remark 1.2. □

REMARK 1.3. Corollary 1.2 is a variant of the classical Rellich's lemma.

REMARK 1.4. It follows from Theorem 4.4 of Chapter I that *property* (1.18) *is invariant under diffeomorphism.*

Let X be a C^∞ manifold of dimension n, countable at infinity, P a pseudodifferential operator of order m on X. If $(\mathcal{O}, x^1, \ldots, x^n)$ is a local chart in X, we may consider the property (1.18) of the symbol $p(x, \xi)$ of P in that chart. According to Remark 1.4 such a property is independent of the choice of local coordinates. Thus Corollary 1.1 (and also Theorem 1.1) remains valid if we replace Ω by the manifold X.

A particular case of foremost importance is that of a compact manifold X. In this case the global Sobolev spaces are well defined, at least as locally convex topological vector spaces (not as Hilbert spaces), and we may state the following corollaries:

COROLLARY 1.3. *Let X be a compact manifold, P a pseudodifferential operator of order m, which satisfies condition* (1.18) *in each local chart in X. Then for any real number s, P induces a compact linear map* $H^s(X) \to H^{s-m}(X)$.

COROLLARY 1.4. *Let X be compact, δ be a number* >0. *For all real s, the natural embedding* $H^s(X) \to H^{s-\delta}(X)$ *is compact.*

These results extend routinely to the case where the operator P, instead of being scalar (that is, acting from complex-valued functions or distributions to the same), is operator-valued, that is to say, valued in the space of linear maps $E \to F$, where E, F are two complex vector spaces, provided their dimensions are finite. (If E and F are, say, Banach spaces, one should then require that the values of P lie in the space of compact linear maps $E \to F$.) Beyond this, the results extend to pseudodifferential operators acting from a vector bundle $\mathbb{E}$ to another $\mathbb{F}$, over the same manifold X (see Chapter I, Section 7). In particular, if we denote by $H^s(X;\mathbb{E})$ the H^s sections over X of the vector bundle $\mathbb{E}$, then for all $s \in \mathbb{R}$ and $\delta > 0$,

(1.19) *the natural injection* $H^s(X;\mathbb{E}) \to H^{s-\delta}(X;\mathbb{E})$ *is compact.*

We conclude this section by deriving a simple and useful consequence of Corollary 1.2.

As usual let Ω be an open subset of $\mathbb{R}^n$. The distributions whose support is the set $\{x_o\}$ (consisting of the single point x_o) are necessarily of the form

$$\sum_\alpha c_\alpha \delta^{(\alpha)}(x - x_o), \tag{1.20}$$

where $\delta^{(\alpha)}$ is the αth derivative of the Dirac measure δ and the c_α are complex numbers. The sum (1.20) is finite.

PROPOSITION 1.2. *Let P be a properly supported pseudodifferential operator in Ω of order m, s a real number such that given any distribution u in Ω whose support is equal to $\{x_o\}$,*

$$Pu \in H^s \quad \textit{implies} \quad u = 0. \tag{1.21}$$

Let s' be an arbitrary number $>s$. To every $\varepsilon > 0$ there is $\varepsilon' > 0$ such that if $B_{\varepsilon'}(x_0)$ denotes the open ball centered at x_o, with radius ε', then

$$\|Pu\|_s \leq \varepsilon \|u\|_{s'+m}, \qquad \forall u \in C_c^\infty(\Omega \cap B_{\varepsilon'}(x_o)). \tag{1.22}$$

PROOF. Suppose that for some $s' > s$ the conclusion does not hold. There would be a sequence of test functions u_j ($j = 1, 2, \ldots$) with supp u_j converging to $\{x_o\}$, such that for all j, $\|Pu_j\|_s = 1$, $\|u_j\|_{s'+m} \leq C < +\infty$. By Corollary 1.2 there is a subsequence, which we may assume to be the sequence u_j itself, that converges in H^{s+m}, say to u_o. We necessarily have supp $u_o = \{x_o\}$. Since the order of P is m, we also have $Pu_o \in H^s$, which implies $u_o = 0$, a contradiction since the H^s norm of $Pu_o = \lim_j Pu_j$ is one. □

COROLLARY 1.5. *If $s' > s > -n/2$, then to every $\varepsilon > 0$ there is $\varepsilon' > 0$ such that*

$$\|u\|_s \leq \varepsilon \|u\|_{s'}, \qquad \forall u \in C_c^\infty(B_{\varepsilon'}(x_o)). \tag{1.23}$$

PROOF. Apply Proposition 1.2 with $P = I$, the identity. The set of s that satisfy the hypothesis is defined by the condition $s > -n/2$. □

COROLLARY 1.6. *Let P be as in Proposition* 1.2 *and s' be a number* $>-n/2$. *Let r, s be two other real numbers such that P defines a continuous linear map* $H^r \to H^s$ *and* $-n/2 < r < s'$. *Then to every* $\varepsilon > 0$ *there is* $\varepsilon' > 0$ *such that*

$$\|Pu\|_s \le \varepsilon \|u\|_{s'}, \qquad \forall u \in C_c^\infty(\Omega \cap B_{\varepsilon'}(x_o)). \tag{1.24}$$

PROOF. Let K be a compact neighborhood of x_o in Ω. We have $\|Pu\|_s \le C\|u\|_r$ for all $u \in C_o^\infty(K)$. Combine this with (1.23). □

COROLLARY 1.7. *Suppose that the pseudodifferential operator P in* Ω *is properly supported and regularizing. Given any real number* $s' > -n/2$ *and any real number s, the conclusion in Corollary* 1.6 *is valid.*

Another well-known application of Corollary 1.4 is the so-called *Korn's lemma*:

PROPOSITION 1.3. *Let* s, s', s'' *be three real numbers such that* $s'' < s' < s$. *To every number* $\varepsilon > 0$ *there is a constant* $C_\varepsilon > 0$ *such that*

$$\|u\|_{s'} \le \varepsilon \|u\|_s + C_\varepsilon \|u\|_{s''}, \qquad \forall u \in H^s(X). \tag{1.25}$$

PROOF. Suppose that the constant C_ε did not exist. It would mean that for some sequence of elements u_j in $H^s(X)$, we would have, for all $j = 1, 2, \ldots$,

$$\varepsilon \|u_j\|_s + j\|u_j\|_{s''} \le \|u_j\|_{s'}.$$

Possibly after replacing u_j by $\|u_j\|_{s'}^{-1} u_j$ we may suppose that $\|u_j\|_{s'} = 1$ for every j. Since $\|u_j\|_s \le 1/\varepsilon$, $\|u_j\|_{s''} \le 1/j$, we would derive that every infinite subset of the sequence $\{u_j\}$ contains a subsequence that converges weakly in $H^s(X)$, necessarily to zero; this in fact means that the sequence u_j itself converges weakly to zero in $H^s(X)$. But by Corollary 1.4 the u_j would converge strongly in $H^{s'}(X)$, to an element whose norm is necessarily equal to one, a contradiction. □

Finally, as an application of Corollary 1.7, let us show that the existence of a parametrix implies the local solvability of a linear partial differential equation. More precisely, we have

PROPOSITION 1.4. *Let P be a linear partial differential operator of order m (with* C^∞ *coefficients), E a pseudodifferential operator of order* $-m'$ *in* Ω, *such that*

$$R = PE - I \tag{1.26}$$

is regularizing in Ω.

Then given any point x_o of Ω and any number $s > -n/2$, there is an open neighborhood U_s of x_o in Ω and a continuous linear operator,

$$K_s : H_c^{-s}(U_s) \to H_{\mathrm{loc}}^{-s+m'}(U_s),$$

such that on $H_c^{-s}(U_s)$ we have

$$PK_s = I. \tag{1.27}$$

PROOF. We may assume that E, and therefore also R, in (1.26), are properly supported. Indicate adjoints by upper stars, as usual; then we have, in $\mathscr{D}'(\Omega)$,

$$I = E^*P^* + R^*. \tag{1.28}$$

By Corollary 1.7 we can determine $\varepsilon > 0$ such that if $U_s = \Omega \cap B_\varepsilon(x_o)$, then

$$\|R^*v\|_s \leq \tfrac{1}{2}\|v\|_s, \qquad \forall v \in C_c^\infty(U_s). \tag{1.29}$$

From (1.29), and from Theorem 2.1 of Chapter I, we derive

$$\|v\|_s \leq \|E^*P^*v\|_s + \|R^*v\|_s \leq C\|P^*v\|_{s-m'} + \tfrac{1}{2}\|v\|_s,$$

whence

$$\|v\|_s \leq 2C\|P^*v\|_{s-m'}, \qquad \forall v \in C_c^\infty(U_s). \tag{1.30}$$

Let M denote the closure of the subspace $P^*C_c^\infty(U_s)$ in $H^{s-m'}(\mathbb{R}^n)$. Estimate (1.30) implies that the map $P^*u \mapsto u$ from M into $H^s(\mathbb{R}^n)$ is well defined and continuous (with norm $\leq 2C$). Let us extend it by zero to the orthogonal $M^\perp$ of M in $H^{s-m'}(\mathbb{R}^n)$ and denote by $\tilde{G}$ the continuous linear map $H^{s-m'}(\mathbb{R}^n) \to H^s(\mathbb{R}^n)$ thus obtained. Note that we have

$$\tilde{G}P^*v = v \qquad \textit{for all } v \in C_c^\infty(U_s). \tag{1.31}$$

We denote by G the adjoint of $\tilde{G}$: G is a continuous linear map $H^{-s}(\mathbb{R}^n) \to H^{m'-s}(\mathbb{R}^n)$. Let f belong to $H_c^{-s}(U_s) \subset H^{-s}(\mathbb{R}^n)$. Then for any $v \in C_c^\infty(U_s)$, we have

$$\langle PGf, \bar{v}\rangle = \langle Gf, \overline{P^*v}\rangle = \langle f, \overline{\tilde{G}P^*v}\rangle = \langle f, \bar{v}\rangle;$$

in other words,

$$PGf = f, \qquad \forall f \in H_c^{-s}(U_s). \tag{1.32}$$

It then suffices to take K_s to be the restriction of G to $H_c^{-s}(U_s)$ regarded as being valued in $H_{\mathrm{loc}}^{-s+m'}(U_s)$. □

2. Fredholm Operators and the Index of Elliptic Pseudodifferential Operators on a Compact Manifold

2.1. Fredholm Operators

Let E and F be two locally convex Hausdorff topological vector spaces (over the complex field), and $L(E; F)$ the space of continuous linear maps $E \to F$. In all the applications in this chapter, E and F are Banach spaces, and $L(E; F)$ are equipped with the operator norm. In Chapter III we shall take E and F to be Fréchet spaces.

DEFINITION 2.1. *A continuous linear map $A: E \to F$ is called a Fredholm operator if the dimension of its kernel (or null space) and the codimension of its image (or range) are both finite.*

Recalling the definition

$$\operatorname{Coker} A = F/A(E), \tag{2.1}$$

we see that when A is Fredholm, dim Ker A and dim Coker A are finite.

DEFINITION 2.2 *Let $A \in L(E; F)$ be Fredholm. The number*

$$\operatorname{Ind} A = \dim \operatorname{Ker} A - \dim \operatorname{Coker} A \tag{2.2}$$

is called the index of A.

Some functorial properties of the index are self-evident. Suppose that we have four spaces E_j, $F_j (j = 1, 2)$ and two continuous linear maps $A_j: E_j \to F_j$. We may form the direct-sum map $A_1 \oplus A_2: x_1 + x_2 \mapsto Ax_1 + Ax_2$ from $E_1 \oplus E_2$ to $F_1 \oplus F_2$. If both A_1 and A_2 are Fredholm, so is $A_1 \oplus A_2$ and we have

$$\operatorname{Ind} A_1 \oplus A_2 = \operatorname{Ind} A_1 + \operatorname{Ind} A_2. \tag{2.3}$$

Let E, F, and G be three spaces, and assume that $\dim G = m < +\infty$. Let us denote by I_G the identity map of G. If $A: E \to F$ is Fredholm, so is $A \otimes I_G: E \otimes G \to F \otimes G$, and we have

$$\operatorname{Ind} A \otimes I_G = m \operatorname{Ind} A. \tag{2.4}$$

Suppose now that G is arbitrary (i.e., not necessarily finite dimensional) and let B be a Fredholm operator $F \to G$. Then $B \circ A$ is Fredholm and we have

$$\operatorname{Ind}(B \circ A) = \operatorname{Ind} A + \operatorname{Ind} B. \tag{2.5}$$

PROPOSITION 2.1. *Let A be a Fredholm operator $E \to F$. The transpose operator ${}^tA: F' \to E'$ is Fredholm* (E', F' *are strong duals*) *and* $\operatorname{Ind} {}^tA = -\operatorname{Ind} A$.

PROOF. First, ${}^tA$ is continuous when E' and F' carry the strong dual topology. But equip the four spaces E, F, E', F' with their weak topologies, for the duality between E and E' and between F and F'. Then $A, {}^tA$ are continuous, and the situation is completely symmetric. The kernel of a continuous linear map is the orthogonal of the image of its transpose; hence

$$\dim \operatorname{Ker} {}^tA = \dim \operatorname{Coker} A, \tag{2.6}$$

$$\dim \operatorname{Coker} {}^tA = \dim \operatorname{Ker} A. \tag{2.7}$$

□

PROPOSITION 2.2. *Suppose that E and F are Fréchet spaces. If $A: E \to F$ is Fredholm, its image and the image of its transpose ${}^tA$ are both closed.*

PROOF. Let G be an algebraic supplementary of $A(E)$: G is a vector subspace of F, $A(E) \cap G = \{0\}$, and $F = A(E) \oplus G$. If A is Fredholm, then G is necessarily finite dimensional. Consider then the direct-sum map $A \oplus I_G: E \oplus G \to F$ (we equip the direct sum with the direct-sum, or product, topology); $A \oplus I_G$ is continuous and surjective. It maps the complement of E onto that of $A(E)$. By the open mapping theorem, the latter is open since the former is, and thus the image of A is closed. To see that $\operatorname{Im} {}^tA$ is closed in E' it suffices to factor A into the sequence

$$E \to E/\operatorname{Ker} A \overset{\tilde{A}}{\to} A(E) \to F,$$

where the two extreme arrows stand for the natural surjection and injection respectively and $\tilde{A}$ is an isomorphism. By transposition we get the sequence

$$E' \leftarrow {}^tA(F') \overset{{}^t\tilde{A}}{\leftarrow} F'/\operatorname{Ker} {}^tA \leftarrow F',$$

and ${}^t\tilde{A}$ is an isomorphism of the complete space $F'/\operatorname{Ker} {}^tA$ onto ${}^tA(F')$, which must therefore be closed. □

PROPOSITION 2.3. *Suppose that E and F are Fréchet spaces and let $A \in L(E; F)$. The following two properties are equivalent:*

(2.8) *A is Fredholm.*

(2.9) *There is $B \in L(F; E)$ such that $A \circ B - I_F$ and $B \circ A - I_E$ both have finite rank.*

If they hold, then the operator B in (2.9) is also Fredholm, and it is possible to select it in such a way that

$$A \circ B \circ A = A, \qquad B \circ A \circ B = B. \tag{2.10}$$

PROOF. First suppose that (2.8) holds. Let $d = \dim \operatorname{Coker} A$. Select a basis $f'_1, \ldots, f'_d$ in $\operatorname{Ker} {}^tA$, and d vectors in F, $f_1, \ldots, f_d$ such that $\langle f'_j, f_k \rangle = \delta_{jk}$, the Kronecker index. Define the following map $F \to F$:

$$Py = y - \sum_{j=1}^{d} \langle f'_j, y \rangle f_j. \tag{2.11}$$

It is clear that $\langle f'_j, Py \rangle = 0$ for all y and all j; therefore the range of P is contained in $A(E)$. The restriction of P to $A(E)$ is the identity; P is a continuous projection of F onto $A(E)$.

Let E_o be a closed vector subspace of E such that E is the direct sum of E_o and $\operatorname{Ker} A$; let A_o be the restriction of A to E_o, regarded as a continuous linear map $E_o \to A(E)$; A_o is an isomorphism, and we denote its "inverse" by $A_o^{-1} : A(E) \to E$. Then set

$$B = A_o^{-1} \circ P. \tag{2.12}$$

The range of B is E_o and therefore $A \circ B = P$. We note that $I_F - P$ is a continuous projection of F onto the linear span of $f_1, \ldots, f_d$ and thus has finite rank. On the other hand $PA = A$, and $BA = A_o^{-1}A$ is a continuous projection of E onto E_o whose kernel is equal to $\operatorname{Ker} A$. Hence $I_E - BA$ is a continuous projection of E onto $\operatorname{Ker} A$.

It can be checked at once that (2.10) holds.

Suppose that (2.9) holds. Since $\operatorname{Ker} A$ is contained in the range of $I_E - BA$, it is finite dimensional. Let F_1 be the range of $I_F - AB$. It is clear that $F_1 + A(E) = F$; therefore $\operatorname{codim} A(E) \le \dim F_1$. □

If E and F are two topological vector spaces, a linear map $E \to F$ is said to be *compact* if there is an open neighborhood of the origin in E whose image under the map has compact closure in F. Such a map is automatically continuous. And the definition agrees with the customary one when E and F are Banach spaces.

Now we denote the space of continuous endomorphisms of E by $L(E)$ (rather than $L(E;E)$), and the identity map of E by I_E.

LEMMA 2.1. *If $K \in L(E)$ is compact, the kernels of $I_E + K$ and $I_{E'} + {}^tK$ are both finite dimensional. If E is a Fréchet space, then $I_E + K$ is Fredholm.*

PROOF. Write $T = I_E + K$. On Ker T we have $x = -Kx$; hence the identity of Ker T is compact, which means that Ker T is locally compact. This is possible only if Ker T is finite dimensional (Treves [2], Theorem 9.2).

In order to prove that $I_{E'} + {}^tK = {}^tT$ has a finite-dimensional kernel it suffices to show that ${}^tK$ is a compact endomorphism of E' when we equip the latter with the topology of uniform convergence on the compact subsets of E. The resulting topological vector space will be denoted by E'_c. Let U be a neighborhood of the origin in E such that $K(U)$ has a compact closure in E. The polar of U,

$$U^0 = \{x' \in E'; \sup_{x\in U} |\langle x', x\rangle| \leq 1\},$$

is a convex compact subset of E'_c, whereas the polar $K(U)^0$ of $K(U)$ is a neighborhood of the origin in E'_c; ${}^tK$ maps $K(U)^0$ into U^0, whence our assertion. (On this see Treves [2], Propositions 32.6, 32.7.) In the case of Banach spaces it is a classical result that ${}^tK$ is compact for the dual norm.

Now suppose that E is a Fréchet space. We are going to show that the range of T is closed, hence equal to the orthogonal of Ker ${}^tT$ and thus of finite codimension. Let U again be a neighborhood of the origin such that $K(U) \subset\subset E$. Select a monotone decreasing basis of convex open neighborhoods of the origin in E, $\{U_j\}_{j=1,2,\ldots}$, all contained in U. It is clear that $K(U_j)$ has a compact closure for every j. Let p_j denote the seminorm associated with U_j; $p_j(x)$ is the infimum of the numbers $\rho > 0$ such that $\rho^{-1}x \in U_j$. We have $p_j \leq p_{j+1}$ for every j.

Let E_o be a closed vector subspace of E such that E is the direct sum of E_o and Ker T. We claim that to every j there is k and $C > 0$ such that

$$p_j(x) \leq Cp_k(Tx), \qquad \forall x \in E_o. \tag{2.13}$$

If this were not the case, there would be an integer $j > 0$ and a sequence $\{x_k\}$ in E_o such that $p_j(x_k) = 1$, $p_k(Tx_k) \leq 1/k$ for each $k = 1, 2, \ldots$, and therefore $Tx_k \to 0$ in E. Since the sequence $\{x_k\}$ is entirely contained in the closure of U_1, the sequence $\{Kx_k\}$ has a compact closure in E. Therefore, by a well-known property of compact subsets of metric spaces, we can extract a subsequence $\{x_{k_\nu}\}$ such that Kx_{k_ν} converges in E, to an element that we denote by $-x_o$. Since Tx_{k_ν} converges to zero, the x_{k_ν} themselves must converge to x_o, which implies the following three facts: (1) $Tx_o = 0$, i.e., $x_o \in \operatorname{Ker} T$; (2) $x_o \in E_o$; (3) $p_j(x_o) = 1$, obviously a contradiction since $E_o \cap \operatorname{Ker} T = \{0\}$.

The inequalities (2.13) imply at once that Im T is closed. □

REMARK 2.1. The reasoning in the proof of Lemma 2.1 shows that when E and F are Fréchet spaces, necessary and sufficient conditions for a continuous linear map $A: E \to F$ to be Fredholm are that its image be closed and that Ker A and Ker ${}^tA$ both be finite dimensional.

LEMMA 2.2. *If $R \in L(E)$ has finite rank, then* $\operatorname{Ind}(I_E + R) = 0$.

PROOF. We shall use a particular case of Lemma 2.2, namely the fact that if $\dim E < +\infty$, then the index of any endomorphism T of E is zero. Indeed, T induces a bijection of $E/\operatorname{Ker} T$ onto Im T, and therefore $\dim \operatorname{Ker} T = \operatorname{codim} \operatorname{Im} T$.

Let E be infinite dimensional, and suppose that the image W of R is finite dimensional. Clearly $I + R$ induces a continuous linear map $J: E/W \to E/W$. The reader can check at once that this map is a bijection; hence it has index zero. But by (2.3) we have

$$\operatorname{Ind}(I + R) = \operatorname{Ind}[(I + R)|_W] + \operatorname{Ind} J,$$

and both terms on the right-hand side are zero. □

THEOREM 2.1. *Let E and F be Fréchet spaces. In order for $A \in L(E; F)$ to be Fredholm it is necessary and sufficient that there be $B \in L(F; E)$ such that both $AB - I_F$ and $BA - I_E$ are compact operators.*

PROOF. Since a linear operator with finite rank is compact, the necessity follows from Proposition 2.3. By Lemma 2.1 we know that if $S = B \circ A - I_E$ and $T = A \circ B - I_F$ are compact operators, then $B \circ A$ and $A \circ B$ are Fredholm. Since

$$\operatorname{Ker} A \subset \operatorname{Ker}(B \circ A) \quad \text{and} \quad \operatorname{Im} A \supset \operatorname{Im}(A \circ B),$$

A is also Fredholm. □

THEOREM 2.2. *Let E and F be Banach spaces. The set of Fredholm operators is open in the Banach space $L(E; F)$, and the index is a locally constant function in this open set.*

PROOF. Let $A \in L(E; F)$ be Fredholm, and let B denote an operator satisfying the conditions in (2.9) and (2.10). Observe that (2.5) and (2.10) imply

$$\text{Ind}\, B = -\text{Ind}\, A. \tag{2.14}$$

Set $\varepsilon = \|B\|^{-1}$ and let T be any operator in $L(E; F)$ such that $\|T\| < \varepsilon$. Then the operators $I_E + B \circ T$ and $I_F + T \circ B$ are bijective; hence their indices are equal to zero. We have

$$\begin{aligned}(A + T) \circ B = A \circ B + T \circ B &= I_F + R + T \circ B \\ &= (I_F + T \circ B)^{-1}(I_F + R'),\end{aligned}$$

where R has finite rank and so does $R' = (I_F + T \circ B)R$. Similarly,

$$B \circ (A + T) = (I_E + B \circ T)^{-1}(I_E + S'),$$

with $S': E \to E$ having finite rank. These formulas show first that $A + T$ is Fredholm, since $\text{Ker}(A + T) \subset \text{Ker}[B \circ (A + T)]$ and $\text{Im}(A + T) \supset \text{Im}[(A + T) \circ B]$, and then that

$$\text{Ind}(A + T) = -\text{Ind}\, B, \tag{2.15}$$

since $\text{Ind}(I_F + R') = \text{Ind}(I_E + S') = 0$, by Lemma 2.2. Combining (2.14) and (2.15) yields $\text{Ind}(A + T) = \text{Ind}\, A$. □

THEOREM 2.3. *Suppose that E and F are Fréchet spaces. If $A \in L(E; F)$ is Fredholm and $K \in L(E; F)$ compact, then $A + K$ is Fredholm. If E and F are Banach spaces we have, moreover,* $\text{Ind}(A + K) = \text{Ind}\, A$.

PROOF. Let B be as in Theorem 2.1. It is clear that $(A + K)B - I_F$ and $B(A + K) - I_E$ are both compact, which shows that $A + K$ is Fredholm. Suppose then that E and F are Banach spaces. Then $A + tK$ is Fredholm for every $t \geq 0$, and therefore A and $A + K$ belong to the same connected component in the open set of Fredholm operators in $L(E; F)$. It suffices then to apply Theorem 2.2. □

COROLLARY 2.1. *Let E be a Banach space. If $K \in L(E)$ is compact,* $\text{Ind}(I_E + K) = 0$.

COROLLARY 2.2. *Let E and F be Banach spaces. The following two properties of an operator $A \in L(E; F)$ are equivalent*:

(2.16) *A is Fredholm and* $\operatorname{Ind} A = 0$.

(2.17) *There is a compact operator $K \in L(E; F)$ such that $A + K$ is invertible.*

PROOF. That (2.17) implies (2.16) is evident, since $A = (A + K) - K$. Suppose that A satisfies (2.16); let E_o be a closed subspace of E, supplementary of Ker A, W a finite-dimensional subspace of F supplementary of $A(E)$; of course $\dim W = \dim(\operatorname{Ker} A)$. Select a linear, bijective map $R: \operatorname{Ker} A \to W$. Extend R to the whole of E by letting it be zero on E_o; then $A + R$ is a bijection of E onto F. □

REMARK 2.2. The meaning of Corollary 2.2 is that the Fredholm operators $E \to F$ having index zero make up the connected component of the bijective operators (in the Fredholm set). When $E = F$, this is the connected component of the identity.

2.2. Application to Pseudodifferential Operators on Compact Manifolds

Throughout the remainder of this section X will denote a compact manifold of dimension n: $\mathbb{E}, \mathbb{F}$ will denote two vector bundles over X. The typical fibers of $\mathbb{E}$ and of $\mathbb{F}$, E and F respectively, are finite-dimensional complex vector spaces. In some of the arguments it is convenient to assume that X is equipped with a Riemannian metric and that $\mathbb{E}$ and $\mathbb{F}$ carry hermitian structures:

A *Riemannian metric* on X is, by definition, the datum for each point x of X of a positive-definite (real) symmetric bilinear form on the cotagent space T_x^*X, g_x, depending smoothly on x, in the sense that if α, β are any two one-forms on X, then $g_x(\alpha, \beta)$ is a C^∞ function of x. In a local chart $(\mathcal{O}, x^1, \ldots, x^n)$, by using the associated basis $dx^1, \ldots, dx^n$ in the cotangent spaces, one can represent g_x by the familiar bilinear form $\sum_{j,k} g^{jk}\xi_j\xi_k$.

The *hermitian structure* on the vector bundle $\mathbb{E}$ is the datum for each $x \in X$ of a positive-definite hermitian (sesquilinear) form on the fiber E_x, h_x such that if α, β are any two C^∞ sections of $\mathbb{E}$, then $h_x(\alpha, \beta)$ is a C^∞ function of x.

The Riemannian metric on X enables us to define the *norm* of an arbitrary covector $\xi \in T_x^*X$:

$$|\xi| = g_x(\xi, \xi)^{1/2}. \tag{2.18}$$

Similarly, the hermitian structure on $\mathbb{E}$ enables us to talk of the norm of any element e of E_x, $|e| = h_x(e, e)^{1/2}$.

We shall now deal with pseudodifferential operators from $\mathbb{E}$ to $\mathbb{F}$, a subject on which we refer the reader to Section 7 of Chapter I. For the sake of simplicity we shall impose two restrictions on the scope of our study. The first one, which could easily be lifted, is that every pseudodifferential operator P that we shall be dealing with has a principal symbol. This means that for each $(x, \xi) \in T^*X\backslash 0$ we are given a linear map $p(x, \xi): E_x \to F_x$ having the following properties:

(i) $p(x, \xi)$ is positive-homogeneous of degree m (the *order* of P) with respect to ξ.

(ii) $p(x, \xi)$ is smooth, in the obvious sense (made clear by using local trivializations of T^*X, of $\mathbb{E}$ and of $\mathbb{F}$).

(iii) In any local chart $(\mathcal{O}, x^1, \ldots, x^n)$ of X, $P - p(x, D)$ is of order $\leq m - \delta$, for some $\delta > 0$.

We shall be interested mainly in the case where P is *elliptic*, which has been defined (Definition 7.1, Chapter I) by the property that the linear map $p(x, \xi): E_x \to F_x$ is injective for each (x, ξ). That is where our second hypothesis (which cannot be relaxed) comes into play. We shall assume throughout that

(2.19) $\mathbb{E}$ *and* $\mathbb{F}$ *have the same fiber dimension.*

Then, to say that P is elliptic is to say that

(2.20) $\forall (x, \xi) \in T^*X\backslash 0$, *the principal symbol* $p(x, \xi)$ *of* P *is a bijection of* E_x *onto* F_x.

Under this hypothesis the result at the end of Section 5, Chapter I, extends at once:

THEOREM 2.4. *Suppose that* P *has a principal symbol* $p(x, \xi)$ *and is elliptic of order* m *on* X. *There is an elliptic operator of order* $-m$ *in* X, Q, *with principal symbol* $p(x, \xi)^{-1}$ *(the inverse of* $p(x, \xi)$ *in* $L(E_x; F_x)$*), such that*

$$QP - I_{\mathbb{E}} \qquad \text{and} \qquad PQ - I_{\mathbb{F}}$$

are regularizing (from $\mathbb{E}$ *to* $\mathbb{E}$, *and from* $\mathbb{F}$ *to* $\mathbb{F}$ *respectively).*

We have denoted by $I_{\mathbb{E}}$ (resp., $I_{\mathbb{F}}$) the identity operator from distribution sections of the vector bundle $\mathbb{E}$ (resp. $\mathbb{F}$) to the same. The formalism for constructing Q is exactly that described in the proof of Proposition 4.1, Chapter I.

If P is elliptic in X, it is hypoelliptic (Chapter I, Corollary 4.5) and therefore the set of distributions $u \in \mathcal{D}'(X; \mathbb{E})$ such that $Pu = 0$ all belong to $C^\infty(X; \mathbb{E})$. On the other hand, the adjoint P^* of P (for the hermitian structures naturally associated with those on $\mathbb{E}$ and on $\mathbb{F}$) is also an elliptic operator, also of order m, this time from $\mathbb{F}$ to $\mathbb{E}$ (by virtue of Theorem 4.2, Chapter I), and its kernel also only consists of C^∞ sections, this time of $\mathbb{F}$. In the statements that follow, consider that we have equipped each space $H^s(X; \mathbb{E})$ with a Hilbert space structure compatible with its locally convex topology. Actually we may use the structure derived from the Riemannian metric on X and the hermitian vector bundle structure on $\mathbb{E}$.

THEOREM 2.5. *Same hypotheses as in Theorem* 2.4. *Whatever the real number s, P is a Fredholm operator* $H^s(X; \mathbb{E}) \to H^{s-m}(X; \mathbb{F})$ *whose index is independent of s and is equal to that of any other elliptic pseudodifferential operator of order m, from* $\mathbb{E}$ *to* $\mathbb{F}$, *whose principal symbol is equal to that of P.*

PROOF. The first part of the statement follows from Theorems 2.1, 2.4, and the remarks preceding the statement. If P_1 is another pseudodifferential operator of order m from $\mathbb{E}$ to $\mathbb{F}$ having the same principal symbol as P, then $P - P_1$ has order $\leq m - \delta$ for some $\delta > 0$, and therefore (Corollary 1.3) defines a compact linear operator $H^s(X; \mathbb{E}) \to H^{s-m}(X; \mathbb{F})$. It suffices to apply Theorem 2.3. □

NOTATION. With P as in Theorem 2.5, we denote its index by Ind P.

We deal momentarily with an elliptic pseudodifferential operator A of order m, from $\mathbb{E}$ to itself, with principal symbol $a(x, \xi)I_E$ with I_E the identity map of the fiber E_x, and $a(x, \xi)$ a complex C^∞ function, nowhere vanishing, in $T^*X\backslash 0$, positive-homogeneous of degree m with respect to ξ. Let us denote by $\mathcal{K}_a$ the set of values $a(x, \xi)/|\xi|^m$ as (x, ξ) ranges over $T^*X\backslash 0$.

LEMMA 2.3. *Suppose that there is an open neighborhood* $\mathcal{O}$ *of* $\mathcal{K}_a$ *in* $\mathbb{C}\backslash\{0\}$ *and a* C^∞ *map* $f(t, z): [0, 1] \times \mathcal{O} \to \mathbb{C}\backslash\{0\}$ *such that* $f(0, z) = z$ *and* $f(1, z) = z_o \neq 0$ *for all* $z \in \mathcal{O}$. *Then* Ind $A = 0$.

PROOF. By reasoning in finitely many local charts of X, we can construct a scalar pseudodifferential operator of order m in X, B_t, depending smoothly on $t \in [0, 1]$, having principal symbol equal to

$$b_t(x, \xi) = f[t, a(x, \xi/|\xi|)]|\xi|^m. \tag{2.21}$$

Denote also by B_t the corresponding pseudodifferential operator from $\mathbb{E}$ to $\mathbb{E}$. Since B_0 and A have the same principal symbol, they have the same index (Theorem 2.5). By Theorem 2.2 we see that A and B_1 also have the same index. And so do B_1 and its adjoint, B_1^* (they have essentially the same principal symbol, see p. 105). But $\operatorname{Ind} B_1^* = -\operatorname{Ind} B_1$ (Proposition 2.1), and thus all these indices are equal to zero. □

COROLLARY 2.3. *If $a(x, \xi)$ is real valued, then* $\operatorname{Ind} A = 0$.

Let K_m be a pseudodifferential operator on X having the principal symbol $|\xi|^m$. We know that $\operatorname{Ind} K_m = 0$; let $f_1, \ldots, f_d$ be a basis of its kernel, M the orthogonal of $\operatorname{Ker} K_m$, say in $H^0(X) = L^2(X)$, for the natural Hilbert space structure associated with the Riemannian metric on X. The codimension of the image of K_m in $H^{-m}(X)$ is equal to d; and by using the density of $C^\infty(X)$ in $H^{-m}(X)$, we can find d C^∞ functions in X, $g_1, \ldots, g_d$, which make up a supplementary of the range of K_m. Let J_m be the continuous linear map $H^0(X) \to H^{-m}(X)$ equal to K_m on M and such that $J_m(f_j) = g_j$ for each $j = 1, \ldots, d$. Each member of $L^2(X)$ can be written, in a unique manner:

$$f = f_o + \sum_{j=1}^{d} (f, f_j)_0 f_j, \qquad f_o \in M. \tag{2.22}$$

As usual $(\ ,\ )_0$ denotes the hermitian product in $H^0(X)$. Therefore

$$J_m f = K_m f_o + \sum_{j=1}^{d} (f, f_j)_0 g_j, \tag{2.23}$$

which shows that J_m is a pseudodifferential operator in X; J_m differs from K_m by a regularizing operator having rank d.

Of course J_m induces an isomorphism of $H^0(X)$ onto $H^{-m}(X)$, and thus

LEMMA 2.4. *There is an elliptic pseudodifferential operator order m in X, J_m, having principal symbol $|\xi|^m$, which, for every real number s, induces an isomorphism of $H^s(X)$ onto $H^{s-m}(X)$.*

We shall denote by J_m that which should be denoted by $J_m I_{\mathbb{E}}$, the pseudodifferential operator from $\mathbb{E}$ to itself. If P is then any elliptic pseudodifferential operator of order m, from $\mathbb{E}$ to $\mathbb{F}$ (having a principal symbol $p(x, \xi)$ as before), we may form PJ_{-m}; this is an elliptic pseudodifferential operator of order zero in X, having the principal symbol $p(x, \xi)/|\xi|^m$. From (2.5) and Corollary 2.3 we deduce that $\operatorname{Ind}(PJ_{-m}) = \operatorname{Ind} P$. We may therefore limit ourselves to pseudodifferential operators of order zero (or, for that matter, of any order we wish).

Let $\mathfrak{S}^0(\mathbb{E};\mathbb{F})$ denote the set of symbols in $T^*X\backslash 0$, $p(x,\xi)$, from $\mathbb{E}$ to $\mathbb{F}$, which are positive-homogeneous of degree *zero* and *elliptic*; the latter means that for every (x,ξ) in $T^*X\backslash 0$, $p(x,\xi)$ is a linear bijection of E_x onto F_x. As usual $p(x,\xi)$ is a smooth function of (x,ξ) in $T^*X\backslash 0$, and we equip $\mathfrak{S}^0(\mathbb{E};\mathbb{F})$ with the standard C^∞ topology. (It is an open subset of the Fréchet space of all symbols from $\mathbb{E}$ to $\mathbb{F}$, positive-homogeneous of degree zero, whether elliptic or not.) It follows from Theorem 2.5 that the functional $\operatorname{Ind} P$ on the set of elliptic pseudodifferential operators of order zero from $\mathbb{E}$ to $\mathbb{F}$ defines a functional $\operatorname{Ind} p$ on $\mathfrak{S}^0(\mathbb{E};\mathbb{F})$.

THEOREM 2.6. *The function* $\operatorname{Ind} p$ *is locally constant in* $\mathfrak{S}^0(\mathbb{E};\mathbb{F})$.

PROOF. Let p_o be an arbitrary element of $\mathfrak{S}^0(\mathbb{E};\mathbb{F})$, $\mathfrak{U}$ a convex neighborhood of p_o in $\mathfrak{S}^0(\mathbb{E};\mathbb{F})$, p_1 any other element of $\mathfrak{U}$. We shall write $p_t = (1-t)p_o + tp_1$ for $0 \le t \le 1$. Let (U_j, χ_j) be local charts in X ($j = 1, \ldots, r$) such that $U_1 \cup \cdots \cup U_r = X$; let $g_1, \ldots, g_r$ be a C^∞ partition of unity subordinate to the covering $(U_1, \ldots, U_r)$, and for each j, a function $h_j \in C_c^\infty(U_j)$ equal to one in a neighborhood of $\operatorname{supp} g_j$. Let us set, for $t \in [0,1]$,

$$P_t = \sum_{j=1}^{r} h_j(x)p_t(x, D_x)g_j(x). \tag{2.24}$$

It can be checked at once that P_t is a classical pseudodifferential operator of order zero in X with principal symbol equal to $p_t(x,\xi)$. As a matter of fact,

$$P_t u(x) = (2\pi)^{-n} \iint e^{i(x-y)\cdot\xi} p_t^\#(x, y, \xi)u(y)\, dy\, d\xi, \tag{2.25}$$

where $p_t^\#(x,y,\xi) = \sum_{j=1}^{r} h_j(x)g_j(y)p_t(x,\xi)$, dy is the Lebesgue measure in the local coordinates in the chart (U_j, χ_j), and $d\xi$ is the one in the dual coordinates.

Speaking loosely, we can say that $p_t^\#$ is a continuous function of t in $[0,1]$ valued in the space of "amplitudes from $\mathbb{E}$ to $\mathbb{F}$" of degree zero. It follows at once from Proposition 2.2, Chapter I, that $r \mapsto P_t$ is a continuous curve in the Banach space of continuous linear operators $H^0(X;\mathbb{E}) \to H^0(X;\mathbb{F})$. By Theorem 2.2, $\operatorname{Ind} P_t$ is constant. □

EXAMPLE 2.1. Let P denote a scalar differential operator on X, of order $m \in \mathbb{Z}_+$, $P_m(x,\xi)$ its principal symbol. This symbol is homogeneous of degree m, not just positive-homogeneous. We may complexify the cotangent spaces to X: $T_x^{*\mathbb{C}}X = T_x^*X \otimes \mathbb{C}$. We recall that the principal symbol of the *transpose* ${}^tP$ of P is $g(x)P_m(x, -\xi)$ and that of its adjoint P^* is $g(x)\overline{P_m(x, \xi)}$. Here $g(x)$ denotes the volume density defined by the Riemannian structure we are using on X. Since $P^*u = \overline{{}^tP\bar{u}}$, $u \mapsto \bar{u}$ is a bijection of Ker P^* onto Ker ${}^tP$, and of Im P^* onto Im ${}^tP$. Thus, when P is elliptic,

$$\text{Ind}\, P^* = \text{Ind}\, {}^tP. \tag{2.26}$$

It follows then from Proposition 2.1 that

$$\text{Ind}\, P^* = -\text{Ind}\, P. \tag{2.27}$$

Formulas (2.26) and (2.27) are valid for arbitrary elliptic pseudodifferential operators on X. Now we use the fact that the function $P_m(x, \xi)$ is homogeneous, and not just positive-homogeneous, with respect to ξ: $P_m(x, -\xi) = \pm P_m(x, \xi)$; therefore, by Theorem 2.5, the index of ${}^tP$ is equal to that of $\pm P$, hence to that of P. By (2.26), (2.27) we conclude:

PROPOSITION 2.4. *If P is a scalar elliptic differential operator on a compact manifold X,* Ind $P = 0$. *More generally this is true if P is an elliptic differential operator from a complex vector bundle $\mathbb{E}$ over X to $\mathbb{E}$ whose principal symbol is a scalar multiple of the identity.*

It is not true that scalar elliptic *pseudo*differential operators always have index zero, as the next example shows:

EXAMPLE 2.2. We refer the reader to Example 5.1 of Chapter I. We consider the projectors P^+, P^- on the unit circle S^1, where the angular coordinate is denoted by θ. They are classical pseudodifferential operators of order zero, with respective principal symbols $\mathcal{Y}(\xi)$, $\mathcal{Y}(-\xi)$ ($\mathcal{Y}$ is Heaviside's function; Chapter I, Theorem 5.3). The operator

$$P = P^+ + P^- e^{i\theta} \tag{2.28}$$

is elliptic, classical, has order zero (cf. last remark in Section 5, Chapter I). Its kernel is one dimensional: it consists of all the functions of the kind const $(1 - e^{-i\theta})$. But P maps $L^2(S^1)$ *onto* itself, as the reader can check at once. Thus Ind $P = +1$.

3. Uniqueness in the Cauchy Problem for Certain Operators with Simple Characteristics

In this section we consider a *differential* operator $P(y, D_y)$ of order m in an open subset Ω of $\mathbb{R}^{n+1}$. We suppose that Ω is subdivided into two parts by a C^∞ hypersurface S; this means that $\Omega = \Omega^+ \cup S \cup \Omega^-$, with Ω^+ and Ω^- connected open subsets, disjoint, and also disjoint from the hypersurface S. Our purpose is to obtain conditions on P and on S sufficient to insure that any function u (endowed with suitable regularity properties), which satisfies

$$P(y, D_y)u = 0 \quad in\ \Omega, \tag{3.1}$$

$$u = 0 \quad in\ \Omega^-, \tag{3.2}$$

must necessarily vanish in a neighborhood of S. This is a version of the so-called *uniqueness in the Cauchy problem.*

Let us denote by $p(y, \eta)$ the *principal symbol* of $P(y, D_y)$. We begin by assuming that the hypersurface S is *noncharacteristic* at any one of its points. Let $y \in S$ be arbitrary, $\eta^o(y)$ a cotangent vector at y to $\mathbb{R}^{n+1}$ orthogonal to the tangent space to S, i.e., conormal to S; we must have $p(y, \eta^o(y)) \neq 0$. Actually we shall make the following stronger hypothesis:

(3.3) *Given any point y in S, any pair of vectors $\eta, \eta^o \in \mathbb{R}_{n+1}\backslash\{0\}$, such that η^o is conormal to S at y and η is not, the polynomial in the complex variable z, $p(y, \eta + z\eta^o)$, has m simple roots.*

REMARK 3.1. When $m = 1$, (3.3) is equivalent to the property that S is noncharacteristic with respect to P. When $m > 1$ it is a stronger property, as shown by the example in $\mathbb{R}^2$,

$$P = (D_{y^1} + \sqrt{-1}\, D_{y^2})^2, \qquad S = \{y \in \mathbb{R}^2;\, y^2 = 0\}.$$

for which (3.3) does not hold.

REMARK 3.2. Let y_o be a point in S, η^o a nonzero covector conormal to S at y_o. Suppose the following:

(3.4) *For any $\eta \in \mathbb{R}_{n+1}$ not conormal to S, $p(y_o, \eta + z\eta^o)$ has m distinct roots.*

It is checked at once, then, that for all points y in a whole neighborhood of y_o in Ω, and all η not collinear to η^o, $p(y, \eta + z\eta^o)$ also has m distinct roots. It follows from this observation that if S' is another smooth hypersurface passing through y_o and tangent to S at this point, property (3.3) holds, in some neighborhood of y_o, for S' in the place of S.

We shall reason in the neighborhood of a point of S, which we take to be the origin of $\mathbb{R}^{n+1}$. We select the coordinates in $\mathbb{R}^{n+1}$ in such a way that S is defined, near the origin, by the vanishing of one of them, which we call t; the remaining ones are called $x^1, \ldots, x^n$. We assume that Ω^- is defined by $t < 0$. The principal symbol of P will henceforth be denoted by $p(x, t, \xi, \tau)$ (the dual coordinates of the x^j are denoted by ξ_j, that of t by τ). The covectors conormal to S are of the kind $(0, \tau)$. Remark 3.2 indicates that hypothesis (3.3) is equivalent, near the origin, to the following:

(3.5) *For all* (x, t) *in a neighborhood* Ω_o *of the origin in* $\mathbb{R}^{n+1}$ *and all* $\xi \in \mathbb{R}_n \backslash \{0\}$, *all* z *in* $\mathbb{C}$,

(3.6) $$p(x, t, \xi, z) = 0 \textit{ implies } p_\tau(x, t, \xi, z) \neq 0.$$

(As usual p_τ denotes the partial derivative of p with respect to τ.)

Thanks to (3.5) we can represent the roots of $p(x, t, \xi, z)$ by C^∞ complex-valued functions $z_j(x, t, \xi)$ in $\Omega_o \times (\mathbb{R}_n \backslash \{0\})$, positive-homogeneous of degree one with respect to ξ $(j = 1, \ldots, m)$.

EXAMPLE 3.1. Let P be the Laplace operator in $\mathbb{R}^{n+1}$, which we write in the fashion $(\partial/\partial t)^2 + \Delta_x$ (Δ_x is the Laplace operator in the x variables). Then the roots of the principal symbol of P, which is $-(\tau^2 + |\xi|^2)$, are the two complex functions $\pm i|\xi|$. If we want them to be smooth they must be positive-homogeneous, not homogeneous, as soon as $n \geq 2$.

Below we assume that $\Omega_o = X \times]-T, T[$, where X is an open neighborhood of the origin in $\mathbb{R}^n$ and T a number >0.

LEMMA 3.1. *Suppose that* (3.5) *holds. There are* m *first-order classical pseudodifferential operators in* X, $Z_j(t)$, *depending smoothly on* $t \in]-T, T[$, *with principal symbol* $z_j(x, t, \xi)$ $(j = 1, \ldots, m)$, *such that*

(3.7) $$P = (D_t - Z_m) \cdots (D_t - Z_1) + R,$$

with $R = R(t)$ *a regularizing operator in* X *whose kernel is a* C^∞ *function in* $X \times X \times]-T, T[$.

PROOF OF LEMMA 3.1. We determine in succession, for $N = 0, 1, \ldots$, first-order classical pseudodifferential operators in X, $Z_{j(N)}$ $(j = 1, \ldots, m)$, having the following properties:

(3.8) *The principal symbol of* $Z_{j(N)}$ *is* $z_j(x, t, \xi)$;

(3.9) $Z_{j,N+1} = Z_{j(N+1)} - Z_{j(N)}$ *has order* $\leq -N$;

(3.10) $R_{(N)} = P - (D_t - Z_{m(N)}) \cdots (D_t - Z_{1(N)})$ *has order* $\leq m - N - 1$.

The $Z_{j(N)}$ and $R_{(N)}$ depend smoothly on t.

The only requirement on $Z_{j(0)}$ is that it verify (3.8); then (3.10) is automatically satisfied. Suppose that we have determined the $Z_{j(N)}$; let us determine $Z_{j,N+1}$ for every j. By virtue of (3.9) and (3.10), and of (3.10) with $N+1$ in place of N (which must be verified), we have

$$\begin{aligned} R_{(N+1)} = R_{(N)} &+ Z_{m,N+1}(D_t - Z_{(m-1)(N+1)}) \cdots (D_t - Z_{1(N+1)}) \\ &+ (D_t - Z_{m(N+1)})Z_{m-1,N+1}(D_t - Z_{(m-2)(N+1)}) \cdots (D_t - Z_{1(N+1)}) \\ &+ \cdots + (D_t - Z_{m(N+1)}) \cdots (D_t - Z_{2(N+1)})Z_{1,N+1}. \end{aligned}$$

We shall write that the principal symbol of $R_{(N+1)}$ *as an operator of order* $m - N - 1$ is identically zero; we denote by $r_{(N)}$ that of R_N, by $z_{j,N+1}$ the principal symbol of $Z_{j,N+1}$ regarded as an operator of order $-N$. Thus we must have

$$\begin{aligned} -r_{(N)} = z_{m,N+1}&(\tau - z_{m-1}) \cdots (\tau - z_1) \\ &+ z_{m-1,N+1}(\tau - z_m)(\tau - z_{m-2}) \cdots (\tau - z_1) \\ &+ \cdots + z_{1,N+1}(\tau - z_m) \cdots (\tau - z_2). \end{aligned} \tag{3.11}$$

The solution of (3.11) is evident; if we write $p(z)$ rather than $p(x, t, \xi, z)$ we have

$$z_{j,N+1} = -r_{(N)}/p'(z_j), \qquad j = 1, \ldots, m. \tag{3.12}$$

Note that (3.12) is indeed a C^∞ function in $X \times]-T, T[\times (\mathbb{R}_n\backslash\{0\})$, positive-homogeneous of degree $-N$ with respect to ξ.

For each j,

$$z_j(x, t, \xi) + \sum_{N=1}^{+\infty} z_{j,N}(x, t, \xi) \tag{3.13}$$

is a classical symbol. It suffices to take for Z_j any pseudodifferential operator having (3.13) as "total" symbol, C^∞ with respect to t, if one is to comply with the statement of Lemma 3.1. □

Before going further we avail ourselves once again of Remark 3.2. In the present setup let S' be the hypersurface $t + |x|^2 = 0$. Since we are trying to prove that the solution u of (3.1), (3.2) vanishes in a neighborhood of the origin, we might as well reason with S' in the place of S, possibly after shrinking slightly the neighborhood Ω_o. After a change of variables, in which S' becomes the hyperplane $t = 0$ and is thereafter called S, this amounts to assuming that in a neighborhood of the kind $X \times]-T, T[$, with X and T suitably small,

(3.14) *the x-projection of* supp u *is contained in a compact subset K of X.*

But then let $g \in C_c^\infty(X)$ be equal to one in a neighborhood of K. We have

$$Pu = gPu = (D_t - gZ_m) \cdots (D_t - gZ_1)u + R^{\#}u, \tag{3.15}$$

where $R^{\#}$ has properties similar to those of R, except that supp $R^{\#}v$ is contained in a compact subset of X independent of t and of $v \in \mathscr{E}'(X)$. To simplify the notation we shall assume the following:

(3.16) *There is a compact subset K' of X such that $Z_j v$ $(j = 1, \dots, m)$ and Rv vanish outside K' whatever $v \in \mathscr{E}'(X)$.*

The next step is to exploit the approximate factorization (3.7) in order to transform equation (3.1) into an approximately diagonal system of first-order differential(in t)–pseudodifferential(in x) equations.

Let us set $u = u^1$ and

$$(D_t - Z_j)u^j = u^{j+1}, \qquad j = 1, \dots, m-1, \tag{3.17}$$

$$(D_t - Z_m)u^m = -Ru^1. \tag{3.18}$$

We then define the following matrix-valued pseudodifferential operator of order one in X (depending smoothly on t).

$$A(t) = \sqrt{-1}\begin{pmatrix} Z_1 & 0 & 0 & \dots & 0 \\ 0 & Z_2 & 0 & \dots & 0 \\ \vdots & & & & \\ 0 & 0 & 0 & \dots & Z_m \end{pmatrix}.$$

We call $\mathscr{N}$ the standard nilpotent $m \times m$ matrix

$$\begin{pmatrix} 0 & 1 & 0 & \dots & 0 \\ 0 & 0 & 1 & \dots & 0 \\ \vdots & & & & \\ 0 & 0 & 0 & \dots & 1 \\ 0 & 0 & 0 & \dots & 0 \end{pmatrix},$$

and $\mathscr{R}(t)$ the $m \times m$ matrix whose entries are all zero, except the one in the lower left corner, equal to $R(t)$. We denote by U the m-vector (written as a column matrix) with components $u^1, \ldots, u^m$. Then (3.17), (3.18) can be rewritten

$$\frac{\partial U}{\partial t} - A(t)U - \sqrt{-1}\,[\mathcal{N} - \mathscr{R}(t)]U = 0. \tag{3.20}$$

We now make some regularity assumptions about u. They will of course reflect on U. A fairly natural condition is the following:

(3.21) *For each $j = 0, 1, \ldots, m$, u is a C^j function of t valued in $H_c^{m-j}(X)$.*

Returning to (3.17) we see that u^k is a C^j function of t with values in $H_c^{m-k+1-j}(X)$ for $j = 1, \ldots, m-k+1$ ($k = 1, \ldots, m$). We conclude:

(3.22) *For $j = 0, 1$, U is a C^j function of t with values in $H_c^{1-j}(X) \otimes \mathbb{C}^m$.*

We are going to prove that under suitable hypotheses about the characteristic roots $z_j(x, t, \xi)$, U will vanish in a neighborhood of the origin. The proof will be based on the exploitation of an inequality of the *Carleman type*:

$$\rho \iint e^{\rho(T-t)^2} |V(x,t)|^2 \, dx\, dt \le \text{const} \iint e^{\rho(T-t)^2} |\mathscr{L}V(x,t)|^2 \, dx\, dt, \tag{3.23}$$

where ρ is a large positive number,

$$\mathscr{L} = I_m \frac{\partial}{\partial t} - A(t) - \sqrt{-1}\,[\mathcal{N} - \mathscr{R}(t)],$$

and

$$V(x,t) = \chi(t)U(x,t), \tag{3.24}$$

with $\chi \in C^\infty(\mathbb{R}^1)$, $\chi(t) = 1$ for $t < 8T/10$, $\chi(t) = 0$ for $t > 9T/10$. Observe that

(3.25) *V has compact support contained in $X \times [0, T[$,*

and because of (3.20)

(3.26) *$\mathscr{L}V$ has compact support contained in $X \times [8T/10, T[$.*

Lemma 3.2. *If (3.23) holds, U vanishes for $t < T/2$.*

Proof. By (3.26) we may restrict the integration on the right-hand side in (3.23) to the region $8T/10 \le t \le T$. If we limit the integration in the left-hand side to $0 \le t \le T/2$, we obtain

$$e^{\rho T^2/4} \iint\limits_{0 \le t \le T/2} |U|^2 \, dx \, dt \le \text{const } e^{\rho T^2/25} \iint |\mathscr{L}V|^2 \, dx \, dt,$$

and we get the result by letting ρ go to infinity. □

The question now is to prove the estimate (3.23). Some additional hypotheses have to be made. Notice that the factor ρ on the left-hand side in (3.23) did not play any role in the proof of Lemma 3.2. Its usefulness is that if we take ρ large enough, we may replace $\mathscr{L}$ by the diagonal "leading part"

$$\mathscr{L}_o = I_m \frac{\partial}{\partial t} - A(t) \tag{3.27}$$

and thus get rid of $\mathcal{N} - \mathscr{R}(t)$. Before stating our hypotheses about the differential operator P (or, equivalently, about the pseudodifferential operator $A(t)$) let us take a look at the standard manner of proving the desired inequality,

$$\rho \iint e^{\rho(T-t)^2} |V|^2 \, dx \, dt \le C \iint e^{\rho(T-t)^2} |\mathscr{L}_o V|^2 \, dx \, dt. \tag{3.28}$$

First one changes function, setting $V = V_o \exp[-\rho(T-t)^2/2]$. What we have to prove then is the inequality

$$\rho \iint |V_o|^2 \, dx \, dt \le C \iint \left| \frac{\partial V_o}{\partial t} - \rho(t-T) V_o - A(t) V_o \right|^2 dx \, dt. \tag{3.29}$$

It suffices to prove this inequality for smooth V_o, with supp V_o compact contained in $X \times [0, T[$. (Then, if needed, go to the limit in $H^1(X \times]-T, T[) \otimes \mathbb{C}^m$.) It is convenient to decompose $A(t)$ into its self-adjoint and anti-self-adjoint parts,

$$A^+ = \tfrac{1}{2}(A + A^*), \qquad A^- = \tfrac{1}{2}(A - A^*), \tag{3.30}$$

where $A^* = A^*(t)$ is the adjoint of A; we assume that it has been multiplied by a C^∞ function with compact support in X, equal to one in a neighborhood of the x-projection of supp V_o. We regard $\partial/\partial t - A^-(t)$ as the anti-self-adjoint part of the operator intervening on the right-hand side of (3.29) and

$-[A^+(t)-\rho(T-t)]$ as its self-adjoint part. We write

$$\text{(3.31)}\qquad J_1 = \iint \left|\frac{\partial V_o}{\partial t} - \rho(t-T)V_o - A(t)V_o\right|^2 dx\,dt$$
$$= \iint \left|\frac{\partial V_o}{\partial t} - A^-(t)V_o\right|^2 dx\,dt$$
$$+ \iint |[A^+(t)-\rho(T-t)]V_o|^2\,dx\,dt + J_2,$$

where

$$\text{(3.32)}\qquad J_2 = -2\,\mathrm{Re}\int \left(\frac{\partial V_o}{\partial t} - A^-V_o, A^+V_o - \rho(T-t)V_o\right)_0 dt$$
$$= -2\,\mathrm{Re}\int\left(\frac{\partial V_o}{\partial t}, A^+V_o + \rho t V_o\right)_0 dt + \int ([A^+, A^-]V_o, V_o)_0\,dt$$
$$= \rho\iint |V_o|^2\,dx\,dt + \iint \bar{V}_o\left(\frac{\partial A^+}{\partial t} + [A^+, A^-]\right)V_o\,dx\,dt,$$

where we have written $(\ ,\ )_0$ for the inner product in $L^2(X)$ and integrated by parts.

If we want to reach the conclusion (3.29) we must show that for some constant $\kappa > 1$, we have

$$\text{(3.33)}\qquad -\kappa\iint \bar{V}_o\left(\frac{\partial A^+}{\partial t} + [A^+, A^-]\right)V_o\,dx\,dt$$
$$\leq \int\left\{\left\|\frac{\partial V_o}{\partial t} - A^-V_o\right\|_0^2 + \|[A^+ - \rho(T-t)]V_o\|_0^2 + \rho\|V_o\|_0^2\right\}dt,$$

where $\|\ \ \|_0$ is the norm in $L^2(X)$.

There are several ways of achieving (3.33), and we shall not aim at great generality. The conditions considered by A. Calderon in his original paper on the subject [1] were of two kinds:

(3.34) $\quad A^+$ *has order zero.*

(3.35) $\quad A^+$ *has order one and is invertible.*

Before reinterpreting these two conditions in terms of the roots $z_j(x, t, \xi)$, let us show how they imply what we want. It is obvious enough in the first case, since we then have, for some $C > 0$,

$$\text{(3.36)}\qquad \left|\iint \bar{V}_o\left(\frac{\partial A^+}{\partial t} + [A^+, A^-]\right)V_o\,dx\,dt\right| \leq C\iint |V_o|^2\,dx\,dt,$$

and it suffices to take $\kappa = \rho/C$, with ρ large enough, in (3.33).

In case (3.35), we have, for a suitable $C > 0$,

(3.37)
$$\left|\iint \bar{V}_o\left(\frac{\partial A^+}{\partial t} + [A^+, A^-]\right) V_o\, dx\, dt\right| \le C \int \|A^+ V_o\|_0 \|V_o\|_0\, dt$$

$$\le C\left\{\int \{\|[A^+ - \rho(T-t)]V_o\|_0 \|V_o\|_0 + T\rho \|V_o\|_0^2\}\, dt\right\}$$

$$\le \tfrac{1}{2}\int \left\|[A^+ - \rho(T-t)]V_o\right\|_0^2 dt + C(T + C/2\rho)\rho \int \|V_o\|_0^2\, dt,$$

and in this case it suffices to select T and $1/\rho$ small enough.

The conditions (3.34), (3.35) bear solely on the principal symbol of A^+ when the latter is regarded as a first-order pseudodifferential operator in X. The principal symbol of A^+ is the diagonal matrix with entries $-\mathrm{Im}\, z_j(x, t, \xi)$ $(j = 1, \ldots, m)$. Consequently, condition (3.34) is equivalent to the following one:

(3.38) *For every $j = 1, \ldots, m$ and every (x, t, ξ) in $X \times]-T, T[\times (\mathbb{R}_n \backslash \{0\})$ the root $z_j(x, t, \xi)$ is real.*

And (3.35) can be restated as follows:

(3.39) *For the same j's and (x, t, ξ) as in (3.38), $z_j(x, t, \xi)$ is nonreal.*

DEFINITION 3.1. *The operator P will be said to be strongly hyperbolic on the hypersurface S if every point y_o of S has an open neighborhood $\mathcal{O}_o$ such that for any y in $\mathcal{O}_o$ and any pair of vectors $\eta, \eta^o \in \mathbb{R}_{n+1}\backslash\{0\}$, where η^o is conormal to S at y_o and η is not, the polynomial in z, $p(y, \eta + z\eta^o)$ has m simple real roots. We shall then also say that the principal symbol of P is strongly hyperbolic on S.*

The meaning of (3.39) is simply that P is elliptic near the origin.

Since the matrix A, and therefore also the matrices A^+, A^- are diagonal, we can handle separately, in the manner described before, the rows corresponding to a wholly real root $z_j(x, t, \xi)$, and those corresponding to nonreal z_j. Thus we obtain the following result:

THEOREM 3.1. *Suppose that the principal symbol $p(y, \eta)$ of P in $\Omega \times (\mathbb{R}_{n+1}\backslash\{0\})$ satisfies* (3.3) *and is equal to $p_1(y, \eta)p_2(y, \eta)$ with p_1 strongly hyperbolic on the hypersurface S and p_2 elliptic. Then every C^m function u in Ω that satisfies* (3.1), (3.2) *vanishes identically in a neighborhood of S.*

Inspection of (3.33) shows that it might be obtainable, even when neither (3.34) nor (3.35) holds. In such a case the principal symbol of A^+ vanishes on some subset of the (x, t, ξ)-space, in fact at some points $(x, 0, \xi)$, but is nonzero at some points (x, t, ξ) for t arbitrarily close to zero (and $t \geq 0$). Then everything would still work out fine if we made the following assumption.

(3.40) *In a neighborhood of the characteristic set of $A^+(t)$ the principal symbol of the following operator is strictly positive*:

$$\frac{\partial A^+}{\partial t} + [A^+, A^-] \tag{3.41}$$

One would then write

$$V_o = g(x, D) V_o + [1 - g(x, D)] V_o, \tag{3.42}$$

where $g(x, \xi)$ is C^∞, positive-homogeneous of degree zero with respect to ξ, equal to one in some neighborhood of Char $A^+(0)$ and to zero outside some larger such neighborhood. Roughly speaking we would reason as in the elliptic case (3.35) in dealing with the second term in (3.42). In dealing with the first term we would use an estimate of the kind

$$-\iint \bar{V}_o\left(\frac{\partial A^+}{\partial t} + [A^+, A^-]\right) g(x, D) V_o \, dx \, dt \leq C \iint |V_o|^2 \, dx \, dt, \tag{3.43}$$

which would follow from hypothesis (3.40). This approach has been carried out in Hörmander [2], Chapter 8, where the geometric interpretation of (3.41) is described. See also Section 5, Chapter XI.

4. The Friedrichs Lemma

The classical Friedrich's lemma concerns the commutator of a differential operator with a Friedrich's mollifier. We shall look at the commutator of such a mollifier with a pseudodifferential operator.

Friedrich's mollifiers are defined as follows: Take any function $\chi \in C_c^\infty = C_c^\infty(\mathbb{R}^n)$ such that $\int \chi(x)\, dx = +1$. For $\varepsilon > 0$ set

$$\chi_\varepsilon(x) = \varepsilon^{-n} \chi(x/\varepsilon).$$

Given u, say in C^∞, form the convolution

$$(\chi_\varepsilon * u)(x) = \int \chi_\varepsilon(y) u(x - y)\, dy = \int \chi(y) u(x - \varepsilon y)\, dy.$$

The word "mollifier" refers to the fact that $u \mapsto \chi_\varepsilon * u$ extends as a continuous linear map $\mathcal{D}' \to C^\infty$. Given any distribution u in $\mathbb{R}^n$ we have, for some suitable constant $C > 0$,

$$\operatorname{supp}(\chi_\varepsilon * u) \subset \operatorname{supp} u + \{x \in \mathbb{R}^n;\, |x| \leq C\varepsilon\}. \tag{4.1}$$

Thus $\chi_\varepsilon *$ maps (continuously) $\mathcal{E}'$ into C_c^∞.

When ε tends to zero, the operators $\chi_\varepsilon *$ converge to the identity as continuous linear operators on a wide class of distribution spaces, such as $\mathcal{D}'$, $\mathcal{E}'$, C^∞, C_c^∞, but also, for instance, $\mathcal{S}$ and the Sobolev spaces $H^s = H^s(\mathbb{R}^n)$; but theirs is the strong convergence, which means that $\chi_\varepsilon * u$ converges to u for each individual u.

PROPOSITION 4.1. *Let s be any real number. The Freidrichs mollifiers $\chi_\varepsilon *$ map continuously H^s into $H^{+\infty}$. Each element u of H^s is the limit, in H^s, of $\chi_\varepsilon * u$ as $\varepsilon \to +0$.*

We have denoted by $H^{+\infty}$ the intersection of the spaces H^s equipped with the projective limit topology. (A set is open in $H^{+\infty}$ if it is equal to the intersection of $H^{+\infty}$ with an open subset of some H^s.)

PROOF. It suffices to note that the Fourier transform of $\chi_\varepsilon * u$ is equal to $\hat{\chi}(\varepsilon\xi)\hat{u}(\xi)$, and that $|\hat{\chi}_\varepsilon| \leq \int |\chi|\, dx$. Then it is evident that $\chi_\varepsilon *$ maps H^s continuously into $H^{+\infty}$. On the other hand, as ε varies, say from 1 to 0, the $\chi_\varepsilon *$ form a bounded set of linear operators on H^s. Since $\chi_\varepsilon * u$ converges to $u \in \mathcal{S}$ (in the space $\mathcal{S}$, which is dense in every H^s), the same is true for an arbitrary $u \in H^s$. □

We now consider a (standard) pseudodifferential operator A on $\mathbb{R}^n$, of order m ($\in \mathbb{R}$). We must make sure that A operates from a global Sobolev space, say H^s, to another one, H^{s-m} (Chapter I, Theorem 2.1). This is the case if we assume that

$$Au(x) = (2\pi)^{-n} \iint e^{i(x-y)\cdot\xi} a(x, y, \xi) u(y)\, dy\, d\xi, \tag{4.2}$$

with an amplitude $a \in S^m(\mathbb{R}^n, \mathbb{R}^n)$, having the following property (which can be greatly relaxed):

(4.3) *$a(x, y, \xi)$ vanishes identically for (x, y) outside some compact subset of $\mathbb{R}^n \times \mathbb{R}^n$.*

We note that $A(\chi_\varepsilon * u)$ converges to Au in H^{s-m} for any $u \in H^s$, by virtue of Theorem 2.1, Chapter I, and of Proposition 4.1. By the latter, $\chi_\varepsilon * Au$ also converges to Au. Thus

$$[A, \chi_\varepsilon *]u = A(\chi_\varepsilon * u) - \chi_\varepsilon * Au \tag{4.4}$$

converges to zero in H^{s-m}. It is the contention of Friedrich's lemma that it does better, that it converges to zero in H^{s-m+1}:

THEOREM 4.1. *Let* $A \in \Psi^m(\mathbb{R}^n)$ *be given by* (4.2) *with* $a \in S^m(\mathbb{R}^n, \mathbb{R}^n)$ *satisfying* (4.3).

The commutators $[A, \chi_\varepsilon *]$, $0 \le \varepsilon \le 1$, *form a bounded set of pseudo-differential operators of order* $m - 1$ *in* $\mathbb{R}^n$.

Given any $u \in H^s$, $[A, \chi_\varepsilon *]u$ *converges to zero in* H^{s-m+1} *as* ε *goes to* 0.

PROOF. It suffices to prove the first part. The preceding arguments render obvious the fact that $[A, \chi_\varepsilon *]u \to 0$ in H^t, for any $t \in \mathbb{R}$, when $u \in \mathcal{S}$. By equicontinuity and the density of $\mathcal{S}$ in H^s, the same is true when $t = s - m + 1$, and u is any element of H^s.

Consider then

$$C_\varepsilon^1 u(x) = A(\chi_\varepsilon * u)(x)$$

$$= (2\pi)^{-n} \iiint e^{i(x-z)\cdot\xi} a(x, z, \xi)\chi_\varepsilon(z - y)u(y)\, dy\, dz\, d\xi,$$

$$C_\varepsilon^2 u(x) = [\chi_\varepsilon * (Au)](x)$$

$$= (2\pi)^{-n} \iiint e^{i(z-y)\cdot\xi} \chi_\varepsilon(x - z)a(z, y, \xi)u(y)\, dy\, dz\, d\xi.$$

Set $z = y + w$ in the integral representing $C_\varepsilon^1 u$, $z = x - w$ in that representing $C_\varepsilon^2 u$. We derive

$$[A, \chi_\varepsilon *]u(x) = (2\pi)^{-n} \iint e^{i(x-y)\cdot\xi} c_\varepsilon(x, y, \xi)u(y)\, dy\, d\xi, \tag{4.5}$$

$$c_\varepsilon(x, y, \xi) = \int e^{-iw\cdot\xi} \chi_\varepsilon(w)[a(x, y + w, \xi) - a(x - w, y, \xi)]\, dw. \tag{4.6}$$

Theorem 4.1 will be proved if we show that (as ε varies from 0 to 1) the c_ε stay in a bounded subset of $S^{m-1}(\mathbb{R}^n, \mathbb{R}^n)$. Formula (4.6) implies at once that $c_\varepsilon(x, y, \xi)$ vanishes identically for (x, y) outside a suitable compact subset of

$\mathbb{R}^n \times \mathbb{R}^n$ independent of ε. We observe that

$$a(x, y + w, \xi) - a(x - w, y, \xi) = \int_0^1 w \cdot [a_x(x - tw, y, \xi) + a_y(x, y + tw, \xi)]\, dt.$$

Let us set

$$b_j(x, y, w, \xi) = \int_0^1 \left[\frac{\partial a}{\partial x^j}(x - tw, y, \xi) + \frac{\partial a}{\partial y^j}(x, y + tw, \xi)\right] dt.$$

It is easily checked that b_j is a C^∞ function of w in $\mathbb{R}^n$, valued in $S^m(\mathbb{R}^n, \mathbb{R}^n)$. And

$$c_\varepsilon(x, y, \xi) = \sum_{j=1}^n \int e^{-iw\cdot\xi}\chi_\varepsilon(w) w^j b_j(x, y, w, \xi)\, dw. \tag{4.7}$$

We are going to look at the jth term in the sum on the right in (4.7), a term that we call $c_{\varepsilon,j}$. We restrict the variation of ξ to an open cone Γ in $\mathbb{R}_n$, with vertex at the origin, strictly convex, in fact to a cone

$$|\xi| < C\xi_n. \tag{4.8}$$

We shall estimate the derivatives of $c_{\varepsilon,j}$ for ξ in Γ. The same estimates will obviously be valid in each rotate of Γ and therefore everywhere in $\mathbb{R}_n$. The differentiations with respect to x, y are passed immediately to b_j and therefore only entail a replacement of the latter. We shall reason as if there were no such differentiations. The differentiations with respect to ξ must also "hit" the exponential $e^{-iw\cdot\xi}$, and thus

$$\begin{aligned} D_\xi^\alpha c_{\varepsilon,j} &= \int e^{-iw\cdot\xi}\chi_\varepsilon(w) w^j (D_\xi - w)^\alpha b_j\, dw \\ &= \sum_{\beta\le\alpha} \binom{\alpha}{\beta} \int e^{-iw\cdot\xi}\chi_\varepsilon(w) w^j (-w)^\beta D_\xi^{\alpha-\beta} b_j\, dw. \end{aligned} \tag{4.9}$$

Consider then an integral

$$\begin{aligned} I_{\alpha,\beta} &= \int e^{-iw\cdot\xi}\chi_\varepsilon(w) w^j w^\beta D_\xi^{\alpha-\beta} b_j\, dw \\ &= \int \left[\left(1 + \frac{\partial}{\partial w^n}\right)^{1+|\beta|} (e^{-iw\cdot\xi})\right] \chi_\varepsilon(w) w^j w^\beta (1 + i\xi_n)^{-1-|\beta|} D_\xi^{\alpha-\beta} b_j\, dw \\ &= \int e^{-iw\cdot\xi}\left(1 - \frac{\partial}{\partial w^n}\right)^{1+|\beta|} [\chi_\varepsilon(w) w^j w^\beta D_\xi^{\alpha-\beta} b_j](1 + i\xi_n)^{-1-|\beta|}\, dw. \end{aligned}$$

At this point we apply the following, which is evident by the change of variables $w = \varepsilon x$:

(4.10) *Given any pair $p, q \in \mathbb{Z}_+^n$, there is $C_{p,q} > 0$ such that, for all $\varepsilon \in]0, 1]$,*

$$\int |D_w^q[w^p \chi_\varepsilon(w)]| \, dw \leq C_{p,q} \varepsilon^{|p-q|}.$$

By using the fact that $b_j \in S^m(\mathbb{R}^n, \mathbb{R}^n)$ (and depends smoothly on w), and by (4.8) we obtain, with $C > 0$ independent of ε,

$$|D^\alpha c_{\varepsilon,j}(x, y, \xi)| \leq C(1 + |\xi|)^{m-1-|\alpha|}. \tag{4.11}$$

We have similar estimates for $D_x^\lambda D_y^\mu c_{\varepsilon,j}$ substituted for $c_{\varepsilon,j}$, and finally with $D_x^\lambda D_y^\mu c_\varepsilon$ substituted. □

The traditional version of the Friedrichs lemma follows:

COROLLARY 4.1. *Let $a(x)$ be a C^∞ function in $\mathbb{R}^n$. Assume that every derivative of a is bounded in $\mathbb{R}^n$. Let $L(D)$ be a differential operator with constant coefficients in $\mathbb{R}^n$, of order one. Given any real number s, any $u \in H^s$,*

$$a[\chi_\varepsilon * L(D)u] - \chi_\varepsilon * [aL(D)u] \tag{4.12}$$

converges to zero in H^s.

PROOF. Let $g, h \in C_c^\infty(\mathbb{R}^n)$, $g = 1$ in the cube $\{x; |x^j| \leq 1, j = 1, \ldots, n\}$, $h = 1$ in a neighborhood of supp g. Set $A_\alpha = \text{Op}\, a_\alpha$, $\alpha \in \mathbb{Z}^n$, with

$$a_\alpha(x, y, \xi) = a(x)h(x - \alpha)L(\xi)g(x - \alpha). \tag{4.13}$$

The (differential) operator A_α is of the kind considered in Theorem 4.1. Because of our hypothesis on a, the amplitudes a_α form a bounded set in $S^1(\mathbb{R}^n, \mathbb{R}^n)$. The distribution (4.12) is equal to

$$\sum_{\alpha \in \mathbb{Z}^n} [A_\alpha, \chi_\varepsilon *]u,$$

whence the result. □

REMARK 4.1. We have used the obvious strengthening of Theorem 4.1, according to which the commutators $[A, \chi_\varepsilon *]$ remain in a bounded set of pseudodifferential operators of order $m - 1$ if ε varies in $[0, 1]$ and also if A varies in a bounded set of pseudodifferential operators of order m.

COROLLARY 4.2. *Let A be a pseudodifferential operator of order m in $\mathbb{R}^n$ satisfying the requirements in Theorem* 4.1. *Let B be an arbitrary pseudo-differential operator of order m' in $\mathbb{R}^n$. Then the commutators*

$$[B, [A, \chi_\varepsilon *]] \qquad (0 \leq \varepsilon \leq 1) \tag{4.14}$$

form a bounded set of pseudodifferential operators of order $m + m' - 2$ on $\mathbb{R}^n$.

Given any $s \in \mathbb{R}$, any $u \in H^s$, $[B, [A, \chi_\varepsilon *]]u$ converges to 0 *in $H^{s+m+m'-2}$.*

5. The Theorem on "Sum of Squares"

Let Ω be an open subset of $\mathbb{R}^n$, where the coordinates are denoted by $x^1, \ldots, x^n$. We consider $m + 1$ *real* C^∞ vector fields in Ω,

$$X_j = \sum_{k=1}^{n} a_j^k(x) \frac{\partial}{\partial x^k}, \qquad j = 0, 1, \ldots, m, \tag{5.1}$$

with $a_j^k \in C^\infty(\Omega)$, real-valued. We shall be interested in the hypoellipticity of the second-order differential operator

$$P = P(x, D) = \sum_{j=1}^{m} X_j^2 + X_0 + c(x), \tag{5.2}$$

where $c(x)$ is a complex-valued C^∞ function in Ω. We recall that the operator P *is said to be hypoelliptic in Ω if, given any open subset U of Ω, any distribution u in U, $Pu \in C^\infty(U)$ demands $u \in C^\infty(U)$.*

It is quite clear that (5.2) will not be hypoelliptic unless some stringent conditions are put on the X_j. The condition we shall now describe has been found by Hörmander who has proved its sufficiency and, under a "constant rank" hypothesis, its necessity. We shall prove the sufficiency following the exposition of Kohn [3]. For an alternative method see Oleinik–Radkevitch [1].

If X, Y are any two real C^∞ vector fields in Ω, let $[X, Y]$ denote their commutation bracket: $[X, Y] = XY - YX$. It is also a real C^∞ vector field in Ω, and thus the set of all such vector fields is a Lie algebra for that bracket. (Here we take the scalars to be the real numbers, not the real-valued C^∞ functions in Ω.) We shall denote by $\mathfrak{g}(\{X_j\}_{0 \leq j \leq m})$, or $\mathfrak{g}$ for short, the Lie algebra (over $\mathbb{R}$) generated by the X_j, that is, the smallest Lie subalgebra of the Lie algebra of all real smooth vector fields in Ω that contains the X_j. It is the real vector space spanned by all the successive brackets of the X_j's.

An arbitrary element of $\mathfrak{g}$ is a real vector field in Ω, with smooth coefficients which we may freeze at any given point x_o of Ω. We thus obtain a first-order operator with constant coefficients (but without a zero-order term) on $\mathbb{R}^n$. When X ranges all over $\mathfrak{g}$, we obtain a real vector space of such first-order operators. Actually, by considering their symbols, we can equate these operators to *linear forms* on $\mathbb{R}_n$. In other words, the freezing of the coefficients at x_o yields, from the Lie algebra $\mathfrak{g}$, a linear subspace of the dual of $\mathbb{R}_n$, i.e., of $\mathbb{R}^n$. The dimension of this subspace is called the *rank* of $\mathfrak{g}$ at the point x_o.

We may now state the Hörmander theorem on sum of squares:

THEOREM 5.1. *Suppose that the rank of* $\mathfrak{g}(\{X_j\}_{0\le j\le m})$ *is constant* (but not zero!) *throughout* Ω. *Then, in order for the operator* (5.2) *to be hypoelliptic in* Ω, *it is necessary and sufficient for the rank of* $\mathfrak{g}$ *to be equal to* n.

In his proof Hörmander established certain estimates involving P, of the so-called subelliptic kind, which imply hypoellipticity. The latter follows from less precise estimates, which are the ones proved later. The best possible such estimates were proved in Rothschild–Stein [1].

PROOF. It suffices to reason locally, in the neighborhood of a point which we take to be the origin of $\mathbb{R}^n$. If the rank of $\mathfrak{g}$ is constant throughout Ω, it follows from the classical Frobenius theorem (see Chapter VII, Theorem 2.1) that we may choose the coordinates in the neighborhood of the origin, in such a way that every vector field X in $\mathfrak{G}$ will be of the form

$$X = \sum_{k=1}^{r} c^k(x)\frac{\partial}{\partial x^k}, \qquad r = \operatorname{rank} \mathfrak{g},$$

where the c^k are C^∞ and real. We may thus regard P as a differential operator with respect to the variables $x' = (x^1, \ldots, x^r)$, depending smoothly on the parameter $x'' = (x^{r+1}, \ldots, x^n)$. A simple argument shows that there are points arbitrarily near the origin where the operator P (in $\mathbb{R}^r$) is *elliptic*, and a classical argument shows that in the neighborhood of such a point x'_o, there is a nowhere zero continuous solution h of the equation $Ph = 0$. Suppose then that $r < n$ and that h is defined in a neighborhood of $x_o = (x'_o, x''_o)$. Define $\tilde{h} = h$ for $x^n > x^n_o$, $\tilde{h} = 0$ for $x^n < x^n_o$. Since $P\tilde{h} = 0$ we see that P is not hypoelliptic.

The remainder of this section is devoted to proving that if rank $\mathfrak{g} = n$, then P is hypoelliptic.

LEMMA 5.1. *Let P be given by* (5.2). *To every real number s, every compact subset K of Ω, every $\varepsilon > 0$ there is $C = C(s, K, \varepsilon) > 0$ such that*

$$(5.3)_s \quad (1-\varepsilon)\sum_{j=1}^{m}\|X_j u\|_s^2 \le -\mathrm{Re}(Pu, u)_s + C\|u\|_s^2, \qquad \forall u \in C_c^\infty(K).$$

As usual we denote by $(\ ,\ )_s$ and $\|\ \|_s$ the inner product and the norm, respectively, in $H^s(\mathbb{R}^n)$ $(=H^s)$.

PROOF OF LEMMA 5.1. Set $b_j = X_j^* - X_j$. We have, for any $u \in C_c^\infty(\Omega)$,

$$\sum_{j=1}^{m}(X_j^* X_j u, u)_0 = -(Pu, u)_0 + (Lu, u)_0 + (cu, u)_0,$$

where $L = \sum_{j=1}^{m} b_j X_j + X_0$ is a real vector field. If we integrate by parts, we see that for some real C^∞ function b,

$$\mathrm{Re}(\{L + c\}u, u)_0 = (bu, u)_0,$$

and consequently

$$(5.4) \qquad \sum_{j=1}^{m}\|X_j u\|_0^2 = -\mathrm{Re}(Pu, u)_0 + (bu, u)_0,$$

which at once implies (5.3) when $s = 0$.

Let $g \in C_c^\infty(\Omega)$ equal one in a neighborhood of K and let us use the notation $G^s = g(1-\Delta)^{s/2}$. We observe that

$$(5.5) \quad X_j^2 G^s = X_j G^s X_j + X_j[X_j, G^s] = 2[X_j, G^s]X_j + [X_j, [X_j, G^s]].$$

Of course G^s is a pseudodifferential operator of order s. By Corollary 4.2, Chapter I, we know that the order of $[X_j, G^s]$ and of $[X_j, [X_j, G^s]]$ is equal to s. Thus

$$(5.6) \qquad PG^s = G^s P + \sum_{j=1}^{m} T_j^s X_j + T_0^s,$$

where the T_j^s $(0 \le j \le m)$ have order s; furthermore, for any $v \in \mathscr{E}'$, supp $T_j^s v$ is contained in $K' = \operatorname{supp} g$.

We substitute $G^s u$ for u in (5.4), and we take (5.6) into account. We note, moreover, that $G^s - (1-\Delta)^{s/2}$ is regularizing when it acts on distributions supported in K. Therefore

$$(5.7)\qquad \sum_{j=1}^{m} \|X_jG^s u\|_0^2 = -\mathrm{Re}(Pu, u)_s + \sum_{j=1}^{m} (T_j^s X_j u, G^s u)_0 + (T_0'^s u, G^s u)_0$$

$$\le -\mathrm{Re}(Pu, u)_s + C\left(\left\{\sum_{j=1}^{m} \|X_j u\|_s^2\right\}^{1/2} + \|u\|_s\right)\|u\|_s.$$

But by the same token

$$(5.8)\qquad \|X_j u\|_s \le \|G^s X_j u\|_0 + C\|u\|_s \le \|X_j G^s u\|_0 + C'\|u\|_s^2,$$

whence we easily get (5.3) if we combine (5.7) and (5.8). □

LEMMA 5.2. *Let P be given by* (5.2). *Suppose that there is a number $\delta > 0$ such that given any compact subset K of Ω, there is $C = C(K) > 0$ such that*

$$(5.9)_0\qquad \|u\|_\delta \le C(\|Pu\|_0 + \|u\|_0), \qquad \forall u \in C_c^\infty(K).$$

Then given any real number s, there is $C(s, K) > 0$ such that

$$(5.9)_s\qquad \|u\|_{s+\delta} \le C(s, K)(\|Pu\|_s + \|u\|_s), \qquad \forall u \in C_c^\infty(K).$$

PROOF OF LEMMA 5.2. We use the same notation as in the proof of Lemma 5.1. Possibly after increasing C we may substitute $G^s u$ for u in $(5.9)_0$. We avail ourselves of (5.6), and again of the fact that $(1-\Delta)^{s/2} - G^s$ is regularizing on the distributions supported in K. We obtain at once

$$\|u\|_{s+\delta} \le C\left(\|Pu\|_s + \sum_{j=1}^{m} \|X_j u\|_s + \|u\|_s\right),$$

from which we get $(5.9)_s$ if we apply (5.3). □

LEMMA 5.3. *Let P be given by* (5.2) *and suppose, furthermore, that given any real number s and any compact subset K of Ω, $(5.9)_s$ holds.*

Then for all distributions u in Ω, whatever the real number s,

$$(5.10)_s\qquad Pu \in H^s_{\mathrm{loc}}(\Omega) \text{ implies } u \in H^{s+\delta}_{\mathrm{loc}}(\Omega),\ X_j u \in H^{s+\frac{1}{2}\delta}_{\mathrm{loc}}(\Omega).$$

PROOF. We begin by proving (5.10) when the support of the distribution u is compact. Then there is a number σ such that $u \in H^\sigma$. We shall suppose that $\sigma \le s + \delta$. Let $t = \inf(s, \sigma)$ so that both u and Pu belong to H^t. We shall use the Friedrichs mollifiers $\chi_\epsilon *$ (Section 4). We shall apply $(5.3)_t$ and $(5.9)_t$ with a choice of the compact set K large enough to contain the supports of $\chi_\varepsilon * u$ for the relevant ε (suitably small). We have

$$(5.11)\quad P(\chi_\varepsilon * u) = \sum_{j=1}^{n} \{X_j[\chi_\varepsilon * (X_j u)] + X_j[X_j, \chi_\varepsilon *]\}u$$

$$+ [(X_0 + c), \chi_\varepsilon *]u + \chi_\varepsilon *(X_0 + c)u$$

$$= \chi_\varepsilon * Pu + \sum_{j=1}^{n} \{[X_j, \chi_\varepsilon *]X_j u + X_j[X_j, \chi_\varepsilon *]u\}$$

$$+ [(X_0 + c), \chi_\varepsilon *]u$$

$$= \chi_\varepsilon * Pu + \sum_{j=1}^{n} \{2X_j[X_j, \chi_\varepsilon *]u + [[X_j, \chi_\varepsilon *], X_j]u\}$$

$$+ [(X_0 + c), \chi_\varepsilon *]u.$$

We take the inner product in H^t of $P(\chi_\varepsilon * u)$ with $\chi_\varepsilon * u$. We observe that the commutator of X_j with $(1 - \Delta)^t$ is a pseudodifferential operator of order $2t$ in Ω. We apply Corollaries 4.1, 4.2 and obtain

$$(5.12)\quad (P(\chi_\varepsilon * u), \chi_\varepsilon * u)_t = (\chi_\varepsilon * Pu, \chi_\varepsilon * u)_t$$

$$- 2 \sum_{j=1}^{n} ([X_j, \chi_\varepsilon *]u, X_j(\chi_\varepsilon * u))_t$$

$$+ (T_\varepsilon u, \chi_\varepsilon * u)_t,$$

where T_ε remains in a bounded set of zero-order pseudodifferential operators for any ε, $0 \le \varepsilon \le \varepsilon_o$, with $\varepsilon_o > 0$ suitably small. We put (5.12) into $(5.3)_t$. By applying the Schwarz inequality to the right-hand side, we obtain at once

$$(5.13)\qquad \sum_{j=1}^{n} \|X_j(\chi_\varepsilon * u)\|_t \le C\{\|\chi_\varepsilon * Pu\|_t + \|u\|_t\}.$$

Letting ε go to zero, we reach the conclusion that $X_j u \in H^t$, $1 \le j \le m$.

We return to (5.11) from which we extract

$$(5.14)\quad P(\chi_\varepsilon * u) = \chi_\varepsilon * Pu + \sum_{j=1}^{n} \{2[X_j, \chi_\varepsilon *]X_j u + [X_j, [X_j, \chi_\varepsilon *]]u\}$$

$$+ [(X_0 + c), \chi_\varepsilon *]u.$$

We put this into $(5.9)_t$, using the fact that $X_j u \in H^t$ and Corollaries 4.1, 4.2. By letting ε go to zero we conclude that $u \in H^{t+\delta}$.

We go back to (5.12) where we substitute $t + \delta/2$ for t. We note that

$$|(\chi_\varepsilon * Pu, \chi_\varepsilon * u)_{t+\delta/2}| \le \|\chi_\varepsilon * Pu\|_t \|\chi_\varepsilon * u\|_{t+\delta}$$
$$\le \|\chi_\varepsilon * Pu\|_t \|u\|_{t+\delta}.$$

We apply $(5.3)_{t+\delta/2}$ and obtain

$$\sum_{j=1}^{n} \|X_j(\chi_\varepsilon * u)\|_{t+\delta/2} \le C\{\|\chi_\varepsilon * Pu\|_t + \|u\|_{t+\delta}\}. \tag{5.15}$$

Once again letting ε tend to zero we conclude that $X_j u \in H^{t+\delta/2}$. Thus we have proved that if $u \in \mathscr{E}'(\Omega)$,

(5.16) *$Pu \in H^t$, $u \in H^t$ implies $u \in H^{t+\delta}$, $X_j u \in H^{t+\delta/2}$* $(1 \le j \le m)$.

We may now repeat this reasoning with $\inf(s, t + \delta)$ in the place of t and reach the analogous conclusion; we may then repeat the same reasoning with $\inf[\inf(s, t + \delta) + \delta, s]$, etc. Finally one reaches the desired conclusion, namely $(5.10)_s$ for $u \in \mathscr{E}'(\Omega)$.

Let Ω' be a relatively compact open subset of Ω, and let u denote a distribution in Ω, possibly not compactly supported, whose restriction to Ω' belongs to $H^\sigma_{\text{loc}}(\Omega')$. Consider a sequence of C^∞ functions $\{g_j\}$ $(j = 0, 1, \ldots)$, with compact support in Ω', such that $g_{j+1} = 1$ in a neighborhood of supp g_j. We write

$$P(g_j u) = g_j Pu + 2\sum_{k=1}^{n} [X_k, g_j]X_k(g_{j+1}u) + [X_0, g_j]u + \sum_{j=1}^{n} [X_k, [X_k, g_j]]u. \tag{5.17$_j$}$$

We assume as before that $Pu \in H^s_{\text{loc}}(\Omega)$; let us set $t = \inf(s, \sigma)$. Then the right-hand side in $(5.17)_{j+1}$ belongs to H^{t-1}, i.e., $P(g_{j+1}u) \in H^{t-1}$. We conclude from this that $X_k(g_{j+1}u) \in H^{t-1+\delta/2}$ (we shall tacitly suppose that $\delta \le 2$). Hence the right-hand side in $(5.17)_j$ belongs to $H^{t-1+\delta/2}$; so does the left-hand side, of course, and we conclude, by the first part of the proof, that $X_k(g_j u) \in H^{t-1+\delta}$, $g_j u \in H^{t-1+3\delta/2}$. We may repeat this argument as many times as we wish, provided that j is large enough, until we reach the conclusion that $P(g_0 u) \in H^t$, from which we derive again, by the first part of the proof, that $g_0 u \in H^{t+\delta}$, $X_k(g_0 u) \in H^{t+\delta/2}$. Since $g_0 \in C^\infty_c(\Omega')$ is arbitrary we have shown that whatever the distribution u in Ω,

(5.18) *$u \in H^t_{\text{loc}}(\Omega')$, $Pu \in H^t_{\text{loc}}(\Omega')$ implies $u \in H^{t+\delta}_{\text{loc}}(\Omega')$,*
$X_k u \in H^{t+\delta/2}_{\text{loc}}(\Omega')$ $(1 \le k \le m)$.

By repeating the obvious argument until $t = s$, we obtain the desired conclusion. □

COROLLARY 5.1. *Under the hypotheses of Lemma 5.3, P is hypoelliptic in Ω.*

Indeed, if the hypotheses are valid in Ω, they are valid in any open subset of Ω. By letting s go to $+\infty$, $(5.10)_s$ (in such a subset) implies that u is C^∞ there.

REMARK 5.1. The hypothesis that P is of the form (5.2) is essential. Hypothesis $(5.9)_s$ does not suffice: for instance, the wave operator $D_1^2 + \cdots + D_{n-1}^2 - D_n^2$ $(D_j = -\sqrt{-1}\,(\partial/\partial x^j))$ verifies it with $\delta = 1$ and is not hypoelliptic.

The effect of the preceding lemmas is to reduce the proof of Theorem 5.1 to that of the inequality $(5.9)_0$ or $(5.9)_s$. Notice that such inequalities are valid for P if and only if they are valid (of course, possibly with increased constants) for $P - c$, and thus we may and shall take $c \equiv 0$. Also observe that we may replace Ω by one of its relatively compact open subsets, arbitrarily chosen, and thus assume that the Lie algebra $\mathfrak{g}$ has a finite set of generators in Ω.

Without our repeating it, all pseudodifferential operators in Ω that we are now going to use will be properly supported (Chapter I, Definition 3.1). We shall denote by $\mathcal{P}_s$ the set of all such operators A of order zero with the property that to every compact subset K of Ω there is a constant $C > 0$ such that

$$\|Au\|_s \leq C(\|Pu\|_0 + \|u\|_0), \qquad \forall u \in C_c^\infty(K). \tag{5.19}$$

Estimate (5.9), and consequently Theorem 5.1, will be proved if we show that the identity I belongs to the union $\mathcal{P}$ of the sets $\mathcal{P}_s$, $s > 0$. This fact will be a consequence of the following properties:

(5.20) *If $A \in \Psi^{-s}(\Omega)$, then A belongs to $\mathcal{P}_s$.*

(5.21) *If A, A' belong to $\mathcal{P}_s$ and $B \in \Psi^0(\Omega)$, then $A + A'$ and BA belong to $\mathcal{P}_s$; and if $s \leq \frac{1}{2}$, so also does the adjoint A^* of A.*

(5.22) *If $S \in \Psi^{-1}(\Omega)$, then $SX_j \in \mathcal{P}_1$ if $j = 1, \ldots, m$, and $SX_0 \in \mathcal{P}_{1/2}$.*

(5.23) *If $s \leq \frac{1}{2}$ and if $A \in \mathcal{P}_s$, then $[X_j, A] \in \mathcal{P}_{s/2}$ if $j = 1, \ldots, m$, and $[X_0, A] \in \mathcal{P}_{s/4}$.*

Property (5.20) is trivial, and so are the statements in (5.21) relative $A + A'$ and to BA. To prove the statement for A^*, let T^{2s} be a properly supported pseudodifferential operator in Ω, equivalent to $(1-\Delta)^s$ modulo regularizing operators. It suffices then to note that if $s \le \frac{1}{2}$,

$$A^*T^{2s}A - AT^{2s}A^* = [A^*, T^{2s}]A + T^{2s}[A^*, A] + [T^{2s}, A]A^*$$

has order $2s - 1 \le 0$.

The part of statement (5.22) relative to X_j for $j \ge 1$ is an immediate consequence of (5.3). Let us prove the part of statement (5.23) relative to the same X_j. We write

$$\begin{aligned}[X_j, A]^*T^s[X_j, A] &= M(X_jA - AX_j) \\ &= -X_j^*MA + [M, X_j]A + (X_j + X_j^*)\, MA \\ &\quad - AMX_j - [M, A]X_j.\end{aligned}$$

Note that $M = [X_j, A]^*T^s$ has order s; therefore $[M, X_j]$ also does, and $[M, A]$ has order $s-1$ (which we are assuming $\le -\frac{1}{2}$). The order of $X_j + X_j^*$ is zero and thus

$$\begin{aligned}(5.24)\qquad \|[X_j, A]u\|_{s/2}^2 &+ (MAu, X_ju)_0 \\ &+ (X_ju, M^*A^*u)_0 \le \text{const}(\|Au\|_s + \|X_ju\|_{s-1})\|u\|_0.\end{aligned}$$

We apply (5.3) and the statement about the adjoint A^* in (5.21) and conclude at once that $[X_j, A] \in \mathscr{P}_{s/2}$.

It remains to prove the analogous conclusions about X_0.

Let $S \in \Psi^{-1}(\Omega)$; set $B = S^*TSX_0$ (where $T \sim (1-\Delta)^{1/2}$); B has order zero. We have $X_0 = P - (X_1^2 + \cdots + X_m^2)$; hence

$$B^*X_0 = B^*P + \sum_{j=1}^{m} \{X_j^*B^*X_j - (X_j + X_j^*)B^*X_j + [B^*, X_j]X_j\},$$

and therefore

$$|(X_0u, Bu)_0| \le \text{const}\left(\|Pu\|_0^2 + \sum_{j=1}^{m} \|X_ju\|_0^2 + \|u\|_0^2\right).$$

Replacing B by its expression and using (5.3) implies at once that $SX_0 \in \mathscr{P}_{1/2}$.

Last, we prove the part in statement (5.23) relative to X_0. We note that the inequality (5.24) is also valid when $j = 0$. However, when $j = 0$ we shall apply it with $\frac{1}{2}s$ substituted for s. Observe then that

$$\|X_0u\|_{s/2-1} \le \text{const}\,\|T^{-1}X_0u\|_{1/2}$$

and that $T^{-1}X_0 \in \mathcal{P}_{1/2}$ by (5.22). It remains therefore to estimate the second and third terms on the left-hand side of (5.24). Since MA and M^*A^* are of the same nature, we shall limit ourselves to estimating

$$(MAu, X_0u)_0 = (MAu, Pu)_0 - \sum_{j=1}^{m} \{(MAu, (X_j + X_j^*)X_ju)_0 - (MAu, X_j^*X_ju)_0\}.$$

Once again using the fact that $X_j + X_j^*$ has order zero, and the inequalities (5.3), we see that everything reduces to estimating $\|X_j(MAu)\|_0^2$. We recall that here the order of M is $\frac{1}{2}s$ and that $j \geq 1$. We apply the full strength of (5.3):

$$\begin{aligned}\|X_j(MAu)\|_0^2 &\leq C(|(PMAu, MAu)_0| + \|MAu\|_0^2)\\ &\leq C(|([P, MA]u, MAu)_0| + |(Pu, A^*M^*MAu)_0|\\ &\quad + |(u, A^*M^*MAu)_0|).\end{aligned}$$

Noting that A^*M^*M has order s and recalling that $A \in \mathcal{P}_s$, we see that everything is reduced to estimating $([P, MA]u, MAu)_0$. We have

$$[P, MA] = \sum_{j=1}^{m} [X_j^2, MA] + [X_0, MA],$$

$$[X_j^2, MA] = 2[X_j, MA]X_j + [X_j, [X_j, MA]],$$

and therefore

$$[P, MA] = \sum_{j=1}^{m} K_jX_j + K_0, \qquad \text{order } K_j \leq \tfrac{1}{2}s \ (j = 0, \ldots, m).$$

From this the conclusion is easily reached:

$$\begin{aligned}|([P, MA]u, MAu)_0| &= \left|\sum_{j=1}^{m} (X_ju, K_j^*MAu)_0 + (u, K_0^*MAu)_0\right|\\ &\leq C\|Au\|_s\left(\sum_{j=1}^{m} \|X_ju\|_0 + \|u\|_0\right),\end{aligned}$$

from which we get the result, by applying once again (5.3).

END OF THE PROOF OF THEOREM 5.1. Let us call C_p a commutator of length p of the vector fields X_j:

$$C_p = [X_{j_p}, C_{p-1}], \qquad C_1 = X_{j_1} \qquad (j_1, \ldots, j_p \in \{0, 1, \ldots, m\}).$$

We shall prove the following assertion:

$$\text{(5.25)}_p \qquad \textit{If } S \in \Psi^{-1}(\Omega), \textit{ then } SC_p \in \mathcal{P}\Big(= \bigcup_{s>0} \mathcal{P}_s\Big).$$

For $p = 1$ this follows from (5.22). We reason by induction on $p > 1$.

$$SC_p = [X_{j_p}, SC_{p-1}] + [X_{j_p}, S]C_{p-1}.$$

The first term on the right belongs to $\mathcal{P}$ since SC_{p-1} does, and we may therefore apply (5.23). The second term belongs to $\mathcal{P}$ simply because $S' = [X_{j_p}, S]$ has order -1, and therefore $S'C_{p-1} \in \mathcal{P}$. Thus $(5.25)_p$ is proved.

If rank $\mathfrak{g} = n$, then each operator $\partial/\partial x^j$ can be expressed as a linear combination (with C^∞ coefficients) of commutators C_p, and therefore we can say:

$$\text{(5.26)} \qquad \textit{If } S \in \Psi^{-1}(\Omega), \textit{ then } S\frac{\partial}{\partial x^j} \in \mathcal{P}.$$

It suffices to write

$$I = (1 - \Delta)^{-1} - \sum_{j=1}^{n} S_j\frac{\partial}{\partial x^j}, \qquad S_j = (1 - \Delta)^{-1}\frac{\partial}{\partial x^j} \in \Psi^{-1},$$

to see that $I \in \mathcal{P}$. □

III

Application to Boundary Problems for Elliptic Equations

It is now common practice to study boundary value problems for an elliptic equation (say, in a domain of Euclidean space or, more generally, on a manifold with boundary) by transferring them to the boundary. For this purpose one makes use of the Poisson kernel for the Dirichlet problem, relative to the equation under study, and of the regularity results and estimates, now well established, in the latter problem. If appropriate results have been established for the problem on the boundary, it is then possible to reach the desired conclusions about the original problem, most often the regularity up to the boundary of its solutions, the finite dimensionality of its kernel and cokernel, etc. Of course, in an expository text, the author is still left with the task of establishing the classical properties of the Dirichlet problem.

In this book I have chosen to follow a more direct path: to deal with all boundary problems for elliptic equations through a unified approach, which does not distinguish between those of the Dirichlet (or, more generally, of the Lopatinski–Shapiro) class and the other usual types, of which the foremost examples are perhaps the oblique derivative problem and the $\bar{\partial}$-Neumann problem. From the start the transfer to the boundary is effected in general, and with it the reduction of the study to that of the operator on the boundary, referred to as the *Calderon operator* (A. Calderon was perhaps the first one to use it systematically; see [3]). The properties of the Calderon operator $\mathscr{B}$ reflect faithfully those of the original problem. For instance, $\mathscr{B}$ is elliptic if and only if the problem is of the Lopatinski–Shapiro type, a fact first pointed out by Calderon himself (Theorem 6.1). $\mathscr{B}$ is hypoelliptic if and only if this is true of the boundary problem (in an obvious sense; Theorem

4.1); and globally hypoelliptic can be substituted for hypoelliptic (Theorem 5.1; global hypoellipticity is a much more common property than hypoellipticity when applied to pseudodifferential operators on a compact manifold). Finally, the boundary problem is Fredholm in the spaces of C^∞ functions if and only if the Calderon operator on the analogous spaces on the boundary is Fredholm (Theorem 5.6). In the oblique derivative problems the Calderon operator is of principal type and, under suitable hypotheses, subelliptic (Definition 7.5), which implies hypoellipticity. In the $\bar{\partial}$-Neumann problem, under hypotheses of the strong pseudoconvexity kind, $\mathscr{B}$ is hypoelliptic with loss of one derivative, and the same is true of its adjoint (Section 8).

Incidentally, we give the name "coercive" to the boundary problems of the Lopatinski–Shapiro type, because of the coercive nature of the concomitant estimates (Theorem 6.2). The traditional meaning of coercive for the variational problems is never used in this book, and perhaps ought to be regarded as outdated, because of the relative "trivialization" of the whole theory of Lopatinski–Shapiro problems.

It is worthwhile to give an idea of our method of transfer to the boundary on a simple example. Let Ω be a bounded open subset of Euclidean space, having a smooth boundary X and lying everywhere on only one side of it. We denote by $-\Delta$ the Laplace–Beltrami operator on X for some Riemannian structure, e.g., the one induced by the surrounding space ($-\Delta$ is a positive operator). We study an elliptic partial differential operator P which, in the neighborhood of X, is given by

$$P = \partial_t^2 + \Delta, \tag{0.1}$$

where $\partial_t = \partial/\partial t$. Here t is a coordinate transversal to the boundary X; we suppose that $t = 0$ defines X and $t > 0$ defines Ω in an open tubular neighborhood $\mathscr{T}$ of X. The boundary problem under study is

$$Pu = f \qquad in\ \Omega, \tag{0.2}$$

$$Lu + c\,\partial_t u = u_o \qquad on\ X. \tag{0.3}$$

In (0.3) L is a smooth vector field and c is a smooth function on X. Let us denote by A the positive square root of $-\Delta$. It is a positive self-adjoint classical pseudodifferential operator of order one in X. Equation (0.1) can be rewritten

$$P = (\partial_t - A)(\partial_t + A) \qquad (\text{in } \mathscr{T}). \tag{0.4}$$

And thus in $\mathscr{T}$ it is natural to look at the following problem:

$$\partial_t u + Au = v, \tag{0.5}$$

(0.6) $$\partial_t v - Av = f,$$

under the boundary condition derived from (0.3) and (0.5):

(0.7) $$\mathscr{B}u = Lu - cAu = u_o - cv \qquad in\, X.$$

In order to solve equations (0.5) and (0.6), or to study their solutions, we use the continuous semigroup of operators on $L^2(X)$, e^{-tA}. Indeed, we have

(0.8) $$u(t) = e^{-tA}u(0) + \int_0^t e^{-(t-t')A}v(t')\,dt',$$

(0.9) $$v(t) = e^{-(T-t)A}v(T) - \int_t^T e^{-(t'-t)A}f(t')\,dt'.$$

We are omitting mention of the variable x on X and thus regarding u, v, f as functions of t valued in the space of distributions on X. We have selected a number $T > 0$, small enough that $X \times [-T, T]$ is a compact neighborhood of X contained in $\mathscr{T}$.

The crucial fact on which our entire approach is based is then the following: e^{-tA} *is a standard pseudodifferential operator on* X, *depending smoothly on* $t \geq 0$ (*in an obvious sense*), *regularizing for* $t > 0$. Observe, however, that under no circumstance is it a classical pseudodifferential operator, that is, an operator whose symbol is an asymptotic series of homogeneous terms whose degrees decrease by integers. Thus the full strength of the theory developed in Chapter I is needed.

Once this is obtained, the exploitation of formulas (0.8) and (0.9) is fairly easy. To give an example, suppose that we wish to study the regularity up to the boundary of the solutions of (0.2)–(0.3). Let $\mathcal{O}$ be any open subset of X and suppose that $f(x, t)$ is a C^∞ function in $\mathcal{O} \times [0, T]$, and $u_o(x)$ is one in $\mathcal{O}$. Define v by (0.9), choosing $v(T)$ arbitrarily, say $v(T) = 0$. Then v is also a C^∞ function in $\mathcal{O} \times [0, T]$, like f, and $v(0)$ is a C^∞ function in $\mathcal{O}$, and so is the right-hand side in (0.7). If $\mathscr{B}$ is hypoelliptic, then $u(0)$ is also C^∞ in $\mathcal{O}$, and by virtue of (0.8), u is a C^∞ function of (x, t) in $\mathcal{O} \times [0, T]$. The reader might object that equation (0.8) does not quite define the solution of (0.2)–(0.3), because of our arbitrary choice of v. But let us look at how it differs from the true solution. In order to obtain the latter we should have taken, according to (0.5),

$$v(T) = (\partial_t u + Au)|_{t=T} \qquad (= v_o(T)).$$

The difference is then "measured" by $e^{-(T-t)}v_o(T)$. But if we keep $t < T$, this is a smooth function of (x, t), whatever the distribution $v_0(T)$, and its presence or absence has no effect on the regularity of u!

The two properties that make this treatment possible remain valid if we go from particular cases such as (0.1) to general elliptic PDEs. Indeed, the analogue of the operator e^{-tA} exists in general. In fact, if one disregards regularizing operators, it even possesses the semigroup properties of e^{-tA}. It is a standard pseudodifferential operator of order zero on X, $U(t)$, depending smoothly on $t \geq 0$ and regularizing for $t > 0$. It is possible that this general property has been known before, but I have not found it emphasized in other treatments of the subject. The significant properties of $U(t)$ are established in Section 1.

The second fundamental property has been often used, in some form, namely that elliptic operators of the kind commonly studied can be factorized into products of the type (0.4), although, of course, the roots will not generally be as symmetric as $+A$ and $-A$ but rather of the form A_1 and $-A_2$, with both A_1 and A_2 positive elliptic, classical, of order one, and depending smoothly on t. And the factorization will not quite be exact; it will only be true modulo regularizing operators.

Once the properties of $U(t)$ are secured, the consequences unfold rather mechanically. Most of this chapter is thus devoted to transforming the equations under study. The nature of the reasoning is algebraic, despite the analytic trimmings. The proof of the regularity up to the boundary, in Lopatinski–Shapiro problems, "evaporates"; it is a direct consequence of the twin facts that $\mathcal{B}$ is elliptic and that standard pseudodifferential operators are pseudolocal. Note also that all the regularity properties are now valid in the microlocal sense, not only in the local one.

For a different approach to boundary problems for elliptic pseudodifferential equations (satisfying the so-called *transmission condition* at the boundary), see Boutet de Monvel [1].

1. The Generalized Heat Equation and Its Parametrix

Throughout this chapter X will be a C^∞ manifold countable at infinity; $n = \dim X$; t will be the variable in the real line $\mathbb{R}$, most often, in the closed half-line $\bar{\mathbb{R}}_+$. By m we denote a strictly positive number, which, in the most significant applications, is equal either to one or to two; T will be some number >0.

We shall deal with functions and distributions valued in a Hilbert space H (over $\mathbb{C}$); H will be finite dimensional; most often there would be not added complication by letting it be infinite dimensional. The norm in H will be denoted by $| \ |_H$, whereas the operator norm in $L(H)$, the space of (bounded) linear operators in H, will be denoted by $\| \quad \|$. The inner product in H will be denoted by $(\quad , \quad)_H$.

Our basic ingredient is a pseudodifferential operator of order m in X, $A(t)$, valued in $L(H)$, depending smoothly on t in $[0, T[$. If one uses a basis in H, it means that $A(t)$ is a matrix whose entries are scalar pseudodifferential operators in X. And this means that in every local chart $(\Omega, x_1, \ldots, x_n)$, $A(t)$ is *congruent modulo regularizing operators which are* C^∞ *functions of* t to an operator

$$(1.1) \qquad A_\Omega(t)u(x) = (2\pi)^{-n} \int e^{ix\cdot\xi} a_\Omega(x, t, \xi)\hat{u}(\xi)\, d\xi, \qquad u \in C_c^\infty(\Omega; H),$$

where

$$(1.2) \qquad a_\Omega(x, t, \xi) \textit{ is a } C^\infty \textit{ function of } t \in [0, T[\textit{ valued in } S^m(\Omega; L(H)).$$

(We shall always shorten this expression about congruence to "equivalent" and symbolize it by $\sim$; here the regularizing operators must be valued in $L(H)$.) We have denoted by $S^m(\Omega; L(H))$ the space of symbols with values in $L(H)$. In the forthcoming we shall often drop the subscript Ω and refer to $a(x, t, \xi)$ as the symbol of $A(t)$ in the chart $(\Omega, x_1, \ldots, x_n)$, although this name should be reserved for the class of $a(x, t, \xi)$ mod $S^{-\infty}(\Omega; L(H))$.

1.1. Existence and "Uniqueness" of the Parametrix

We are interested in solving the following initial value problem:

$$(1.3) \qquad \frac{dU}{dt} - A(t) \circ U \sim 0 \qquad \textit{in } X \times [0, T[,$$

$$(1.4) \qquad U|_{t=0} = I, \quad \textit{the identity of } H, \qquad \textit{in } X.$$

In principle the solution $U(t)$ should be an equivalence class, modulo regularizing operators in X depending smoothly on t, of continuous linear operators $\mathscr{E}'(X; H) \to \mathscr{D}'(X; H)$ depending smoothly on t (in $[0, T[$). But without additional hypotheses about $A(t)$ there is no reason that such a solution should exist. We are going to make a hypothesis on $A(t)$ which will insure not only that U exists but also that it possesses a convenient integral representation. Observe that when $X = \mathbb{R}^n$ and $A(t) = \Delta_x$ (the Laplace operator in n variables, in which case the symbol of $A(t)$ is $-|\xi|^2$), then equations (1.3)–(1.4) define the parametrix in the forward Cauchy problem for the heat equation $\partial U/\partial t - \Delta_x U = 0$. In general we shall make the following hypothesis:

(1.5) *Let* $(\Omega, x_1, \ldots, x_n)$ *be any local chart in* X. *There is a symbol* $a(x, t, \xi)$ *satisfying* (1.2) *and defining via* (1.1) *the operator* $A_\Omega(t)$ *congruent to* $A(t)$ *modulo regularizing operators in* Ω *depending smoothly on* $t \in [0, T[$, *such that*

(1.6) *to every compact subset* K *of* $\Omega \times [0, T[$ *there is a compact subset* K' *of the open half-plane* $\mathbb{C}_- = \{z \in \mathbb{C}; \operatorname{Re} z < 0\}$ *such that*

(1.7) $$zI - a(x, t, \xi)/(1 + |\xi|^2)^{m/2} : H \to H$$

is a bijection (hence also a homeomorphism), for all (x, t) *in* K, ξ *in* $\mathbb{R}_n$, z *in* $\mathbb{C}\backslash K'$.

We may now state the main result of this section:

THEOREM 1.1. *Under hypothesis* (1.5) *the problem* (1.3)–(1.4) *has a solution* $U(t)$ *which is a function of* $t \in [0, T[$ *valued in* $\dot{\Psi}^0(X; L(H))$ (cf. *Chapter I, Definition* 4.2). *There is a representative of the equivalence class* $U(t)$ *with the following property:*

In each local chart $(\Omega, x_1, \ldots, x_n)$ *of* X *the representative in question is equivalent to an element* $U_\Omega(t)$ *of* $\Psi^0(\Omega; L(H))$ *given by*

(1.8) $$U_\Omega(t)u(x) = (2\pi)^{-n} \int e^{ix\cdot\xi} \mathcal{U}_\Omega(x, t, \xi)\hat{u}(\xi)\, d\xi, \qquad u \in C_c^\infty(\Omega; H),$$

whose symbol $\mathcal{U}_\Omega$ *has the following properties*:

(1.9) $\mathcal{U}_\Omega$ *is a* C^∞ *map* $\Omega \times [0, T[\times \mathbb{R}^n \to L(H)$.

(1.10) *To every compact subset* $\mathcal{K}$ *[of* $\Omega \times [0, T[$, *to every pair of n-tuples* $\alpha, \beta \in \mathbb{Z}_+^n$, *and to every pair of integers* r, $N \geq 0$, *there is a constant* $C > 0$ *such that for all* (x, t) *in* $\mathcal{K}$, $\xi \in \mathbb{R}_n$,

(1.11) $$\|\partial_x^\alpha \partial_\xi^\beta \partial_t^r \mathcal{U}_\Omega(x, t, \xi)\| \leq Ct^{-N}(1 + |\xi|)^{rm - |\beta| - Nm}$$

Any C^∞ *function of* t *in* $[0, T[$ *valued in the space of continuous linear mappings* $\mathcal{E}'(X; H) \to \mathcal{D}'(X; H)$ *which satisfies* (1.3), (1.4) *belongs to the equivalence class* $U(t)$.

It follows from (1.11) that for $t > 0$, $\mathcal{U}_\Omega(x, t, \xi)$ belongs to $S^{-\infty}(\Omega; L(H))$, i.e., the operator (1.8) is regularizing; in other words, the equivalence class $U(t)$ is zero. This generalizes the well-known property of the parametrix of the heat equation.

PROOF OF THEOREM 1.1. A: EXISTENCE OF THE PARAMETRIX $U(t)$. It suffices to reason in the (generic) local chart $(\Omega, x_1, \ldots, x_n)$ and patch the $U_\Omega(t)$ together afterwards, by means of a smooth partition of unity in X. Thus we construct the symbol $\mathcal{U}_\Omega$; actually we construct a formal symbol (see Chapter I, Section 4)

$$\mathcal{U}(x, t, \xi) = \sum_{j=0}^{+\infty} \mathcal{U}_j(x, t, \xi), \tag{1.12}$$

from which a true symbol can later be constructed, by using cutoffs as indicated in Chapter I. We take the operator (1.1) to be the operator $A_\Omega(t)$ in (1.5) and omit the subscripts Ω; we no longer distinguish between $A(t)$ and $A_\Omega(t)$, which we also denote by $a(x, t, D_x)$ $\big((a(x, t, \xi)$ is its symbol$\big)$. Reasoning formally, we write

(1.13)

$$\left[\frac{d}{dt} - A(t)\right] U(t)u = (2\pi)^{-n} \int e^{ix\cdot\xi}\left[\frac{\partial}{\partial t} - a(x, t, D_x + \xi)\right]\mathcal{U}(x, t, \xi)\hat{u}(\xi)\, d\xi,$$

and we require, for $0 \le t < T$,

$$\frac{\partial \mathcal{U}}{\partial t} - a(x, t, D_x + \xi)\mathcal{U} = \frac{\partial \mathcal{U}}{\partial t} - \sum_{\alpha \in Z_+^n} \frac{1}{\alpha!}\partial_\xi^\alpha a(x, t, \xi) D_x^\alpha \mathcal{U} = 0, \tag{1.14}$$

which may be rewritten, with the notation (4.24) of Chapter I,

$$\frac{\partial \mathcal{U}}{\partial t} - a(x, t, \xi) \odot \mathcal{U} = 0, \quad 0 \le t < T. \tag{1.15}$$

Equation (1.15) is the "translation" of (1.3); as for (1.4) it translates into

$$\mathcal{U}(x, 0, \xi) = I \text{ (the identity of } H). \tag{1.16}$$

By availing ourselves of the basic hypothesis, (1.5), we are going to obtain $\mathcal{U}(x, t, \xi)$ in the form

$$\mathcal{U}(x, t, \xi) = (2\pi i)^{-1} \oint_\gamma e^{\rho t z} k(x, t, \xi; z)\, dz, \tag{1.17}$$

where k is a suitable formal symbol of degree zero, valued in $L(H)$, depending holomorphically on the complex variable z in an open neighborhood of the integration contour γ provided (x, t) remains in a given compact subset of $\Omega \times [0, T[$. We have used the notation $\rho = \rho(\xi) = (1 + |\xi|^2)^{m/2}$ and shall continue to use it.

We select arbitrarily a relatively compact open subset Ω_o of Ω, a number T_o, $0 < T_o < T$, and take the compact set K in (1.6) to be the closure of $\mathcal{O} = \Omega_o \times [0, T_o[$. We take the compact subset K' of $\mathbb{C}_-$ in (1.6) accordingly, and denote by M the maximum norm of the inverse of the mapping (1.7) as (x, t) ranges over K, ξ over $\mathbb{R}_n$, and z over a simple closed smooth curve γ winding around K' in $\mathbb{C}_-\backslash K'$.

Since $a \odot (e^{\rho t z} k) = e^{\rho t z}(a \odot k)$ we may rewrite equation (1.15) as

$$\oint_\gamma e^{\rho t z} \mathcal{L} k(x, t, \xi; z)\, dz = 0, \tag{1.18}$$

where

$$\mathcal{L} k = \frac{\partial k}{\partial t} + \rho z k - a(x, t, \xi) \odot k. \tag{1.19}$$

We are going to solve (in the sense of formal symbols) the equation

$$\mathcal{L} k = \rho I, \tag{1.20}$$

which implies at once (1.18). It turns out that the (unique) formal symbol k satisfying (1.20) will also satisfy

$$(2\pi i)^{-1} \oint_\gamma k(x, t, \xi; z)\, dz = I, \qquad \forall (x, t) \in \mathcal{O}, \xi \in \mathbb{R}_n, \tag{1.21}$$

which, for $t = 0$, is nothing but (1.16).

Solution of (1.20). We rewrite (1.20) as follows:

$$k = E\left[I - \rho^{-1}\left(\frac{\partial k}{\partial t} - a \odot k + ak\right)\right], \tag{1.22}$$

setting $E = [zI - \rho^{-1} a(x, t, \xi)]^{-1}$ (inverse in $L(H)$; cf. (1.7)). We note that

$$\rho^{-1}(a \odot k - ak) = \sum_{\alpha \neq 0} \frac{1}{\alpha!} \rho^{-1} \partial_\xi^\alpha a D_x^\alpha k \tag{1.23}$$

has degree $\leq \deg k - 1$. We solve (1.22) by taking $k = \sum_{j=0}^{+\infty} k_j$ and requiring

$$k_0 = E, \tag{1.24$_0$}$$

$$k_j = -E\rho^{-1}\left[\frac{\partial}{\partial t} k_{j-1} - \sum_{1 \leq |\alpha| \leq j} \frac{1}{\alpha!} \partial_\xi^\alpha a D_x^\alpha k_{j-|\alpha|}\right]. \tag{1.24$_{j>0}$}$$

By induction on j we see easily that

$$m_j = \deg k_j \leq -j \inf(1, m), \tag{1.25}$$

which implies that $\sum_j k_j$ indeed defines a formal symbol (since $m > 0$). Furthermore,

(1.26) *If $j \geq 1$, k_j is a finite sum of terms of the form $Eb_1E \cdots b_rE$ with r varying from term to term but always remaining ≥ 2, and with each b_i a C^∞ function of t in $[0, T[$ valued in $S^{d_i}(\Omega; L(H))$ independent of z ($i = 1, \dots, r$) and, moreover, such that $d_1 + \cdots + d_r \leq m_j$.*

According to (1.26), therefore, we have the following:

(1.27) *To every relatively compact open subset Ω_o of Ω and to every number T_o, $0 < T_o < T$, there is a compact subset K' of $\mathbb{C}_-$ such that for each $j = 0, 1, \dots,$ $k_j(x, t, \xi, z)$ is a C^∞ function of (t, z) in $[0, T_o[\times (\mathbb{C}\backslash K')$, holomorphic with respect to z, valued in $S^{m_j}(\Omega_o; L(H))$.*

Proof of (1.21). Fix arbitrarily (x, t) in $\mathcal{O}$ and ξ in $\mathbb{R}_n$; then $a_0 = \rho^{-1}a(x, t, \xi)$ is a (bounded) linear operator $H \to H$ and so are the b_i in (1.26). Writing $E(z) = (zI - a_0)^{-1}$ and keeping in mind that γ winds around the spectrum of a_0, we get

$$(2\pi i)^{-1}\oint_\gamma E(z)\,dz = I, \qquad \oint_\gamma E(z)b_1E(z)\cdots b_rE(z)\,dz = 0 \quad (\text{if } r \geq 1),$$

whence we get (1.21) by using (1.26).

Estimate of the Symbols $\mathcal{U}_j(x, t, \xi) = (2\pi i)^{-1}\oint_\gamma e^{\rho t z}k_j(x, t, \xi; z)\,dz$: By (1.27) we see that this formula defines $\mathcal{U}_j(x, t, \xi)$ for *all* (x, t) in $\Omega \times [0, T[$; indeed, it does for (x, t) in $\mathcal{O}$. But if we replace $\mathcal{O}$ by a larger open set $\mathcal{O}_1$ we might replace γ by a different contour. If we then restrict (x, t) to $\mathcal{O}$, it follows from the Cauchy integral theorem that we recover the same value as before. We note that if $z \in \gamma$, then

$$\begin{aligned} |\partial_\xi^\beta \partial_t^r(e^{\rho t z})| &\leq \text{const}(1 + |\xi|)^{-|\beta|}\rho^r \sum_{l=0}^{|\beta|} (t\rho)^l e^{\rho t \operatorname{Re} z} \\ &\leq \text{const}(1 + |\xi|)^{-|\beta|}\rho^{r-N}t^{-N} \sum_{l=N}^{|\beta|+N} (\rho t)^l e^{-c_o \rho t} \\ &\leq \text{const}\, t^{-N}(1 + |\xi|)^{-|\beta|}\rho^{r-N}. \end{aligned} \tag{1.28}$$

We have availed ourselves of the fact that $\operatorname{Re} z \le -c_o < 0$ on γ. On the other hand, we derive from (1.27), for (x, t) in $\mathcal{O}$ and z in γ,

$$\|\partial_x^\alpha \partial_\xi^\beta \partial_t^r k_j(x, t, \xi; z)\| \le \text{const}(1 + |\xi|)^{m_j - |\beta|}. \tag{1.29}$$

By combining (1.28) and (1.29) and applying Leibniz's formula, we get

$$\|\partial_x^\alpha \partial_\xi^\beta \partial_t^r \mathcal{U}_j(x, t, \xi)\| \le Ct^{-N}(1 + |\xi|)^{m_j + (r-N)m - |\beta|} \tag{1.30}$$

for all (x, t) in $\mathcal{O}$, ξ in $\mathbb{R}_n$. This implies (1.11).

PROOF OF THEOREM 1.1. B: UNIQUENESS OF THE PARAMETRIX. The uniqueness of the parametrix, needless to say in the sense of equivalence class modulo regularizing operators, follows from various standard considerations which are of interest in their own right, and which we now go into rapidly.

First, we did not have to solve equation (1.3) while prescribing the value of the solution at time $t = 0$. We could have solved

$$\frac{dU}{dt} - A(t) \circ U \sim 0 \quad \text{in } X \times [t', T[, \qquad U|_{t=t'} = I \quad \text{in } X, \tag{1.31}$$

where t' is any number such that $0 \le t' < T$. By the same procedure as in part A we can find a solution $U(t, t')$ having a representative which, in any local chart $(\Omega, x_1, \dots, x_n)$, is equivalent to an operator $U_\Omega(t, t')$ defined by

$$U_\Omega(t, t')u(x) = (2\pi)^{-n} \int e^{ix\cdot\xi} \mathcal{U}_\Omega(x, t, t', \xi)\hat{u}(\xi)\, d\xi, \tag{1.32}$$

with

$$\mathcal{U}_\Omega(x, t, t', \xi) = (2\pi i)^{-1} \oint_\gamma e^{(t-t')\rho z} k_\Omega(x, t, \xi; z)\, dz, \tag{1.33}$$

where k_Ω is the same symbol as in (1.17); in particular it is independent of t'. This is due to the validity of (1.21) where we may take $t = t'$. The contour of integration γ may also be taken to be the same as in part A.

The solution of (1.31) enables us to solve the *inhomogeneous Cauchy problem*:

$$\frac{\partial u}{\partial t} - A(t)u = f \quad \text{in } X \times [0, T[, \qquad u|_{t=0} = u_0 \quad \text{in } X. \tag{1.34}$$

Here f is an H-valued function or distribution in $X \times [0, T[$, u_0 an element of $\mathcal{D}'(X; H)$ (in all rigor we must reason modulo $C^\infty(X)$). If f is sufficiently regular with respect to t, say continuous, we may write

$$u(t) = U(t)u_0 + \int_0^t U(t, t')f(t')\,dt'. \tag{1.35}$$

Next we look at the *backward Cauchy problem for the adjoint equation.* For each t in $[0, T]$ we denote by $A^*(t)$ the adjoint of the operator $A(t)$ as an $L(H)$-valued pseudodifferential operator in X. (In order to define the adjoint of $A(t)$ we make use of a strictly positive density ϖ in X.) Then in any local chart $(\Omega, x_1, \ldots, x_n)$, in which the symbol of $A(t)$ is $a(x, t, \xi)$, a formal symbol or $A^*(t)$ is given by

$$\sum_{\alpha \in \mathbf{Z}_+^n} \frac{1}{\alpha!} D_x^\alpha\, \partial_\xi^\alpha[\varpi(x)a(x, t, \xi)^*], \tag{1.36}$$

where $a(x, t, \xi)^*$ stands for the adjoint of $a(x, t, \xi)$ as a bounded linear operator on H. If we assume that $a(x, t, \xi)$ satisfies the basic hypothesis (1.6) so will any reasonable true symbol constructed from (1.36). By duplicating the construction in part A we can now construct an operator $V(t, t')$ solution to

$$\frac{dV}{dt} + A^*(t)V \sim 0 \quad \text{in } X \times]0, t'], \qquad V|_{t=t'} \sim I \quad \text{in } X, \tag{1.37}$$

where t' is any number such that $0 < t' < T$. As a matter of fact, our hypothesis that $A(t)$ is smooth up to $t = 0$ implies that we can construct a representative of $V(t, t')$ which is C^∞ in the *closed* interval $[0, t]$. This is important in the applications.

Let $V^*(t, t')$ denote the adjoint of $V(t, t')$ or, rather, a properly supported representative of this equivalence class. By transposing (1.37) we get

$$\frac{dV^*}{dt'}(t', t) + V^*(t', t) \circ A(t') = R(t, t') \qquad \text{in } X \times [0, t], \tag{1.38}$$

$$V^*(t, t) - I = R_o(t) \qquad \text{in } X, \tag{1.39}$$

with R, R_o regularizing and depending smoothly on t, t'. If we therefore define the operator G by the formula

$$Gf(t) = \int_0^t V^*(t', t)f(t')\,dt', \tag{1.40}$$

and suppose that (1.34) holds, then we obtain, by integration by parts,

$$\begin{aligned} Gf(t) = {} & V^*(t, t)u(t) - V^*(0, t)u_0 \\ & - \int_0^t \left[\frac{dV^*}{dt'}(t', t) + V^*(t', t) \circ A(t')\right] u(t')\,dt'. \end{aligned} \tag{1.41}$$

At this point it is advisable to use a properly supported representative of $A(t)$ so that both R and R_o in (1.38), (1.39) are properly supported. Equation (1.41) can be rewritten as

(1.42)
$$u(t) = G\left[\frac{\partial u}{\partial t} - A(t)u\right] + V^*(0, t)u(0) - R_o(t)u(t) + \int_0^t R(t, t')u(t')\,dt'.$$

We may now easily prove the uniqueness of the parametrix. Let $U(t)$ be a representative of the parametrix constructed in part A, $U_1(t)$ a representative of another solution of (1.3)–(1.4); we assume that $U_1(t)$ is a C^∞ function of t in $[0, T[$ valued in the space of continuous linear mappings $\mathscr{E}'(X; H) \to \mathscr{D}'(X; H)$. It suffices to put $u(t) = [U(t) - U_1(t)]w_o$ in (1.42), with $w_o \in \mathscr{E}'(X; H)$ arbitrary, to conclude that $U(t) - U_1(t)$ is a C^∞ function of t in $[0, T[$ valued in $\Psi^{-\infty}(X; L(H))$.

The proof of Theorem 1.1 is complete. □

REMARK 1.1. The uniqueness of the parametrix $U(t)$ and the similar property for $U(t, t')$ have the consequence that

(1.43) $$U(t, t'') \sim U(t, t')U(t', t'') \qquad \text{if } 0 \le t'' \le t' \le t < T;$$

(1.44) $$U(t, t') \sim V^*(t', t) \qquad \text{if } 0 \le t' \le t.$$

Thus formula (1.42) can be rewritten in the form

(1.45)
$$u(t) = U(t)u(0) + \int_0^t U(t, t')\left[\frac{\partial u}{\partial t}(t') - A(t')u(t')\right] dt'$$
$$- R_o(t)u(t) + \int_0^t R(t, t')u(t')\,dt',$$

for every $u \in C^\infty([0, T[; \mathscr{D}'(X; H))$. In turn the expression (1.45) implies the *hypoellipticity of the Cauchy problem* (1.34):

THEOREM 1.2. *Let $\mathcal{O}$ be an open subset of X, u a C^∞ function of t in $[0, T[$ valued in $\mathscr{D}'(X; H)$.*

Suppose that $u(0) \in C^\infty(\mathcal{O}; H)$ and that

$$\frac{\partial u}{\partial t} - A(t)u \in C^\infty(\mathcal{O} \times [0, T[; H).$$

Then $u \in C^\infty(\mathcal{O} \times [0, T[; H)$.

Since pseudodifferential operators decrease wave-front sets (Chapter I, Corollary 6.6) we could have replaced the open set $\mathcal{O} \subset X$, in Theorem 1.2, by any conic open subset of $T^*X \backslash 0$.

REMARK 1.2. An observation often used in the present context is that the requirement that u be smooth with respect to t, with values in $\mathscr{D}'(X; H)$, is no restriction at all, if we assume that $\partial u/\partial t - A(t)u = f$ has that same property. Indeed, if we were to assume that u is merely a distribution with respect to t, hence to $(x, t) \in X \times [0, T]$ (valued in H), we could write $u = \partial_t^m v$ with $v \in C^0([0, T]; \mathscr{D}'(X; H))$ and m an integer suitably large. We have, by the transposed Leibniz formula (Notation and Background, (0.3))

$$\begin{aligned}\partial_t^{m+1} v &= A(t)\, \partial_t^m v + f \\ &= \sum_{0 \le p \le m} (-1)^{m-p} \binom{m}{p}\, \partial_t^p [A^{(m-p)}(t) v] + f.\end{aligned}$$

Let us define ∂_t^{-1} as integration from 0 to t. We obtain

$$v = -\sum_{0 \le p \le m} \binom{m}{p} (-\partial_t)^{p-m-1} (A^{(m-p)} v) + \partial_t^{-m-1} f.$$

This shows that, whatever $j = 0, 1, \ldots$, if we have proved that v is a C^j function of $t \in [0, T]$ valued in $\mathscr{D}'(X; H)$, it is automatically a C^{j+1} such function, and therefore $u \in C^\infty([0, T]; \mathscr{D}'(X; H))$.

1.2. Reduced Symbol of the Parametrix. Operator $U^* U$. Estimates. "Orthogonal Projections" on the Kernel and Cokernel

Reduced Symbol

Modulo regularizing operators (depending smoothly on t), the parametrix $U(t)$ constructed before is equivalent to an operator $U^{\#}(t)$ which is defined for all $t \ge 0$ and forms an "approximate continuous semigroup," in the sense that $U^{\#}(t) U^{\#}(t') \sim U^{\#}(t + t')$. Note that the composition is defined only if we replace $U^{\#}(t')$ by an equivalent properly supported pseudodifferential operator. In particular, this has the effect that

$$U(t, t') \sim U^{\#}(t - t'), \qquad 0 \le t' \le t \le T.$$

In many applications it is preferable, as we show later in this section, to use $U^{\#}(t)$ in the place of $U(t)$.

Let $k(x, t, \xi; z)$ be the symbol in (1.17) and set

$$k^{\#}(x, \xi; z) = \sum_{j=0}^{+\infty} \frac{1}{j!} (-\rho)^{-j} \frac{\partial^{2j} k}{\partial t^j\, \partial z^j}(x, 0, \xi; z), \tag{1.46}$$

$$\mathscr{U}^{\#}(x, t, \xi) = (2\pi i)^{-1} \oint_\gamma e^{\rho t z} k^{\#}(x, \xi; z)\, dz, \tag{1.47}$$

where γ is the same contour as in (1.17). Note that $k^{\#}$ and $\mathcal{U}^{\#}$ are bona fide formal symbols. From $(1.24)_0$ we derive that the principal symbol in $k^{\#}$ is equal to

$$k_0(x, 0, \xi; z) = [zI - \rho^{-1}a(x, 0, \xi)]^{-1}.$$

It is easy to check that $k^{\#}$, as a function of z, is the Laplace transform of $\mathcal{U}^{\#}$, regarded as a function of t, in the sense that

$$k^{\#}(x, \xi; z) = \rho \int_0^{+\infty} e^{-\rho z t}\mathcal{U}^{\#}(x, y, \xi)\, dt. \tag{1.48}$$

For us the important thing is that

(1.49) *$\mathcal{U}(x, t, \xi) - \mathcal{U}^{\#}(x, t, \xi)$ is a C^∞ function of t in* $[0, T[$ *with values in $S^{-\infty}(\Omega; L(H))$.*

The motivation for this is that (reasoning formally)

$$\begin{aligned}\mathcal{U}(x, t, \xi) &\equiv (2\pi i)^{-1} \sum_{j=0}^{+\infty} \oint_\gamma e^{\rho t z} \frac{t^j}{j!}\left(\frac{\partial}{\partial t}\right)^j k(x, 0, \xi; z)\, dz \\ &\equiv (2\pi i)^{-1} \sum_{j=0}^{+\infty} \frac{1}{j!} \oint_\gamma \left(\frac{1}{\rho}\frac{\partial}{\partial z}\right)^j (e^{\rho t z}) \frac{\partial^j k}{\partial t^j}(x, 0, \xi; z)\, dz \\ &\equiv \mathcal{U}^{\#}(x, t, \xi) \qquad \textit{by integrating by parts with respect to } z.\end{aligned}$$

One could have gone directly to $k^{\#}$ by performing a Laplace transform with respect to t on the operator $\partial/\partial t - \sum_j t^j/j![A^{(j)}(0)]$, and obtained $\mathcal{U}^{\#}$ by inverse Laplace transform. By means of $\mathcal{U}^{\#}$ we obtain a representative $U_\Omega^{\#}(t)$ in the local chart $(\Omega, x_1, \ldots, x_n)$ of the parametrix $U(t)$ which happens to be a smooth function of t for *all* positive values of t, not just those $<T$: simply set $U(t) = 0$ (as an equivalence class) for all $t > 0$. This is not so surprising since we know that we can extend $U(t)$ to such values. The symbol (1.47) is now going to be used, in the study of the operator U^*U. Our assertion that $U^{\#}(t) = \operatorname{Op} \mathcal{U}^{\#}(x, t, \xi)$ forms an approximate continuous semigroup is a direct consequence of the fact that on the right-hand side in (1.47), t appears only in the exponent of the exponential and not in $k^{\#}(x, \xi; z)$.

The Operator U^*U

We assume that X is equipped with a density $\varpi > 0$ and use a properly supported representative of $U(t)$, which we may and shall assume defined for all $t > 0$, in accordance with the last remark. We then regard $u_o \mapsto U(t)u_o$ as a continuous linear mapping $\mathcal{D}'(X; H) \to C^\infty(\bar{\mathbb{R}}_+; \mathcal{D}'(X; H))$, which we call U. It has an adjoint U^*, which we may regard as a continuous linear map $C_c^\infty(\bar{\mathbb{R}}_+; \mathscr{D}'(X; H)) \to \mathscr{D}'(X; H)$, and which is given by

$$(1.50)\quad (U^*v)(x) = \int_0^{+\infty} U(t)^* v(x, t)\, dt, \qquad v \in C_c^\infty(X \times \mathbb{R}_+; H),$$

where, for each $t \geq 0$, $U(t)^*$ is the adjoint of the pseudodifferential operator in X, $U(t)$; it is also properly supported. Let $\zeta \in C_c^\infty(\bar{\mathbb{R}}_+)$, $\zeta(t) = 1$ for $t < R$ $(R > 0)$ and let us set

$$(1.51)\qquad K_\zeta = \int_0^{+\infty} \zeta(t) U(t)^* U(t)\, dt.$$

From the fact that $U(t)$ is regularizing in X for all $t > 0$ we derive at once that if $\zeta_1 \in C_c^\infty(\bar{\mathbb{R}}_+)$, $\zeta_1(t) = 1$ for $t < R_1$ ($R_1 > 0$ also) the difference $K_\zeta - K_{\zeta_1}$ is regularizing in X. We denote by U^*U the equivalence class of K_ζ modulo regularizing operators in X.

THEOREM 1.3. *The compose U^*U is an elliptic self-adjoint pseudo-differential operator of order $-m$ in X, valued in $L(H)$.*

Actually, in proving Theorem 1.3, one can obtain detailed information about the symbol of U^*U in an arbitrary local chart $(\Omega, x_1, \ldots, x_n)$ of X. It is convenient to use the symbol (1.47), although we shall omit the super-scripts $\#$. A formal symbol for $U(t)^*U(t) \in \Psi^0(X; L(H))$ in the local chart under consideration is given by

$$(1.52)\qquad \sum_{\alpha,\beta \in \mathbf{Z}_+^n} \frac{1}{\alpha!\beta!} \partial_\xi^{\alpha+\beta} D_x^\alpha[\varpi(x)\mathscr{U}(x, t, \xi)^*] D_x^\beta \mathscr{U}(x, t, \xi),$$

and a formal symbol for U^*U is then obtained by integrating (1.52) from 0 to $+\infty$. This shows easily that U^*U is formally self-adjoint, i.e., its asymptotic symbol is self-adjoint, in an obvious sense. By taking into account (1.46), (1.47), and the subsequent remark concerning the principal symbol of $k^\#$, we see that modulo symbols of degree $< -m$, that of U^*U is equal to

$$(2i\pi)^{-2}\varpi(x) \int_0^{+\infty} \oint_{\bar\gamma}\oint_\gamma e^{\rho t(\bar z + z')} k(x, \xi; z)^* k(x, \xi; z')\, d\bar z\, dz'\, dt,$$

and therefore to

$$(2i\pi)^{-2}\rho^{-1}\varpi(x) \oint_{\bar\gamma}\oint_\gamma [\bar z I - \rho^{-1} a(x, 0, \xi)^*]^{-1} [z' I - \rho^{-1} a(x, 0, \xi)]^{-1} \frac{d\bar z\, dz'}{\bar z + z'}.$$

But since z' lies on γ, $-z'$ lies in the exterior of $\bar\gamma$ and therefore the preceding integral is equal to

$$-(2i\pi\rho)^{-1}\varpi(x)\oint_\gamma [zI+\rho^{-1}a(x,0,\xi)^*]^{-1}[zI-\rho^{-1}a(x,0,\xi)]^{-1}\,dz.$$

We deform the contour of integration γ into the contour defined as follows (for $R>0$ large). We follow up the veritical line segment $\operatorname{Re} z=0$, $-R\le \operatorname{Im} z\le R$, by the semicircle $\operatorname{Re} z<0$, $|z|=R$; this contour is oriented counterclockwise. As R goes to $+\infty$, the (operator) norm of the integrand decays like $1/R^2$ on the semicircular arc, and therefore the contribution from that arc decays like $1/R$. We reach the conclusion that the symbol of U^*U differs from the following symbol

$$\mathcal{K}(x,\xi)=\frac{\varpi(x)}{2\pi}\int_{-\infty}^{+\infty}([a(x,0,\xi)-i\tau I]^{-1})^*[a(x,0,\xi)-i\tau I]^{-1}\,d\tau, \tag{1.53}$$

by a symbol of order $<-m$. Thus (1.53) plays a role analogous to that of principal symbol. Notice that

(1.54) *$\mathcal{K}(x,\xi)$ is elliptic (of order $-m$) positive-definite*;

more precisely, if h is an arbitrary element of H, $x\in\Omega$, $\xi\in\mathbb{R}_n$,

$$(\mathcal{K}(x,\xi)h,h)_H\ge\varpi(x)|h|_H^2/[\pi\|a(x,0,\xi)\|]. \tag{1.55}$$

Furthermore, if $a(x,0,\xi)$ and $a(x,0,\xi)^*$ commute in $L(H)$, then

$$\mathcal{K}(x,\xi)=-\varpi(x)[a(x,0,\xi)+a(x,0,\xi)^*]^{-1}. \tag{1.56}$$

It should be emphasized that if $a(x,0,\xi)$ and $a(x,0,\xi)^*$ do not commute, then the equality (1.56) might not hold; in fact, the right-hand side might not even exist, as the example,

$$a(x,t,\xi)=\begin{pmatrix}-\frac{1}{2} & -1\\ 0 & -\frac{1}{2}\end{pmatrix}|\xi|,\ H=\mathbb{C}^2,$$

shows.

Instead of dealing with $U(t)=U(t,0)$ we could have dealt with $U(t,t')$, $0\le t'\le t<+\infty$ and looked at the class of operators that can be denoted by $\int_{t'}^{+\infty}U(t,t')^*U(t,t')\,dt$. But the expression (1.47) shows that $U(t,t')\sim U(t-t')$; hence this class is equivalent to U^*U modulo regularizing operators.

These results will now be exploited.

The Local Estimates

Let us denote by $U(t)$, $U(t,t')$ properly supported representatives of the equivalence classes so denoted; we may also assume that $U(t)$ is defined for all $t \geq 0$, and $U(t, t')$ for all $t \geq t'$ ($t' \geq 0$). By means of a positive density in X, or else by restricting the argument to a local chart $(\Omega, x_1, \ldots, x_n)$, we may use Sobolev norms on X. For each s in $\mathbb{R}$ we denote the corresponding norm $\| \ \|_s$ and the inner product by $(\ , \ )_s$. In what follows $\mathcal{O}$ will denote a *relatively compact* open subset of X.

First, if u_o is an arbitrary element of $C_c^\infty(\mathcal{O}; H)$, then we have

$$\int_0^T \|U(t)u_o\|_0^2 \, dt = (K_T u_o, u_o)_0, \qquad K_T = \int_0^T U(t)^* U(t) \, dt. \tag{1.57}$$

But of course K_T is a representative of the class U^*U, and thus by Theorem 1.3 we derive that

$$\int_0^T \|U(t)u_o\|_0^2 \, dt \leq \text{const}\|u_o\|_{-m/2}^2. \tag{1.58}$$

Next, let f be an arbitrary element of $C_c^\infty(\mathcal{O} \times \bar{\mathbb{R}}_+; H)$. We have

$$\int_0^T \left\| \int_0^t U(t, t') f(t') \, dt' \right\|_0^2 dt \leq \text{const} \int_0^T \|f(t)\|_{-m}^2 \, dt. \tag{1.59}$$

PROOF OF (1.59). The left-hand side in (1.59) is equal to

$$J = \int_0^T \int_0^t \int_0^t (U(t, t')f(t'), U(t, t'')f(t''))_0 \, dt \, dt' \, dt''.$$

We can subdivide the domain of integration with respect to (t', t'') into two regions, one where $t' \leq t''$, the other one where $t'' \leq t'$. But because of the symmetry with respect to t' and t'', we have

$$J = 2 \int_0^T \int_0^t \int_0^{t'} (U(t, t')f(t'), U(t, t'')f(t''))_0 \, dt \, dt' \, dt''.$$

If we neglect any contribution from regularizing operators, we may apply (1.43) and deal with

$$J_1 = 2 \int_0^T \int_0^{t'} \left[\int_{t'}^T (U(t, t')^* U(t, t') f(t'), U(t', t'') f(t''))_0 \, dt \right] dt' \, dt''.$$

We use the expression (1.47) for the symbol of $U(t)$ and thus $U(t, t') = U(t - t')$. Then in J_1 we have the right to integrate with respect to t from t' to $+\infty$, that is, to integrate with respect to $t - t'$ from 0 to $+\infty$ which yields, modulo negligible terms,

$$J_1 \equiv 2 \int_0^T \left(U^* U f(t), \int_0^t U(t, t') f(t') \, dt' \right)_0 dt.$$

By the Cauchy–Schwarz inequality, this yields

$$J_1^2 \le 4 \int_0^T \|U^* U f(t)\|_0^2 \, dt \int_0^T \left\| \int_0^t U(t,t') f(t') \, dt' \right\|_0^2 dt.$$

It suffices then to apply Theorem 1.3. □

By availing ourselves of (1.58) and (1.59) in conjunction with formula (1.45), we reach the conclusion that if T' is any number such that $0 < T' < T$,

$$(1.60) \quad \int_0^{T'} \|u(t)\|_0^2 \, dt \le C\left\{ \|u(0)\|_{-m/2}^2 + \int_0^{T'} \left\| \frac{\partial u}{\partial t}(t) - A(t)u(t) \right\|_{-m}^2 dt \right.$$

$$\left. + \int_0^{T'} \|u(t)\|_{s'}^2 \, dt \right\} \qquad (s' < 0).$$

All other estimates that one might reasonably expect follow from this basic one. For instance, instead of the zero norm of $u(t)$, one can estimate the sth norm of that function. One can also estimate the sth norm of the derivatives of $u(t)$ with respect to t:

THEOREM 1.4. *Let $\mathcal{O}$ be an arbitrary relatively compact open subset of X, T' any number such that $0 < T' < T$, s and s' any pair of real numbers, k an arbitrary integer ≥ 0. There is a constant $C > 0$ such that for all functions u in $C_c^\infty(\mathcal{O} \times \bar{R}_+; H)$,*

$$(1.61) \quad \int_0^{T'} \|\partial_t^k u\|_s^2 \, dt \le C\left\{ \|u(0)\|_{s+km-\frac{1}{2}m}^2 + \right.$$

$$\left. \sum_{j \le \sup(0,k-1)} \int_0^T \|\partial_t^j(\partial_t u - Au)\|_{s+(k-j-1)m}^2 \, dt + \int_0^T \|u\|_{s'}^2 \, dt \right\}$$

(In practice one takes s' small in comparison to s.) Suitable estimates are also valid for the t-derivatives of negative order (i.e., integrals from 0 to t, iterated a number of times).

Orthogonal Projections on the Kernel and Cokernel

Here we let $A(t)$, $U(t)$, $U(t, t')$ act on C^∞ functions of t valued in the space of microdistributions of x, i.e., on elements of

$$(1.62) \quad C^\infty([0, T[; \dot{\mathcal{D}}'(X; H)) = C^\infty([0, T[; \mathcal{D}'(X; H))/C^\infty(X \times [0, T[; H),$$

and we let U^*U act on $\dot{\mathcal{D}}'(X; H) = \mathcal{D}'(X; H)/C^\infty(X; H)$, by using properly supported representatives of all these classes of operators. Let us then denote by $(U^*U)^{-1}$ the inverse of U^*U; by Theorem 1.3 it is an (elliptic) pseudodifferential operator of order m. We then form

(1.63) $$P_0 = U(U^*U)^{-1}U^* \text{ acting on } C^\infty([0, T[; \dot{\mathcal{D}}'(X; H)).$$

THEOREM 1.5. *In the sense of operators on the space* (1.62), *we have*

(1.64)
$$P_0 = P_0^* = P_0^2; \qquad P_0 = I \text{ on } \operatorname{Ker}\left[\frac{\partial}{\partial t} - A(t)\right]; \qquad \left[\frac{\partial}{\partial t} - A(t)\right] \circ P_0 = 0.$$

One may paraphrase the equations (1.64) by saying that P_0 is an approximate orthogonal projection on the kernel of $\partial/\partial t - A(t)$ in the space of microdistributions, (1.62).

Lastly, for any $\dot{g}$ in (1.62), let us call $E_0\dot{g}$ the class in (1.62) of the distribution $\int_0^t U(t, t')g(t')\,dt'$, for some g in the class $\dot{g}$.

THEOREM 1.6. *With the preceding notation set* $E = (I - P_0)E_0$. *Then, in the sense of linear operators acting on the space* (1.62), *we have*

(1.65) $$\left[\frac{\partial}{\partial t} - A(t)\right]E = I; \qquad E\left[\frac{\partial}{\partial t} - A(t)\right] = I - P_0.$$

1.3. Exact Solution When the Manifold X Is Compact

In the remainder of this section we shall assume that the C^∞ manifold X is compact. This is of course always the case when X is the boundary of a bounded open subset of $\mathbb{R}^{n+1}$ ($n = \dim X$). We shall suppose that all the Sobolev spaces $H^s(X)$, $s \in \mathbb{R}$, are equipped once and for all with a Hilbert space structure. In fact, we can also suppose that X is equipped with a Riemannian structure, and that $H^0(X) = L^2(X)$, equipped with the norm naturally defined by the volume element in X. If $-\Delta$ then denotes the Laplace–Beltrami operator on X, the Hilbert space structure on $H^s(X)$ is that obtained by transferring the structure of $L^2(X)$ via the map $u \mapsto (1 - \Delta)^{-s/2}u$ (see Chapter XII, Section 1).

DEFINITION 1.1. *Let s be a real number, k an integer ≥ 0. We denote by* $\mathbb{E}^{s,k} = \mathbb{E}^{s,k}(X \times [0, T]; H)$ *the space of H-valued distributions in* $X \times [0, T]$ *such that*†

(1.66) $$\frac{\partial^j u}{\partial t^j} \in L^2(0, T; H^{s+(k-j)m}(X; H)), \qquad j = 0, \ldots, k.$$

† In the remainder of this section, for the sake of convenience, we deal with functions and distributions defined in the *closed* interval $[0, T]$ (and generally valued in spaces of distributions on X). We shall assume that $A(t)$ satisfies the basic hypothesis (1.5) in that same closed interval, which can be obtained from (1.5) by decreasing $T > 0$.

We equip $\mathbb{E}^{s,k}$ with the Hilbert norm

$$\|u\|_{s,k} = \left\{ \sum_{j=0}^{k} \int_0^T \left\| \frac{\partial^j u}{\partial t^j} \right\|^2_{s+(k-j)m} dt \right\}^{1/2}. \tag{1.67}$$

We have

$$\mathbb{E}^{s,0} = L^2(0, T; H^s(X; H)).$$

We shall write

$$L = \partial_t - A(t).$$

Since the order of $A(t)$ is m, we have the continuous linear map

$$L : \mathbb{E}^{s,k} \to \mathbb{E}^{s,k-1} \qquad (k \geq 1). \tag{1.68}$$

Let us focus on the case $k = 1$ (s arbitrary). It is clear that the elements of $\mathbb{E}^{s,1}$ are *absolutely continuous* functions of t with values in $H^s(X; H)$.

THEOREM 1.7. *Let s be any real number. To every $f \in \mathbb{E}^{s,0}$, every $u_o \in H^{s+m/2}(X; H)$, there is a unique element u of $\mathbb{E}^{s,1}$ such that*

$$Lu = f \qquad \text{in } X \times]0, T[, \tag{1.69}$$

$$u|_{t=0} = u_o \qquad \text{in } X. \tag{1.70}$$

PROOF. For $\tau > 0$ we define the following "inner product" on $H^{s+m}(X; H)$:

$$a_s(t, \tau; u, v) = \tau(u, v)_{s+m/2} - (A(t)u, v)_{s+m/2}. \tag{1.71}$$

Actually the second term on the right-hand side makes good sense when u and v belong to $C^\infty(X; H)$ and extends "naturally" when they belong to $H^{s+m}(X; H)$. From hypothesis (1.5) we derive the following in a straightforward manner:

(1.72) *There exists $\tau(s) > 0$ and $c_s > 0$ such that for all $u \in H^{s+m}(X; H)$, all t in $[0, T]$, all $\tau \geq \tau(s)$,*

$$c_s\|u\|^2_{s+m} \leq \operatorname{Re} a_s(t, \tau; u, u). \tag{1.73}$$

Consider then the following two sesquilinear forms on $\mathbb{E}^{s,1} \times \mathbb{E}^{s,1}$:

$$\mathfrak{A}_s(\tau; u, v) = \int_0^T [-(u, v_t)_{s+m/2} + a_s(t, \tau; u, v)]\, dt, \tag{1.74}$$

$$\mathfrak{A}'_s(\tau; u, v) = \int_0^T [(u_t, v)_{s+m/2} + a_s(t, \tau; u, v)]\, dt. \tag{1.75}$$

Observe that

$$2\,\mathrm{Re}\int_0^T (u_t, u)_{s+m/2}\,dt = \int_0^T \frac{d}{dt}\|u\|^2_{s+m/2}\,dt$$
$$= \|u(\cdot\,, T)\|^2_{s+m/2} - \|u(\cdot\,, 0)\|^2_{s+m/2},$$

whence, taking (1.73) into account, we get the following.

(1.76) *For all* $\tau > \tau(s)$ *and all* $u \in C^\infty(X \times [0, T]; H)$,

(1.77)

$$c_s \int_0^T \|u\|^2_{s+m}\,dt + \|u(\cdot\,, 0)\|^2_{s+m/2} \leq 2\,\mathrm{Re}\,\mathfrak{A}_s(\tau; u, u) + \|u(\cdot\,, T)\|^2_{s+m/2};$$

(1.78)

$$c_s \int_0^T \|u\|^2_{s+m}\,dt + \|u(\cdot\,, T)\|^2_{s+m/2} \leq 2\,\mathrm{Re}\,\mathfrak{A}'_s(\tau; u, u) + \|u(\cdot\,, 0)\|^2_{s+m/2}.$$

In (1.69) we set $u = e^{\tau t}v$, $f = e^{\tau t}g$, and thus (1.69)–(1.70) is equivalent to

(1.79) $v_t - A(t)v + \tau v = g \quad \text{in } X \times]0, T[, \qquad v|_{t=0} = u_o \quad \text{in } X.$

Existence of the Solution of (1.79)

We apply the following elementary lemma about Hilbert spaces; for a proof see Treves [3], Lemma 41.2.

LEMMA 1.1. *Let* $\mathbb{E}$ *be a Hilbert space,* $\mathfrak{h}$ *a linear subspace of* $\mathbb{E}$, $\mathfrak{A}(w, h)$ *a sesquilinear functional on* $\mathbb{E} \times \mathfrak{h}$ *having the following properties:*

(1.80) *For each fixed* $h \in \mathfrak{h}$, $w \mapsto \mathfrak{A}(w, h)$ *is a continuous linear functional on* $\mathbb{E}$.

(1.81) *For some* $c_o > 0$ *and all* h *in* $\mathfrak{h}$, $c_o\|h\|^2 \leq |\mathfrak{A}(h, h)|$.

Under these circumstances there is a bounded linear map G *of the antidual* $\bar{\mathbb{E}}^*$ *of* $\mathbb{E}$ *into* $\mathbb{E}$, *with norm* $\leq c_o^{-1}$, *such that for every continuous antilinear functional* λ *on* $\mathbb{E}$, *we have*

(1.82) $\mathfrak{A}(G\lambda, h) = \lambda(h) \qquad \text{for all } h \text{ in } \mathfrak{h}.$

We apply Lemma 1.1 with $\mathbb{E}$ the space of pairs (w, w_o) such that

(1.83) $w \in \mathbb{E}^{s+m,0}, \qquad w_o \in H^{s+m/2}(X; H),$

and with $\mathfrak{h}$ the subspace of $\mathbb{E}$ consisting of the pairs (h, h_o) such that

(1.84) $h \in \mathbb{E}^{s,1}, \qquad h_o \in H^{s+m/2}(X; H),$

(1.85) $h(\cdot\,, 0) = h_o, \qquad h(\cdot\,, T) = 0.$

We shall consider the following antilinear functional on $\mathbb{E}$:

$$(1.86)\qquad (w, w_o) \mapsto \int_0^T (g, w)_{s+m/2}\, dt + (u_o, w_o)_{s+m/2}.$$

Once again we are slightly abusing notation: $(g, w)_{s+m/2}$ stands for the extension of the sesquilinear form which would be so denoted if both g and w belonged to $C^\infty(X; H)$; it is clear that the extension is well defined on $H^s(X; H) \times H^{s+m}(X; H)$.

As for the sesquilinear form $\mathfrak{A}$, we take the form $\mathfrak{A}_s$ defined in (1.74): $\mathfrak{A}((w, w_o), (h, h_o)) = \mathfrak{A}_s(\tau; w, h)$.

We reach the conclusion that there is $(v, v_o) \in \mathbb{E}$ such that

$$(1.87)\qquad \mathfrak{A}_s(\tau; v, h) = \int_0^T (g, h)_{s+m/2}\, dt + (u_o, h_o)_{s+m/2}, \qquad \forall (h, h_o) \in \mathfrak{h}.$$

First take $h \in C_c^\infty(X \times]0, T[\ ; H)$ and thus $h_o = 0$ by virtue of (1.85). If we return to the definition (1.74) of $\mathfrak{A}_s$, we see that after integration by parts with respect to t (1.87) reads

$$(1.88)\qquad \int_0^T \big(v_t - A(t)v + \tau v, h\big)_{s+m/2}\, dt = \int_0^T (g, h)_{s+m/2}\, dt,$$

which simply means that the first equation (1.79) is satisfied. But this same equation implies that $v_t \in L^2(0, T; H^s(X; H))$; hence v is an absolutely continuous function in $[0, T]$ valued in $H^s(X; H)$. If h is an arbitrary element of $\mathbb{E}^{s,1}$ such that $h(\cdot, T) = 0$, then integration by parts in t yields

$$\mathfrak{A}_s(\tau; v, h) = \int_0^T \big(v_t - A(t)v + v, h\big)_{s+m/2}\, dt + \big(v(\cdot, 0), h(\cdot, 0)\big)_{s+m/2}.$$

Putting this into (1.87) and taking (1.85) and (1.88) into account yields the second relation (1.79). We are tacitly using the obvious fact that the "traces" $h(\cdot, 0)$, which belong to $H^{s+m/2}(X; H)$, of elements of $\mathbb{E}^{s,1}$, h, such that $h(\cdot, T) = 0$, form a dense subset of $H^{s+m/2}(X; H)$: take $h(x, t) = h_o(x)\cos(\pi t/2T)$ with $h_o \in C^\infty(X)$ arbitrary. Actually it can be shown that those traces make up the entire space $H^{s+m/2}(X; H)$.

As we have already pointed out, we have $v_t \in L^2(0, T; H^s(X; H))$; in other words v belongs to $\mathbb{E}^{s,1}$, as was to be shown.

Uniqueness of the Solution of (1.79)

By subtraction we must show that if $w \in \mathbb{E}^{s,1}$ satisfies

$$(1.89)\qquad w_t - A(t)w + \tau w = 0 \quad \textit{in}\ X \times]0, T[, \qquad w(\cdot, 0) = 0,$$

then we must necessarily have $w \equiv 0$. By (1.71) and (1.75) we have $\mathfrak{A}_s'(\tau; w, w) = 0$, which, by virtue of (1.78) and of the fact that w vanishes identically in X at $t = 0$, implies that the latter is true at all $t \in [0, T]$. □

COROLLARY 1.1. *To every* $u_o \in H^{s+m/2}(X; H)$ *there is a unique* $u \in \mathbb{E}^{s,1}$ *such that*

$$(1.90) \qquad Lu = 0 \qquad \text{in } X \times]0, T[,$$

and that $u|_{t=0} = u_o$ *in* X.

Let us denote by $U_o(t)u_o$ the solution u of (1.90). Since

$$\partial_t U_o(t)u_o = A(t)U_o(t)u_o,$$

we see, by successive differentiation with respect to t, that if $u_o \in H^{s+m/2}(X; H)$, then

$$(1.91) \quad \partial_t^j U_o(t)u_o \in L^2(0, T; H^{s-(j-1)m}(X; H)), \qquad j = 0, 1, \ldots.$$

In particular, if $u_o \in C^\infty(X; H)$ we have $U_o(t)u_o \in C^\infty(X \times [0, T]; H)$.

A restatement of Corollary 1.1 is that $U_o(t)$ is the unique solution of

$$(1.92) \qquad \partial_t U_o(t) - A(t)U_o(t) = 0, \qquad U_o(0) = I.$$

Similarly we could have defined the unique solution $U_o(t, t')$ of

$$(1.93) \qquad \partial_t U_o(t, t') - A(t)U_o(t, t') = 0, \qquad U_o(t', t') = I,$$

for $0 \le t' \le t \le T$. This enables us to represent the solution of (1.69)–(1.70) in the standard manner:

$$(1.94) \qquad u(x, t) = U_o(t)u_o(x) + \int_0^t U_o(t, t')f(x, t')\, dt'.$$

It also enables us to compare the exact solution $U_o(t)$ of (1.92) to the parametrix $U(t)$ constructed in the earlier part of this section. Indeed (cf. (1.3)), $R(t) = \partial_t U(t) - A(t)U(t)$ is a linear operator on $\mathcal{D}'(X; H)$ whose associated kernel is a C^∞ function of (x, t, y) in $X \times [0, T] \times X$; and since, by (1.4), $U(0) = I$, we have (by availing ourselves of the uniqueness of the solution u in Theorem 1.7)

$$(1.95) \qquad U(t) = U_o(t) + \int_0^t U_o(t, t')R(t')\, dt'.$$

It is checked at once that the second term on the right-hand side is a regularizing operator on $\mathcal{D}'(X; H)$ depending smoothly on t in $[0, X]$. We reach the following conclusion:

PROPOSITION 1.1. *For each* $t \in [0, T]$, $U_o(t)$ *is a standard pseudo-differential operator of order zero in X, valued in* $L(H)$.

Let us look at the dependence on t of $U_o(t)$ and of $U_o(t, t')$.

PROPOSITION 1.2. *If k is any integer* ≥ 0, *and if* $f \in \mathbb{E}^{s,k}$ *and* $u_o \in H^{s+km+m/2}(X; H)$, *the solution u of* (1.69)–(1.70) *belongs to* $\mathbb{E}^{s,k+1}$.

PROOF. The result is stated in Theorem 1.7 when $k = 0$. We shall assume henceforth that $k \geq 1$ and use the formula

$$\partial_t^{j+1} u = \partial_t^j(Au) + \partial_t^j f, \qquad j = 0, \ldots, k. \tag{1.96}$$

We know that $\partial_t^j f \in L^2(0, T; H^{s+(k-j)m}(X; H))$; by Theorem 1.7 and by the fact that

$$\mathbb{E}^{s,k} \subset \mathbb{E}^{s+km,0},$$

we also know that $\partial_t^j u \in L^2(0, T; H^{s+(k+1-j)m}(X; H))$ for $j = 0, 1$. Thus we can deduce the same conclusion from (1.96) for j up to and including $k + 1$. □

PROPOSITION 1.3. *Given any real number s, any integer* $j \geq 0$, *any* $u_0 \in H^s(X; H)$, $U_o(t)u_o$ *is a* C^j *function of t in* $[0, T]$ *valued in* $H^{s-jm}(X; H)$.

PROOF. It suffices to prove the result for $j = 0$. After this, equation (1.96) (where $f \equiv 0$) can be used to deduce the continuity of the successive derivatives of $u = U_o(t)u_o$ with respect to t. We know already, by Theorem 1.7, that when $u_o \in H^{s+m/2}(X; H)$, $U_o(t)u_o$ is an absolutely continuous function of t in $[0, T]$ valued in $H^s(X; H)$. When $u_o \in H^s(X; H)$ is arbitrary, we select a sequence $\{u_\nu\}$ in $H^{s+m/2}(X; H)$ converging to u_o in $H^s(X; H)$ and, by using Proposition 1.1, we write that

$$\sup_{0 \leq t \leq T} \|U_o(t)u_o - U_o(t)u_\nu\|_s \leq C\|u_o - u_\nu\|_s$$

and thus see that $U_o(t)u_o$ is a uniform limit of continuous functions.

The result for $U(t, t')$ analogous to Proposition 1.3 is of course valid, and from this fact and from formula (1.94) we derive

PROPOSITION 1.4. *Suppose that* $u_o \in H^s(X; H)$ *and that*

$$f \in C^j([0, T]; H^{s-jm}(X; H)), \qquad j = 0, \ldots, k. \tag{1.97}$$

If then u is the solution of (1.69)–(1.70) *we have*

$$u \in C^j([0, T]; H^{s-jm}(X; H)), \qquad j = 0, \ldots, k+1. \tag{1.98}$$

2. Preliminaries to the Study of Elliptic Boundary Problems: Sobolev Spaces in Bounded Open Subsets of Euclidean Spaces. Traces

Throughout this section Ω will be a bounded open subset of Euclidean space $\mathbb{R}^N$ ($N \geq 1$) whose boundary is a C^∞ hypersurface X; the domain Ω lies on one side of it. We write $N = n + 1$; thus $\dim X = n$, as before. We shall always assume that X is equipped with the Riemannian structure induced by $\mathbb{R}^N$ and that this structure defines the Sobolev spaces $H^s(X)$, s real.

We denote by $\mathcal{T}$ a tubular open neighborhood of X in $\mathbb{R}^N$ and assume that there is defined in $\mathcal{T}$ a C^∞ real function t, whose differential does not vanish anywhere in $\mathcal{T}$, such that $t > 0$ in $\mathcal{T} \cap \Omega$, $t < 0$ in $\mathcal{T} \cap (\mathbb{R}^N \backslash \bar{\Omega})$. We shall select a number $T > 0$ such that $|t| \leq T$ defines a compact neighborhood of X contained in $\mathcal{T}$. We denote by x the variable point in X; local coordinates in X will be denoted by $x^1, \ldots, x^n$.

We shall apply some of the results of the preceding pages, always with the choice $m = 1$. Below we use m to denote an arbitrary integer ≥ 0.

DEFINITION 2.1. *By* $H^m(\Omega)$ *one denotes the space of* L^2 *functions in* Ω *whose derivatives of order* $\leq m$ *all belong to* $L^2(\Omega)$.

It is customary to equip $H^m(\Omega)$ with the inner product

$$(u, v)_m = \sum_{|\alpha| \leq m} \int_\Omega D^\alpha u \overline{D^\alpha v}\, dx. \tag{2.1}$$

Then $H^m(\Omega)$ is a Hilbert space: if $\{u_j\}$ is a Cauchy sequence in $H^m(\Omega)$, for each α, $|\alpha| \leq m$, the $D^\alpha u_j$ converge in $L^2(\Omega)$, necessarily to $D^\alpha u$ if u is the limit of the u_j, since they do so in $\mathcal{D}'(\Omega)$.

We have the continuous injection with norm one $H^m(\Omega) \hookrightarrow H^{m'}(\Omega)$ if $m' \leq m$. Of course $H^0(\Omega) = L^2(\Omega)$. If we relax the constraint that Ω be bounded and allow Ω to be equal to $\mathbb{R}^N$, the space $H^m(\Omega)$ is equal to the standard Sobolev space $H^m(\mathbb{R}^N)$, though not normwise when $m \geq 2$.

It is also customary to denote by $H_o^m(\Omega)$ the closure in $H^m(\Omega)$ of the space of test functions, $C_c^\infty(\Omega)$. The dual of $H_o^m(\Omega)$ can be identified with a space of distributions in Ω, denoted by $H^{-m}(\Omega)$, and always equipped with the dual Hilbert space structure. Here again, when $\Omega = \mathbb{R}^N$, one finds the standard Sobolev space.

The restriction to $\mathcal{T} \cap \Omega$ of a function $u \in H^m(\Omega)$ has the property that

$$\partial_t^j u \in L^2(0, T; H^{m-j}(X)), \qquad j = 0, \ldots, m. \tag{2.2}$$

Compare with Definition 1.1; keep in mind that here $H = \mathbb{C}$, that the number m in Definition 1.1 must now be taken equal to one, and that what is called $s + km$ there is now called m.

Let us suppose that $m \geq 1$. Then

(2.3) *$\partial_t^j u$ is an absolutely continuous function of t in $[0, T]$, valued in $H^{m-1-j}(X)$ for $j = 0, \ldots, m - 1$.*

We can therefore define the first m *traces* of u on the boundary X of Ω:

$$\gamma_j u = \partial_t^j u|_X, \qquad j = 0, \ldots, m - 1. \tag{2.4}$$

Of course, if $u \in C^\infty(\bar{\Omega})$, we have $\gamma_j u \in C^\infty(X)$ for every j.

Let $-\Delta_x$ denote the Laplace–Beltrami operator on X, Λ the positive square root of $1 - \Delta_x$. We shall set

$$L = \partial_t + \Lambda. \tag{2.5}$$

We denote by $U_o(t)$ the solution of

$$LU_o = 0 \quad in\, X \times [0, +\infty[, \qquad U_o(0) = I. \tag{2.6}$$

We could write $U_o(t) = e^{-t\Lambda}$.

Then let $\zeta \in C^\infty(\mathbb{R}^1)$, $\zeta(t) = 1$ for $|t| < \frac{1}{4}T$, $\zeta(t) = 0$ for $|t| > \frac{1}{2}T$. We can and shall regard ζ as a function of (x, t), independent of x, i.e., as an element of $C_c^\infty(\mathcal{T})$, and therefore as an element of $C_c^\infty(\mathbb{R}^{n+1})$. Let $\psi = 1 - \zeta$ in Ω, $\psi = 0$ elsewhere. For any $u \in \mathcal{D}'(\Omega)$ we may write

$$u = \zeta u + \psi u \qquad (in\ \Omega). \tag{2.7}$$

We recall that $C^\infty(\bar{\Omega})$, the space of C^∞ functions in Ω whose derivatives of all orders extend continuously to the closure of Ω, is also equal to the space of restrictions to Ω of C^∞ functions (with compact support, if you like) in the whole space $\mathbb{R}^{n+1}$.

THEOREM 2.1. *$C^\infty(\bar{\Omega})$ is dense in $H^m(\Omega)$.*

PROOF. We may restrict ourselves to the case $m \geq 1$, since $C_c^\infty(\Omega)$ is dense in $L^2(\Omega)$.

Let $u \in H^m(\Omega)$ and consider the decomposition (2.7). The second term, ψu, which belongs to $H_c^m(K)$, with K a compact neighborhood of supp ψ in Ω, is the limit of a sequence of functions belonging to $C_c^\infty(K)$. Let us set

$$v_j(x, t) = \zeta(t) U_o(1/j) u(x, t), \qquad j = 1, 2, \ldots.$$

For each t in $[0, \frac{1}{2}T]$ we regard $u(x, t)$ as a function on X on which we let the regularizing operator $U_o(1/j)$ act. All the traces of v_j on $t = 0$ are C^∞ functions in X, and when $j \to +\infty$, v_j converges to ζu in $H^m(\Omega)$, as one can check at once. We then define a function $\tilde{v}_j$ in $\mathbb{R}^{n+1}$ as follows: $\tilde{v}_j$ is equal to v_j in Ω, and in the complement of Ω,

$$\tilde{v}_j(x, t) = \zeta(t) \sum_{k=0}^{m-1} \frac{t^k}{k!} (\gamma_k v_j)(x);$$

thus here $t \leq 0$. It can also be checked at once that $\tilde{v}_j \in H^m(\mathbb{R}^{n+1})$; of course, supp $\tilde{v}_j$ is a compact subset of $\Omega \cup \mathcal{T}$. By the density of C^∞ in H^m we can approximate $\tilde{v}_j$ by C^∞ functions, from which we get Theorem 2.1 by the diagonal process. □

THEOREM 2.2. *Let m be an integer* ≥ 1. *The map*

$$u \mapsto (\gamma_j u)_{j=0,\ldots,m-1} \tag{2.8}$$

extends from $C^\infty(\bar{\Omega})$ *as a continuous linear map of* $H^m(\Omega)$ *onto* $\prod_{j=0}^{m-1} H^{m-j-1/2}(X)$. *The kernel of the map* (2.8) *is exactly equal to* $H_o^m(\Omega)$.

PROOF. Let $u \in C^\infty(\bar{\Omega})$. We have

$$\|u(0)\|_{m-1/2}^2 = -2 \operatorname{Re} \int_0^T \left(\frac{\partial}{\partial t}(\zeta u), \zeta u\right)_{m-1/2} dt \leq 2 \int_0^T \left\|\frac{\partial}{\partial t}(\zeta u)\right\|_{m-1} \|u\|_m \, dt,$$

whence, by Cauchy–Schwarz and Leibniz formulas,

$$\|u(0)\|_{m-1/2}^2 \leq C \int_0^T (\|u(t)\|_m^2 + \|\partial_t u(t)\|_{m-1}^2) \, dt. \tag{2.9}$$

In all this, Sobolev inner products and norms are those on X. Replacing u by $\partial_t^j u$ for $j = 1, \ldots, m-1$ (if $m \geq 2$) we get immediately that (2.8) extends from the dense subspace $C^\infty(\bar{\Omega})$ (Theorem 2.1) as a continuous linear map *into*.

In order to show that it is onto, let $u_j \in H^{m-j-1/2}(X)$ be arbitrary, for $j = 0, \ldots, m-1$. Using the Vandermonde determinant, we can select m

elements of $H^{m-1/2}(X)$, $v_0, \ldots, v_{m-1}$, such that

$$\sum_{j=1}^{m} j^k v_{j-1} = u_k, \qquad k = 0, \ldots, m-1. \tag{2.10}$$

Take

$$u(x,t) = \zeta(t) \sum_{j=1}^{m} U_o(jt) v_{j-1}(x).$$

We have $v \in H^m(\Omega)$ and $\gamma u = u_k$ for each $k = 0, \ldots, m-1$.

Finally we prove the assertion about the kernel of the map (2.8). It is evident that $H_o^m(\Omega)$, which is the closure of $C_c^\infty(\Omega)$ in $H^m(\Omega)$, is contained in that kernel. On the other hand, if all the traces $\gamma_j u$ $(j = 0, \ldots, m-1)$ vanish, the functions $(\zeta u)(x, t-\varepsilon)$ (assumed to vanish identically for $0 \le t \le \varepsilon$) belong to $H^m(\Omega)$, hence to $H_o^m(\Omega)$, and converge to ζu in $H^m(\Omega)$. Since ψu belongs in any case to $H_o^m(\Omega)$, so does u (see (2.7)). □

COROLLARY 2.1. *The restriction to* Ω *is a continuous linear surjection of* $H_c^m(\Omega \cup \mathcal{T})$ *onto* $H^m(\Omega)$.

PROOF. If $u \in H^m(\Omega)$, define $\tilde{u} \in H_c^m(\Omega \cup \mathcal{T})$ by setting $\tilde{u} = u$ in Ω and

$$\tilde{u}(x,t) = \zeta(t) \sum_{k=0}^{m-1} \frac{t^k}{k!} (\gamma_k u)(x) \tag{2.11}$$

for $t \le 0$. □

COROLLARY 2.2. *Continue to suppose that* $m \ge 1$. *The natural injection*

$$H^m(\Omega) \to H^{m-1}(\Omega) \tag{2.12}$$

is compact.

PROOF. Factorize the map (2.8) into the sequence

$$H^m(\Omega) \to H_c^m(\Omega \cup \mathcal{T}) \to H^{m-1}(\Omega \cup \mathcal{T}) \to H^{m-1}(\Omega),$$

where the first arrow stands for the extension defined in the proof of Corollary 2.1, the second arrow is the natural injection, which is compact according to Corollary 1.2 of Chapter II, and the third arrow is the restriction to Ω map. □

Corollary 2.2 is a variant of Rellich's lemma.

Let $\mathcal{N}^m(\Omega)$ denote the orthogonal of $H_o^m(\Omega)$ in $H^m(\Omega)$. It follows at once from the definition (2.1) of the inner product in the latter space that $\mathcal{N}^m(\Omega)$ is the space of elements $h \in H^m(\Omega)$ satisfying

$$\sum_{|\alpha| \leq m} D^{2\alpha} h = 0. \tag{2.13}$$

The map (2.8) induces a bijection $\mathcal{N}^m(\Omega) \to \prod_{j=0}^{m-1} H^{m-j-1/2}(X)$. This yields a representation of the continuous linear functionals on $H^m(\Omega)$:

$$(H^m(\Omega))' \cong H^{-m}(\Omega) \times \prod_{j=0}^{m-1} H^{-m+j+1/2}(X). \tag{2.14}$$

Let us also note that

$$u \mapsto \sum_{|\alpha| \leq m} D^{2\alpha} \bar{u} \tag{2.15}$$

is the natural antilinear isometry of $H_o^m(\Omega)$ onto its dual $H^{-m}(\Omega)$. It follows at once that every element f of $H^{-m}(\Omega)$ can be represented (in an infinity of manners) as a sum

$$f = \sum_{|\alpha| \leq m} D^\alpha f_\alpha, \qquad f_\alpha \in L^2(\Omega), |\alpha| \leq m. \tag{2.16}$$

All that precedes extends routinely to the case where functions and distributions, instead of being scalar, are valued in a finite-dimensional Hilbert space, as will be the case in the applications.

3. Approximate Triangulization of Boundary Problems for Elliptic Equations

We use the same notation as in Section 2, in particular with regard to the meaning of Ω, X, $\mathcal{T}$, and t. We propose to study a boundary problem for an elliptic linear partial differential operator P whose coefficients are C^∞ functions in $\Omega \cup \mathcal{T}$, which is an open neighborhood of the closure $\bar{\Omega}$ of Ω, valued in $L(H)$. We shall modify the meaning of m; henceforth it will be the *order* of P, not that of $A(t)$, which will be assumed to be equal to one whenever we apply the results of Section 1.

We shall restrict our study mainly to the region $X \times [0, T]$. Throughout the sequel we assume that in this region P has the form

$$P = I\,\partial_t^m + \sum_{j=1}^{m} P_j(x, t, D_x)\,\partial_t^{m-j}, \tag{3.1}$$

where, for each $j = 1, \ldots, m$, $P_j(x, t, D_x)$ is a linear partial differential operator of order j in X whose coefficients are C^∞ functions of (x, t) in $X \times [0, T[$ valued in $L(H)$; thus P is a determined system of linear PDOs. Actually we do not lose anything by taking the P_j to be *classical pseudodifferential operators* of order j respectively, in X, with values in $L(H)$, which we then denote by $P_j(t)$. For each j the principal symbol of $P_j(t)$ can be regarded as a C^∞ function of (x, ξ, t) in $(T^*X\backslash 0) \times [0, T[$ valued in $L(H)$, positive-homogeneous of degree j with respect to the fiber variable ξ. We are now going to make a truly restrictive hypothesis about our operator P, namely that its principal part is *scalar*; this will cover the main cases that we shall be interested in.

(3.2) *For every $j = 1, \ldots, m$, the principal symbol $P_{j,0}(x, t, \xi)$ is equal to $P_j^0(x, t, \xi)I$, where $P_j^0(x, t, \xi)$ is complex-valued.*

Let us then select a scalar pseudodifferential operator of order j in X, depending smoothly on $t \in [0, T[$, with principal symbol $P_j^0(x, t, \xi)$; moreover, it is classical. Let it be P_j^0 $(1 \leq j \leq m)$ and set

$$P = IP^0 + P', \qquad P^0 = \partial_t^m + \sum_{j=1}^{m} P_j^0(t)\, \partial_t^{m-j}, \tag{3.3}$$

where P' is not necessarily scalar but has order $\leq m - 1$ (with respect to both x and t). Of course, if P is a differential operator, we may also take each $P_j^0(t)$ to be a differential operator. Note that

$$P' = \sum_{j=1}^{m} P_j'(t)\, \partial_t^{m-j}, \qquad \deg P_j'(t) \leq j - 1,\ 1 \leq j \leq m. \tag{3.4}$$

We now state the *ellipticity* hypothesis; it concerns the principal symbol of P, or rather that of P^0:

$$\sigma(P^0) = z^m + \sum_{j=1}^{m} P_j^0(x, t, \xi) z^{m-j}. \tag{3.5}$$

(3.6) *There are two integers, m^+ and m^-, such that $m^+ + m^- = m$, $m^- \geq 1$, and for all $(x, \xi) \in T^*X\backslash 0$, $t \in [0, T[$, the polynomial $\sigma(P^0)$ with respect to z has exactly m^+ roots with real part >0 and m^- roots with real part <0.*

REMARK 3.1. When P is a differential operator or, more generally, an antipodal pseudodifferential operator, which means that $\sigma(P^0)(x, t, \xi; z) =$

$\pm\sigma(P^0)(x, t, -\xi; -z)$, and provided that $\dim X > 1$, then hypothesis (3.6) requires that $m^+ = m^- = m/2$, and then m is necessarily *even*. This follows by a standard continuity argument and the fact that the complement of the origin in ξ-space is connected.

We shall use the ellipticity hypothesis (3.6) to factorize the principal symbol of P^0 as follows:

$$(3.7)\quad \sigma(P^0) = \sigma(M^{+0})\sigma(M^{-0}), \qquad \sigma(M^{\pm 0}) = \prod_{k=1}^{m^\pm} [z - z_k^\pm(x, t, \xi)],$$

where the z_k^+ (resp., the z_k^-) are the roots with real part >0 (resp., <0). We wish to show that the factorization (3.7) leads to a factorization of the operator P itself, of the kind

$$(3.8)\qquad P \sim M^+M^-,$$

with $\sigma(M^\pm) = I\sigma(M^{\pm 0})$. If $m^+ = 0$ this is obvious: we may take $M^+ = I$. Thus, in the remainder, we shall assume $m^+ \geq 1$. Let us write

$$(3.9)\qquad \sigma(M^{\pm 0})(x, t, \xi, z) = z^{m^\pm} + \sum_{j=1,\dots,m^\pm} M_j^{\pm 0}(x, t, \xi)z^{m^\pm - j}.$$

Although the roots $z_k^\pm$ cannot, in general, be represented as continuous functions of (x, t, ξ), for each j the coefficient $M_j^{+0}(x, t, \xi)$ (resp., $M_j^{-0}(x, t, \xi)$) is a C^∞ function in $(T^*X\backslash 0) \times [0, T[$, positive-homogeneous of degree j with respect to ξ. This follows from the fact that these coefficients are *symmetric functions* of the roots z_k^+ (resp., z_k^-) and that the two sets of roots z_k^+ and z_k^- stay apart, as (x, t, ξ) varies.

LEMMA 3.1. *Let s denote the variable point in a C^∞ manifold S and let $f(s; z)$ be a monic polynomial in z, of degree m, whose coefficients are complex-valued C^∞ functions of s in S. Suppose that the roots of $f(s; z)$ can be labeled in such a manner that for some integer d, $1 \leq d \leq m$, independent of s, the distance between the sets*

$$\{z_1(s), \dots, z_d(s)\} \quad \text{and} \quad \{z_{d+1}(s), \dots, z_m(s)\}$$

is strictly positive. Then the coefficients of

$$g(s; z) = \prod_{j=1}^{d} [z - z_j(s)]$$

are C^∞ functions of s in S.

PROOF. If s remains in a suitably small subset U of S we can find a smooth closed curve γ in the complex plane whose interior contains all the roots $z_j(s)$ for $j \leq d$ and none for $j > d$, and itself does not contain any root, for any s in U. Then apply Rouché's theorem:

$$\sum_{j=1}^{d} z_j(s)^k = (2i\pi)^{-1} \oint_\gamma z^k \frac{f_z(s;z)}{f(s;z)}\,dz, \qquad k = 0, 1, \ldots.$$

The right-hand side, and hence the left-hand side, are C^∞ functions of s in U. The coefficients of $g(s;z)$ are "universal" polynomial functions of the basic symmetric functions $\sum_{j=1}^{d} z_j^k$. □

Now for each j we may select a classical pseudodifferential operator in X, scalar, depending smoothly on $t \in [0, T[$, $M_j^{\pm 0}(t)$, with symbol $M_j^{\pm 0}(x, t, \xi)$, and we set

$$M^{\pm 0} = \partial_t^{m^\pm} + \sum_{j=1}^{m^\pm} M_j^{\pm 0}(t)\,\partial_t^{m^\pm - j}. \tag{3.10}$$

Then we have

$$P = IM^{+0}M^{-0} + P'', \tag{3.11}$$

with P'' of the same type as P' in (3.3).

We shall reason by induction. Let us assume that we have found, for $N \geq 1$, operators $M^\pm_{(N)}$ and $R_{(N)}$ such that

$$M^\pm_{(N)} = \partial_t^{m^\pm} + \sum_{j=1,\ldots,m^\pm} M^\pm_{(N)j}(t)\,\partial_t^{m^\pm - j}; \tag{3.12}$$

(3.13) $\sigma(M^\pm_{(N)j}(t)) = IM_j^{\pm 0}(x, t, \xi)(j = 1, \ldots, m^\pm)$ *and the* $M^\pm_{(N)j}$ *are classical pseudodifferential operators in* X, *valued in* $L(H)$, *depending smoothly on* t *in* $[0, T[$;

$$R_{(N)} = \sum_{j=1,\ldots,m} R_{(N)j}(t)\,\partial_t^{m-j}; \tag{3.14}$$

$$R_{(N)j}(t) \in C^\infty([0, T[; \Psi^{j-N}(X; L(H))), \qquad j = 1, \ldots, m; \tag{3.15}$$

$$P = M^+_{(N)}M^-_{(N)} + R_{(N)}. \tag{3.16}$$

We shall obtain all this with $N + 1$ in the place of N. In order to do this we regard $A_N^\pm = M^\pm_{(N+1)} - M^\pm_{(N)}$ as unknowns and write that

$$-R_{(N+1)} = M^+_{(N)}A_N^- + A_N^+M^-_{(N)} + A_N^+A_N^- - R_{(N)}. \tag{3.17}$$

We seek $A_N^\pm$ in the form

$$A_N^\pm = \sum_{j=1,\ldots,m^\pm} A_{N,j}^\pm(t)\,\partial_t^{m^\pm - j},$$

with $A^{\pm}_{N,j}(t)$ a classical pseudodifferential operator of order $j - N$ in X, valued in $L(H)$, depending smoothly on $t \in [0, T[$. We proceed as follows. We write that the principal symbol of the right-hand side in (3.17), regarded as an operator of order $m - N$, vanishes identically. Observe that the order of $A^{+}_{N}A^{-}_{N}$ is $\leq m^{+} + m^{-} - 2N < m - N$. Consequently we require that

$$\sigma(M^{+}_{(N)})\sigma(A^{-}_{N}) + \sigma(A^{+}_{N})\sigma(M^{-}_{(N)}) = \sigma(R_{(N)}), \tag{3.18}$$

that is, by virtue of (3.13),

$$\sigma(M^{+0})\sigma(A^{-}_{N}) + \sigma(M^{-0})\sigma(A^{+}_{N}) = \sigma(R_{(N)}). \tag{3.19}$$

By their definition the polynomials $\sigma(M^{\pm 0})(x, t, \xi, z)$ (with respect to z) are *coprime*, and uniformly so as (x, ξ, t) remains in compact subsets of $(T^*X\backslash 0) \times [0, T[$. Observe that equation (3.19) is homogeneous with respect to (ξ, z).

We are now going to apply the classical Bezout theorem. Let us denote by $\mathcal{P}^d$ the space of complex polynomials in one variable, of degree $\leq d$.

LEMMA 3.2. *Let $p \in \mathcal{P}^{m^+}$, $q \in \mathcal{P}^{m^-}$ have no common root. Then $(f, g) \mapsto pf + qg$ is a bijection of $\mathcal{P}^{m^- -1} \times \mathcal{P}^{m^+ -1}$ onto $\mathcal{P}^{m^+ + m^- -1}$.*

PROOF. Note $\dim \mathcal{P}^d = d + 1$; therefore the dimension of the source and target spaces of our map are the same. It suffices to show that the map is injective. If $pf = -qg$, each root of q (with its multiplicity) must be a root of f since it is not one of p; but $\deg f \leq \deg q - 1$, and hence this is possible only if f and g vanish identically. □

COROLLARY 3.1. *Let p, q be as in Lemma 3.2. Then $(f, g) \mapsto pf + qg$ is a bijection of $[\mathcal{P}^{m^- -1} \otimes L(H)] \times [\mathcal{P}^{m^+ -1} \otimes L(H)]$ onto $\mathcal{P}^{m^+ + m^- -1} \otimes L(H)$.*

PROOF. Use a basis in H and represent the elements of $\mathcal{P}^d \otimes L(H)$, which is the space of polynomials of degree $\leq d$ with coefficients in $L(H)$, as matrices with polynomial entries. Apply Lemma 3.2 to each entry. □

Suppose now that p, q depend smoothly on some parameter θ, for us (x, t, ξ), which varies in some smooth manifold Θ, and that the set of roots of p and q stay at a distance from each other bounded from below on every compact subset of Θ. Let $r \in \mathcal{P}^{m^+ + m^- -1} \otimes L(H)$ also depend smoothly on θ. The same will be true of the solutions f, g of $pf + qg = r$, since the coefficients of f and g are linear functions of those of r (linear functions that depend smoothly on θ; one could also apply the closed graph theorem to reach the same conclusion).

Thus, by virtue of Lemma 3.2, $\sigma(A_N^+)$ and $\sigma(A_N^-)$ are uniquely determined polynomials in z, of degrees $\leq m^+ - 1$, $m^- - 1$ respectively, with coefficients that are smooth functions of (x, ξ, t), the degree of homogeneity of these symbols with respect to (ξ, z) is $m^+ - N$ and $m^- - N$ respectively. It remains only to select for $A_{N,j}^\pm(t)$ classical pseudodifferential operators in X, valued in $L(H)$, depending smoothly on t, with the appropriate principal symbols (specifically $(1/j!)(\partial^j/\partial z^j)\sigma(A_N^\pm)|_{z=0}$). If (3.17) is taken as definition of $R_{(N+1)}$, we shall have all the properties (3.12)–(3.16) satisfied with $N + 1$ instead of N. Since we know, by (3.11), that they are satisfied when $N = 1$, we can go to the limit for N arbitrarily large, namely $N = +\infty$; indeed, the operators $M_{(N)}^\pm$ obviously converge. At the limit, rather than writing $N = +\infty$, we shall omit the subscripts N. Thus (3.18) is satisfied. Actually we shall go one step further and divide R by M^-:

(3.20) $$R = QM^- + R', \qquad R' = \sum_{j=1,\dots,m^-} R_j'(t)\, \partial_t^{m^- - j};$$

(3.21) $$R_j'(t) \in C^\infty([0, T[; \Psi^{-\infty}(X; L(H))), \qquad j = 1, \dots, m^-.$$

This poses no problem: it is just a question of replacing

$$\partial_t^{m^-} \text{ by } M^- - \sum_{j=1,\dots,m^-} M_j^-(t)\, \partial_t^{m^- - j}$$

a large enough number of times. Finally, after dropping primes in the notation for R' and substituting M^+ for $M^+ + Q$, we obtain the decomposition

(3.22) $$P = M^+M^- + R,$$

with $M^\pm$ of the same kind as $M^{\pm 0}$ (see (3.10)) and R the same as R' in (3.20)–(3.21).

Now suppose that we are dealing with the equation

(3.23) $$Pu = f, \qquad u, f \in C^\infty([0, T[; \mathscr{D}'(X; H)).$$

By (3.22) we see that it is equivalent to the system of (two) equations

(3.24) $$M^-u = v,$$

(3.25) $$M^+v + Ru = f.$$

Let us consider a typical equation of the kind

(3.26) $$Mw = g, \qquad M = \partial_t^r + \sum_{j=1,\dots,r} C_j(t)\partial_t^{r-j},$$

where, for each j, $C_j(t)$ is a pseudodifferential operator in X of order j with values in $L(H)$, depending smoothly on t. Let us select once and for all an elliptic pseudodifferential operator Λ of order one in X, scalar; we take Λ to be *classical*, and *properly supported*. We shall furthermore require that there be a (classical elliptic) pseudodifferential operator of order -1 in X, properly supported, which we denote by Λ^{-1} and such that $\Lambda\Lambda^{-1}$ = Identity of $\mathscr{D}'(X)$. Let us not worry whether such an operator exists: When the manifold X is *compact*, which is the only case that truly interests us, it certainly does exist. Equip X with a Riemannian metric and denote by $-\Delta_x$ the Laplace–Beltrami operator on X for that metric (it is an operator ≥ 0). A possible choice will then be $\Lambda = (1 - \Delta_x)^{1/2}$. When X is compact, the requirement "properly supported" is void, and $\mathscr{D}'(X; H) = \mathscr{E}'(X; H)$.

Let us then set, for each $j = 1, \dots, r$, $w^j = \Lambda^{1-j}\, \partial_t^{j-1} w$; we shall denote by $\mathbf{W}$ the r-vector with components $w^1, \dots, w^r$. Each component is a smooth function of t with values in $\mathscr{D}'(X; H)$, hence $\mathbf{W}$ itself is such a function but valued in $\mathscr{D}'(X; H \otimes \mathbb{C}^r)$. Note that we have

$$\partial_t w^j = \Lambda w^{j+1}, \qquad j = 1, \dots, r-1. \tag{3.27}$$

If we multiply both members in equation (3.26) by Λ^{1-r}, we may rewrite it in the following manner:

$$\partial_t w^r + \sum_{j=1}^{r} \Lambda^{1-r} C_j(t) \Lambda^{r-j} w^{r-j+1} = \Lambda^{1-r} g. \tag{3.28}$$

We observe that the "coefficients" $C_j^{\#}(t) = \Lambda^{1-r} C_{r-j+1}(t) \Lambda^{j-1}$ are pseudodifferential operators of order one in X, valued in $L(H)$, depending smoothly on t. We gather equations (3.27) and (3.28) in a single system,

$$\partial_t \mathbf{W} - \mathscr{M}(t)\mathbf{W} = \mathbf{G}, \tag{3.29}$$

with $\mathbf{G}$ the r-vector with components all zero, except the rth one, equal to $\Lambda^{1-r} g$, and where $\mathscr{M}(t)$ is the $r \times r$ matrix

$$\begin{pmatrix} 0 & I\Lambda & 0 & \cdots & 0 \\ 0 & 0 & I\Lambda & \cdots & 0 \\ \vdots & & & & \\ 0 & 0 & 0 & \cdots & I\Lambda \\ -C_1^{\#}(t) & -C_2^{\#}(t) & -C_3^{\#}(t) & \cdots & -C_r^{\#}(t) \end{pmatrix}. \tag{3.30}$$

Of course we view $\mathscr{M}(t)$ as a pseudodifferential operator of order one in X with values in $L(H \otimes \mathbb{C}^r)$. A standard and important remark is that its principal symbol $\sigma(\mathcal{M}(t))$ has the property that

$$\det\{zI - \sigma(\mathcal{M}(t))\} = \sigma(M)(x, t, \xi, z), \tag{3.31}$$

where we view $\sigma(\mathcal{M}(t))$ as a matrix over the *ring* $L(H)$, depending smoothly on $((x, \xi), t) \in (T^*X\backslash 0) \times [0, T[$, and positive-homogeneous of degree one with respect to ξ. Thus the determinant det is computed in that ring. Although the ring $L(H)$ is not commutative if $\dim H > 1$, the computation of that determinant is made easy by the fact that all the rows in $\sigma(\mathcal{M}(t))$, except possibly the last one, are scalar multiples of the identity of H; at any rate in the application of what precedes to equations (3.24) and (3.25) the last row will be a scalar multiple of the identity.

First we apply the preceding transformation to equation (3.24); in this case $r = m^-$; we set $u^j = \Lambda^{1-j}\,\partial_t^{j-1}u$, $j = 1, \ldots, m^-$, and shall denote by $\mathbf{u}$ the m^--vector with components u^j. Let us set right away $v^j = \Lambda^{1-j}\,\partial_t^{j-1}v$, $j = 1, \ldots, m^+$, and call $\mathbf{v}$ the vector with components v^j. We then denote by $J\mathbf{v}$ the m^--vector whose components are all zero, except the last one, equal to $\Lambda^{1-m^-}v$. With this notation equation (3.24) reads

$$\partial_t \mathbf{u} - A^-(t)\mathbf{u} = J\mathbf{v}. \tag{3.32}$$

According to (3.31) we have (cf. (3.10) and (3.22))

$$\det\{zI - \sigma[A^-(t)]\} = \sigma(M^{-0})(x, t, \xi, z), \tag{3.33}$$

where the determinant is now computed in the complex field. We have taken advantage of the fact that the principal symbols of the operators $M_j^-(t)$ are scalar multiples of the identity of H.

On the other hand, we note that according to (3.20),

$$\Lambda^{1-m^+}Ru = \sum_{j=1,\ldots,m^-} \Lambda^{1-m^+}R_j(t)\Lambda^{m^- - j}u^{m^- - j+1}. \tag{3.34}$$

We may then denote by $\mathcal{R}\mathbf{u}$ the m^+-vector with components all zero, except the last one, equal to $\Lambda^{1-m^+}Ru$. It is clear that

(3.35) *$\mathcal{R}$ is a regularizing operator in X, depending smoothly on $t \in [0, T[$, valued in $L(H \otimes \mathbb{C}^{m^-}; H \otimes \mathbb{C}^{m^+})$.*

We now denote by $\mathbf{g}$ the m^+-vector with all components equal to zero, except the last one, equal to $\Lambda^{1-m^+}f$. Thus equation (3.25) reads

$$\partial_t \mathbf{v} - A^+(t)\mathbf{v} = \mathbf{g} - \mathcal{R}\mathbf{u}. \tag{3.36}$$

We have

$$\det\{zI - \sigma[A^+(t)]\} = \sigma(M^{+0})(x, t, \xi, z). \tag{3.37}$$

The relations (3.33) and (3.37) show that the eigenvalues of $A^{-}(t)$ and those of $-A^{+}(t)$ stay in the open half-plane $\mathbb{C}_{-}$. They are the roots z_k^{-}, and the negatives $-z_k^{+}$, of the polynomial $\sigma(P^0)(x, t, \xi, z)$. We ought to emphasize that we are here viewing $\sigma(A^{\pm}(t))$ as matrices of size $m^{\pm} \times m^{\pm}$. To have them as symbols valued in the space of linear mappings $H \otimes \mathbb{C}^{m^{\pm}} \to H \otimes \mathbb{C}^{m^{\pm}}$, one must then tensor them (on the left) with the identity of H.

We may thus state the following:

(3.38) *The basic hypothesis of Section* 1, (1.5), *is verified by* $A^{-}(t)$ *and by* $-A^{+}(t)$.

The next step is to adjoin "initial conditions" to equation (3.23):

$$B_j(x, D_x, \partial_t)u|_{t=0} = h_j \qquad (j = 1, \dots, \nu). \tag{3.39}$$

Here $h_j \in \mathcal{D}'(X; H)$ for each j, and

$$B_j(x, D_x, \partial_t) = \sum_{k=0,\dots,d_j} B_{j,k}(x, D_x)\,\partial_t^k, \tag{3.40}$$

where, for each choice of $j = 1, \dots, \nu$, $k = 1, \dots, d_j$, $B_{j,k}(x, D_x)$ is a pseudodifferential operator in X, valued in $L(H)$.

First we divide each $B_j = B_j(x, D_x, \partial_t)$ by P, using the fact that P is a monic polynomial with respect to ∂_t:

$$B_j = Q'_j P + B'_j. \tag{3.41}$$

Note that the degree of B'_j as a polynomial in ∂_t does not exceed $m - 1$. We then replace the conditions (3.39) by

$$B'_j u|_{t=0} = h_j - (Q'_j f)|_{t=0}. \tag{3.42}$$

Next we divide B'_j by M^{-}:

$$B'_j = Q_j M^{-} + B_j^{\#} \qquad (j = 1, \dots, \nu). \tag{3.43}$$

Here not only is the degree of $B_j^{\#} \leq m^{-} - 1$ (as a polynomial in ∂_t) but also now $\deg Q_j \leq \deg B'_j - \deg M^{-} \leq m - m^{-} - 1 = m^{+} - 1$. By virtue of (3.24) we may replace the conditions (3.42) by

$$B_j^{\#} u|_{t=0} = h_j - (Q'_j f)|_{t=0} - (Q_j v)|_{t=0}. \tag{3.44}$$

Let us denote by $\mathbf{h}_{\#}$ the ν-vector with components $h_j - (Q'_j f)|_{t=0}$ ($j = 1, \dots, \nu$), by $\mathcal{Q}\mathbf{v}(0)$ the one whose components are $(Q_j v)|_{t=0}$. The fact that the degree (in ∂_t) of Q_j is $\leq m^{+} - 1$ implies that $\mathcal{Q}$ may indeed be regarded as a $\nu \times m^{+}$ matrix, since

$$Q_j v = \sum_{k=0,\dots,m^{+}-1} Q_{j,k}(t)\,\partial_t^k v = \sum_{k=1}^{m^{+}} Q_{j,k-1}(t)\Lambda^{k-1} v^k. \tag{3.45}$$

Finally we may rewrite the initial conditions (3.39) in the manner

$$\mathscr{B}\mathbf{u}(0) = \mathbf{h}_{\#} - \mathscr{Q}\mathbf{v}(0), \tag{3.46}$$

where $\mathscr{B}$ is the pseudodifferential operator in X, valued in the space of $\nu \times m^-$ matrices with entries in $L(H)$ defined as follows. If one writes

$$B_j^{\#} = \sum_{k=0}^{m^- -1} B_{j,k}^{\#}(t)\, \partial_t^k, \tag{3.47}$$

we have

$$B_j^{\#} u = \sum_{k=1}^{m^-} B_{j,k-1}^{\#}(t)\Lambda^{k-1} u^k. \tag{3.48}$$

Consequently, if $\mathscr{B}_{j,k}$ ($j = 1, \ldots, \nu$, $k = 1, \ldots, m^-$) is a generic entry of $\mathscr{B}$, we have

$$\mathscr{B}_{j,k} = B_{j,k-1}^{\#}(0)\Lambda^{k-1}. \tag{3.49}$$

DEFINITION 3.1. *The pseudodifferential operator $\mathscr{B}$ on the boundary X will be called the Calderon operator of the boundary problem* (3.23)–(3.39).

Actually there is much leeway in the definition of $\mathscr{B}$, and we shall apply the name Calderon operator to any operator on the manifold X of the form $\mathscr{A}\mathscr{B}\mathscr{C}$, with $\mathscr{A}$, $\mathscr{C}$ elliptic classical pseudodifferential operators in X, valued in $L(H \otimes \mathbb{C}^\nu)$ and $L(H \otimes \mathbb{C}^{m^-})$ respectively.

Let us summarize what we have done so far in this chapter. The system of equations (3.23), (3.39),

$$(*) \qquad Pu = f, \qquad B_j(x, D_x, \partial_t)u|_{t=0} = h_j \qquad (1 \le j \le \nu),$$

has been transformed into the system (3.32), (3.36), (3.46):

$$(**) \qquad \begin{aligned} &\partial_t \mathbf{u} - A^-(t)\mathbf{u} = J\mathbf{v}, \qquad \mathscr{B}\mathbf{u}(0) = \mathbf{h}_{\#} - \mathscr{Q}\mathbf{v}(0); \\ &\partial_t \mathbf{v} - A^+(t)\mathbf{v} = \mathbf{g} - \mathscr{R}\mathbf{u}. \end{aligned}$$

Let us emphasize the fact that these equations are exact, insofar as $\Lambda\Lambda^{-1} = I$ exactly, which is possible when X is compact. Nevertheless, we have not quite succeeded in "triangulizing" the problem (*); we have only approximately done so, since the third equation in (**) still contains $\mathscr{R}\mathbf{u}$. But we shall see that this transformation still enables us to analyze some important aspects of (*).

Let us emphasize that there is a one-to-one correspondence between solutions of (*) and solutions of (**): the argument in the preceding pages has shown how to go from

$$u \in C^\infty([0, T[; \mathscr{D}'(X; H))$$

to

$$\mathbf{u} \in C^\infty([0, T[; \mathscr{D}'(X; H \otimes \mathbb{C}^{m^-})), \qquad \mathbf{v} \in C^\infty([0, T[; \mathscr{D}'(X; H \otimes \mathbb{C}^{m^+})).$$

Conversely one can go from $\mathbf{u}$ to u simply by taking the latter to be the first component of the former.

Appendix: More General Elliptic Systems

As before let Ω be an open and bounded subset of $\mathbb{R}^N$, whose boundary X is a C^∞ hypersurface; we assume that Ω lies on one side of X. We shall denote by x the variable point in X, by t a coordinate transversal to X, defined in a tubular neighborhood $\mathscr{T}$ of $X (t > 0$ defines $\Omega \cap \mathscr{T}$ in $\mathscr{T})$.

We shall suppose that we are given r^2 linear partial differential operators P_{jk} $(j, k = 1, \ldots, r)$, whose coefficients are C^∞ functions in $\Omega \cap \mathscr{T}$. Let us denote by $\sigma(P_{jk})$ the principal symbol of P_{jk}, and set

$$\text{(3.50)} \qquad P = (P_{jk})_{1 \le j,k \le r}, \qquad \sigma(P) = (\sigma(P_{jk}))_{1 \le j,k \le r}.$$

We shall make the following hypothesis:

(3.51) *There are* $2r$ *integers* d_j, d'_k $(1 \le j, k \le r)$ *such that, for every pair* (j, k),

$$\text{(3.52)} \qquad \deg \sigma(P_{jk}) \;(=\text{order of } P_{jk}) = d_j - d'_k$$

(with the agreement that $P_{jk} \equiv 0$ if $d'_k > d_j$).

It follows immediately from this that $\det \sigma(P)$ is a homogeneous symbol of degree

$$\text{(3.53)} \qquad m = \sum_{j=1}^{r} (d_j - d'_j).$$

The *ellipticity* of the system P is then expressed by the following property:

(3.54) $\det \sigma(P)$ *is elliptic in* $\Omega \cup \mathscr{T}$.

Moreover,

(3.55) $\det \sigma(P)$ *satisfies condition* (3.6),

where, of course, the number $T > 0$ is chosen small enough that the condition $0 \le t \le T$ defines a compact subset of $\mathscr{T}$.

Let $q = (q_{jk})_{1 \le j,k \le r}$ be the *cofactor matrix* of $\sigma(P)$. We have, and this can be taken as a definition of q,

$$\sigma(P)q = q\sigma(P) = \det \sigma(P)\, I_r, \tag{3.56}$$

where I_r is the $r \times r$ identity matrix. For each (j, k) the symbol q_{jk} is a *polynomial* with respect to the fiber variables; let Q_{jk} be the homogeneous differential operator whose (total) symbol is equal to q_{jk}. We set

$$Q = (Q_{jk})_{1 \le j,k \le r}, \tag{3.57}$$

and thus

$$QP = I_r P^0 + R, \qquad PQ = IP^0 + R', \tag{3.58}$$

where P^0 is a *scalar* differential operator of order m with principal symbol $\det \sigma(P)$ (we may for instance take P^0 to be the unique *homogeneous* such operator), and R and R' are matrix-valued operators of order $\le m - 1$. In other words, both QP and PQ satisfy the hypotheses of Section 3, most notably, (3.1), (3.2), (3.6).

Suppose then that the equations $(*)$ hold, where now P is the operator in (3.50). This implies

$$QPu = Qf, \qquad Q(B_j u - h_j)|_{t=0} = 0, \quad j = 1, \ldots, \nu. \tag{3.59}$$

Conversely, suppose that there is a v, say in $C^\infty(\bar{\Omega}; \mathbb{C}^r)$, satisfying

$$PQv = f, \qquad B_j Qv|_{t=0} = h_j, \quad j = 1, \ldots, \nu. \tag{3.60}$$

Then, obviously, $u = Qv$ shall satisfy $(*)$. This shows that it might be of interest to study the boundary operators QB_j or B_jQ, in relation to the operators QP or PQ respectively, a study that enters in the framework of the preceding sections.

For details on the subject of elliptic systems we refer to Volevič [1].

4. Hypoelliptic Boundary Problems

Because of the imperfect "triangulization" in $(**)$, it might be difficult to solve the latter problem, no less difficult than it is to solve, directly and exactly, problem $(*)$. But in a number of questions one can take advantage of the special feature of the triangulization, namely that it is "perfect" if one neglects regularizing operators. One such question is that of regularity up to the boundary of the solutions, which we define locally (and even microlocally, if one wishes). An approach that works is to replace (3.36) by the approximation

$$\partial_t \mathbf{v}_\# - A^+(t)\mathbf{v}_\# = \mathbf{g} \qquad \textit{in } X \times [0, T[, \tag{4.1}$$

and also to modify (3.32) and (3.46) accordingly:

$$\partial_t \mathbf{u}_\# - A^-(t)\mathbf{u}_\# = J\mathbf{v}_\# \qquad \textit{in } X \times [0, T[, \tag{4.2}$$

$$\mathscr{B}\mathbf{u}_\#(0) = \mathbf{h}_\# - \mathscr{Q}\mathbf{v}_\#(0) \qquad \textit{in } X. \tag{4.3}$$

We can easily solve this modified problem, provided that we strengthen our hypotheses on $A^\pm(t)$ and on f and, by way of consequence, on $\mathbf{g}$. Specifically, we assume that all these functions are smooth, with respect to t, in the *closed* interval $[0, T]$. This is of no great importance in the applications; it can be achieved by slightly decreasing T.

Since $-A^+(t)$ satisfies the basic assumption (1.5), we can solve the *backward* Cauchy problem for equation (4.1), starting at $t = T$, and thus write, for an arbitrary choice of $\mathbf{v}_\#(T) \in \mathscr{D}'(X; H)$,

$$\mathbf{v}_\#(t) = U^+(t, T)\mathbf{v}_\#(T) - \int_t^T U^+(t, t')\mathbf{g}(t')\, dt', \tag{4.4}$$

where $U^+(t, t')$ is the relevant parametrix. We can then put the solution $\mathbf{v}_\#$ of (4.1) thus obtained into (4.2)–(4.3) and solve the *forward* Cauchy problem, starting at $t = 0$ (cf. (1.45)):

$$\mathbf{u}_\#(t) = U^-(t, 0)\mathbf{u}_\#(0) + \int_0^t U^-(t, t')J\mathbf{v}_\#(t')\, dt', \tag{4.5}$$

where $U^-(t, t')$ is the parametrix for (4.2)–(4.3). In doing this we have taken advantage of the fact that $A^-(t)$ satisfies (1.5).

Of course, once we use the parametrices for these Cauchy problems, we obtain only approximate solutions, modulo errors that involve regularizing operators acting on the data. The question is then to go from these solutions to solutions of (**), and from there to solutions of our original problem (*). Here we shall focus on the regularity of these solutions and show how the regularity of one set of solutions determines that of the other.

LEMMA 4.1. *Let* $\mathbf{u}$, $\mathbf{v}$ *be solutions of* (**), *where we assume that* $\mathbf{g}$ *belongs to* $C^\infty([0, T[; \mathscr{D}'(X; H \otimes \mathbb{C}^{m^+}))$, *and let* $\mathbf{u}_\#$, $\mathbf{v}_\#$ *be defined by* (4.4)–(4.5). *Then*

$$\mathbf{v} - \mathbf{v}_\# \in C^\infty(X \times [0, T[; H \otimes \mathbb{C}^{m^+}). \tag{4.6}$$

If moreover we assume that $\mathbf{u}_\#(0) - \mathbf{u}(0) \in C^\infty(X; H \otimes \mathbb{C}^{m^-})$, *then*

$$\mathbf{u} - \mathbf{u}_\# \in C^\infty(X \times [0, T[; H \times \mathbb{C}^{m^-}). \tag{4.7}$$

PROOF. It is very simple: it suffices to observe that $\mathbf{u}_1 = \mathbf{u} - \mathbf{u}_\#$, $\mathbf{v}_1 = \mathbf{v} - \mathbf{v}_\#$ satisfy the equations

$$[\partial_t - A^-(t)]\mathbf{u}_1 = J\mathbf{v}_1 + \boldsymbol{\phi}, \tag{4.8}$$

$$[\partial_t - A^+(t)]\mathbf{v}_1 = \boldsymbol{\psi}, \tag{4.9}$$

where $\boldsymbol{\phi}$ and $\boldsymbol{\psi}$ are linear combinations of $\mathbf{u}$, $\mathbf{v}$, $\mathbf{u}_1$, $\mathbf{v}_1$ with coefficients that are matrices, of the appropriate size, with regularizing operators as entries. We apply then formula (1.45) to $\mathbf{u}_1$ and the analogue for equation (4.9) to $\mathbf{v}_1$:

$$\mathbf{u}_1(t) = U^-(t, 0)\mathbf{u}_1(0) + \int_0^t U^-(t, t')[J\mathbf{v}_1(t') + \boldsymbol{\phi}(t')]\, dt' \\ -R_o^-(t)\mathbf{u}_1 + \int_0^t R^-(t, t')\mathbf{u}_1(t')\, dt', \tag{4.10}$$

$$\mathbf{v}_1(t) = U^+(t, T)\mathbf{v}_1(T) - \int_t^T U^+(t, t')\boldsymbol{\psi}(t')\, dt' - R_o^+(t)\mathbf{v}_1(t) \\ + \int_t^T R^+(t, t')\mathbf{v}_1(t')\, dt', \tag{4.11}$$

where the R's stand for regularizing operators, depending smoothly on t or on (t, t'). First (4.6) follows at once from (4.11) if we recall that $U^+(t, T)$ is regularizing as soon as $t < T$. Taking this fact into account in (4.10) we deduce (4.7) if we assume that $\mathbf{u}_1(0)$ is C^∞ in X. □

DEFINITION 4.1. *The problem* (**) *is said to be hypoelliptic if given any open subset* $\mathcal{O}$ *of* X *and any data* $\mathbf{g} \in C^\infty([0, T]; \mathcal{D}'(X; H \otimes \mathbb{C}^{m^+}))$, $\mathbf{h}_\# \in \mathcal{D}'(X; H \otimes \mathbb{C}^\nu)$ *whose restrictions to* $\mathcal{O}$ *are smooth, i.e.,*

$$\mathbf{g} \in C^\infty(\mathcal{O} \times [0, T]; H \otimes \mathbb{C}^{m^+}), \qquad \mathbf{h}_\# \in C^\infty(\mathcal{O}; H \otimes \mathbb{C}^\nu), \tag{4.12}$$

then every solution $(\mathbf{u}, \mathbf{v})$ *of* (**) *such that*

$$\mathbf{u} \in C^\infty([0, T]; \mathcal{D}'(X; H \otimes \mathbb{C}^{m^-})), \qquad \mathbf{v} \in C^\infty([0, T]; \mathcal{D}'(X; H \otimes \mathbb{C}^{m^+})) \tag{4.13}$$

is in fact smooth in $\mathcal{O}$ *for* $t < T$, *i.e.,*

$$\mathbf{u} \in C^\infty(\mathcal{O} \times [0, T[; H \otimes \mathbb{C}^{m^-}), \qquad \mathbf{v} \in C^\infty(\mathcal{O} \times [0, T[; H \otimes \mathbb{C}^{m^+}). \tag{4.14}$$

REMARK 4.1. There would be no gain in generality in relaxing condition (4.13) and allowing **u** and **v** to be distributions in t. Indeed (see Remark 1.2), our hypotheses on **g** and the equations (**) themselves would automatically imply (4.13). One could allow **g** not to be smooth with respect to t outside $\mathcal{O}$, but what counts is its smoothness in $\mathcal{O}$.

REMARK 4.2. The hypoellipticity of problem (**) is obviously equivalent to that of our original problem (*), defined in evident manner.

REMARK 4.3. If one prefers to consider "wave-front sets" hypoellipticity, namely the property that the operator under study preserves the wave-front sets, and not merely the singular supports, it should be said that the counterparts of all the statements in the present section, in particular Theorem 4.1, are valid for this concept.

Finally we recall that the operator $\mathcal{B}$ in X is said to be hypoelliptic if it preserves the singular supports.

THEOREM 4.1. *Problem* (*) (*or, equivalently, problem* (**)) *is hypoelliptic if and only if the Calderon operator* $\mathcal{B}$ *is hypoelliptic.*

PROOF. Suppose first that $\mathcal{B}$ is hypoelliptic. By the fact that pseudodifferential operators, here $U^+(t, t')$, are pseudolocal, we derive from (4.4) that $\mathbf{v}_\#$ is smooth in $\mathcal{O} \times [0, T[$; remember that $U^+(t, T)$ is regularizing for $t < T$. By the first part of Lemma 4.1 we conclude that **v** is also smooth in $\mathcal{O} \times [0, T[$; in particular $\mathbf{v}(0) \in C^\infty(\mathcal{O}; H \otimes \mathbb{C}^{m^+})$. We take this into account in the relation $\mathcal{B}\mathbf{u}(0) = \mathbf{h}_\# - \mathcal{Q}\mathbf{v}(0)$ and derive that $\mathbf{u}(0) \in C^\infty(\mathcal{O}; H \otimes \mathbb{C}^{m^-})$. It then suffices to apply the last part of Lemma 4.1 with $\mathbf{u}_\#(0) = \mathbf{u}(0)$.

Suppose now that $\mathcal{B}$ is *not* hypoelliptic: there is then a distribution $\mathbf{u}_o$ in X, valued in $H \otimes \mathbb{C}^{m^-}$, whose restriction to some open set $\mathcal{O} \subset X$ is not C^∞ but such that the one of $\mathcal{B}\mathbf{u}_o$ is C^∞. Set $\mathbf{w}(t) = U^-(t, 0)\mathbf{u}_o$. Then

$$\text{(4.15)} \qquad \mathbf{v}_* = \left[\frac{\partial}{\partial t} - A^-(t)\right]\mathbf{w} \textit{ is a } C^\infty \textit{ function } X \times [0, T] \to H \otimes \mathbb{C}^{m^-}.$$

Note that $v_*^j = \partial_t w^j - \Lambda w^{j+1}$ if $j < m^-$. Let us define inductively

$$\text{(4.16)} \qquad u^{m^-} = w^{m^-},$$

$$\text{(4.17)} \quad u^j = w^j + \int_0^t [\Lambda(u^{j+1} - w^{j+1})(t') - v_*^j(t')]\, dt' \qquad \textit{if } j < m^-.$$

By (descending) induction on j and by (4.15) we see that $\mathbf{u}-\mathbf{w}$ is C^∞ in $X \times [0, T]$ with values in $H \otimes \mathbb{C}^{m^-}$. We have

(4.18) $$\partial_t u^j = \Lambda u^{j+1} \qquad \textit{if } j < m^-.$$

Let χ denote the last (i.e., the m^-th) component of $\partial_t \mathbf{u} - A^-(t)\mathbf{u}$, and define $v = \Lambda^{m^- -1}\chi$, and the m^+-vector $\mathbf{v} = (v^1, \ldots, v^{m^+})$ as before, by setting $v^j = \Lambda^{1-j}\,\partial_t^{j-1} v$. Equation (3.32) is automatically satisfied. Since

$$[\partial_t - A^-(t)]\mathbf{u} = \mathbf{v}_* + [\partial_t - A^-(t)](\mathbf{u} - \mathbf{w}),$$

we see that $\mathbf{v} \in C^\infty(X \times [0, T]; H \otimes \mathbb{C}^{m^+})$. We then set $\mathbf{g} = [\partial_t - A^+(t)]\mathbf{v} + \mathscr{R}\mathbf{u}$; we have $\mathbf{g} \in C^\infty(X \times [0, T]; H \otimes \mathbb{C}^{m^+})$. On the other hand, $\mathbf{u}(0) = \mathbf{w}(0) = \mathbf{u}_o$. By virtue of our hypothesis about $\mathscr{B}\mathbf{u}_o$, $\mathbf{h}_\# = \mathscr{B}\mathbf{u}_o + \mathscr{Q}\mathbf{v}(0)$ is a C^∞ mapping $\mathscr{O} \to H \otimes \mathbb{C}^\nu$. Thus (4.12) and, of course, (4.13) are true, but (4.14) is not. □

5. Globally Hypoelliptic Boundary Problems. Fredholm Boundary Problems

Throughout the present section we suppose that the C^∞ manifold X is the boundary of a bounded open subset Ω of $\mathbb{R}^{n+1}$. The function t used in the preceding sections is defined and has a nonvanishing differential in a tubular neighborhood $\mathscr{T}$ of X. The equation $t = 0$ defines X, while $t > 0$ defines $\Omega \cup \mathscr{T}$. We assume that P is an elliptic pseudodifferential operator in $\Omega \cap \mathscr{T}$, classical, of order m, valued in $L(H)$, having the form (3.1) and satisfying conditions (3.2) and (3.6) in the set $\mathscr{T}$. It is imperative, in what we are going to say, that the dimension of the Hilbert space H be *finite*. Our purpose is to take a closer look at the boundary problem (*) of Section 3.

We shall use the Sobolev spaces of distributions in X (and in Ω, see Section 2) valued in various finite-dimensional vector spaces. As usual, we denote by $\| \ \|_s$ the sth Sobolev norm. In X we may assume that it is defined by the Riemannian structure induced by $\mathbb{R}^{n+1}$.

DEFINITION 5.1. *The Calderon operator* $\mathscr{B}$ (*Definition* 3.1) *is said to be globally hypoelliptic in* X *if, given any distribution* u *in* X, *valued in* $H \otimes \mathbb{C}^{m^-}$,

(5.1) $$\mathscr{B}u \in C^\infty(X; H \otimes \mathbb{C}^\nu) \textit{ implies } u \in C^\infty(X; H \otimes \mathbb{C}^{m^-}).$$

If $\mathscr{B}$ is hypoelliptic it is globally hypoelliptic. The converse is not true (see Hörmander [18]).

PROPOSITION 5.1. *Suppose that* $\mathscr{B}$ *is globally hypoelliptic in* X. *Then*

(i) $\operatorname{Ker} \mathscr{B}$ *in* $\mathscr{D}'(X; H \otimes \mathbb{C}^{m^-})$ *is finite dimensional and, of course, contained in* $C^\infty(X; H \otimes C^{m^-})$;

(ii) *to any pair of real numbers,* s, s', *there is a third number,* s'', *and a constant* $C > 0$ *such that*

$$\|u\|_s \leq C(\|u\|_{s'} + \|\mathscr{B}u\|_{s''}), \qquad \forall u \in C^\infty(X; H \otimes \mathbb{C}^{m^-}); \tag{5.2}$$

(iii) *the range of* $\mathscr{B}: C^\infty(X; H \otimes \mathbb{C}^{m^-}) \to C^\infty(X; H \otimes \mathbb{C}^\nu)$ *is closed.*

PROOF. Let us equip $C^\infty(X; H \otimes \mathbb{C}^{m^-})$ with the topology defined by the norms

$$u \mapsto \|u\|_{s'} + \|\mathscr{B}u\|_{s''}, \tag{5.3}$$

where s' is fixed and s'' ranges over $\mathbb{R}$. It is weaker than the standard C^∞ topology. Let us show that the underlying metric space is complete.

Let $\{u_j\}$ be a Cauchy sequence for the norms (5.3). It converges in $H^{s'}(X; H \otimes \mathbb{C}^{m^-})$, to an element u_o, and the $\mathscr{B}u_j$ converge in $C^\infty(X; H \otimes \mathbb{C}^\nu)$, necessarily to $\mathscr{B}u_o$. By (5.1) u_o must belong to $C^\infty(X; H \otimes \mathbb{C}^{m^-})$, and our contention follows.

By the open mapping theorem the standard topology and that defined by the norms (5.3) are identical, which is what (ii) expresses. If we restrict (5.2) to the subspace $\operatorname{Ker} \mathscr{B}$, we see that the topologies induced on the latter by all the spaces $H^s(X; H \otimes \mathbb{C}^{m^-})$ are identical. Since the injection of H^s into $H^{s'}$, when $s' < s$, is compact (Chapter II, Corollary 1.4), $\dim \operatorname{Ker} \mathscr{B} < +\infty$.

Property (iii) is a simple consequence of (ii) and of Corollary 1.4, Chapter II. We leave its proof to the reader (cf. proof of Theorem 5.4). ☐

There is an evident definition of global hypoellipticity for boundary problems, whether they are of the kind (*) or of the kind (**); for instance, concerning the latter, it suffices to modify Definition 4.1; instead of considering any open subset $\mathcal{O}$ of X, put $\mathcal{O} = X$. If (**) is obtained from (*) by the transformation described in Section 3, either both are globally hypoelliptic or neither is (cf. Remark 4.2).

The same modification in the proof of Theorem 4.1 (putting $\mathcal{O} = X$) immediately yields the following theorem.

THEOREM 5.1. *Problem* (*) (*and therefore also problem* (**)) *is globally hypoelliptic if and only if the Calderon operator* $\mathscr{B}$ *is globally hypoelliptic in* X.

Consider now the space of H-valued distributions u in $\Omega \cup \mathcal{T}$ satisfying

$$(5.4) \qquad Pu = 0 \qquad in\ \Omega.$$

Since P is elliptic, u is a C^∞ function in Ω; there is of course no reason, in general, that it should be a C^∞ function in $\bar{\Omega}$. But if we know that its t-derivatives of order $\leq m - 1$ are continuous functions of t in $[0, T]$ (for $T > 0$ suitably small) valued in $\mathcal{D}'(X; H)$, then equation (5.4) implies that the same is true of *all* t-derivatives of u (cf. Remark 1.2). Under these circumstances the boundary conditions

$$(5.5) \qquad B_j(x, D_x, \partial_t)u|_{t=0} = 0, \qquad j = 1, \ldots, \nu,$$

make good sense. We shall denote by $\mathcal{N}$ the space of distributions u of the kind just described that satisfy (5.4)–(5.5).

REMARK 5.1. If u belongs to the Sobolev space $H^m(\Omega; H)$, its t-derivatives of order $\leq m - 1$ are C^0 functions of t in $[0, T]$ valued in $L^2(X; H)$.

THEOREM 5.2. *If the Calderon operator $\mathcal{B}$ is globally hypoelliptic, the "null space" $\mathcal{N}$ is a finite-dimensional subspace of $C^\infty(\bar{\Omega}; H)$.*

PROOF. That $\mathcal{N} \subset C^\infty(\bar{\Omega}; H)$ follows at once from Theorem 5.1. Select an integer $M \geq m$ such that, for each $j = 1, \ldots, \nu$, $u \mapsto B_j(x, D_x, \partial_t)u|_{t=0}$ is a continuous linear map of $H^M(\Omega; H)$ into, say, $H^0(X; H)$ (cf. Theorem 2.2). Then $\mathcal{N}$ is a closed linear subspace of $H^M(\Omega; H)$ and also of $H^{M+1}(\Omega; H)$. Since the latter space injects compactly into the former (Corollary 2.2), $\mathcal{N}$ must be finite dimensional. □

Let m_j denote the (total) order of $B_j(x, D_x, \partial_t)$. By virtue of (3.49) we have

$$(5.6) \qquad \text{Order } \mathcal{B}_{jk} \leq m_j, \qquad k = 1, \ldots, m^-.$$

Let s and s' be two arbitrary real numbers, with $s' < s$. We shall denote by s'' a third real number such that, for some constant $C > 0$,

$$(5.7) \qquad \|\mathbf{w}\|_{s+m-1/2} \leq C\left(\|\mathbf{w}\|_{s'} + \sum_{j=1}^{\nu} \left\| \sum_{k=1}^{m^-} \mathcal{B}_{jk} w^k \right\|_{s''+m-m_j-1/2}\right),$$

$$\textit{for all } \mathbf{w} = (w^1, \ldots, w^{m^-}) \in C^\infty(X; H \otimes \mathbb{C}^{m^-}).$$

If $\mathcal{B}$ is globally hypoelliptic, such a real number s'' certainly exists, according to Proposition 5.1. We may and shall assume that $s'' \geq s$.

REMARK 5.2. The number s' in (5.7) can be replaced by any other real number without modifying s'', but possibly after modifying the constant C. It suffices to apply Proposition 1.3, Chapter II. If s_1' is any real number (most of the time small in comparison to s'), then for some suitable $C' > 0$ one writes

$$\|\mathbf{w}\|_{s'} \le (2C)^{-1}\|\mathbf{w}\|_{s+m-1/2} + C'\|\mathbf{w}\|_{s_1'},$$

which yields (assuming that $C' \ge 1$)

$$(5.7') \qquad \|\mathbf{w}\|_{s+m+1/2} \le 2CC'\left(\|\mathbf{w}\|_{s_1'} + \sum_{j=1}^{\nu}\left\|\sum_{k=1}^{m^-}\mathscr{B}_{jk}w^k\right\|_{s''+m-m_j-1/2}\right).$$

We shall repeatedly take advantage of this leeway in the choice of s' in what follows.

We shall make another hypothesis, one which we can dispense with, but at the cost of more complicated expressions later on:

(5.8) *For all $j = 1, \ldots, \nu$, the degree of $B_j(x, D_x, \partial_t)$ as a polynomial in ∂_t does not exceed $m - 1$.*

When this hypothesis is not satisfied, in order to apply the following conclusions it suffices to replace $B_ju|_{t=0}$ by $B_ju|_{t=0} - Q_j'f|_{t=0}$ (cf. (3.42)). Its effect is that the operators Q_j' of (3.41) vanish identically.

Until otherwise specified, $(\ ,\ )_s$ and $\|\ \|_s$ stand for the inner product and the norm in the Sobolev space on X. In order to distinguish, we shall denote momentarily by $|||\ |||_s$ the norm in the Sobolev space on $\mathbb{R}^{n+1}$.

We select a C^∞ function χ on the real line, $\chi(t) = 1$ for $|t| < \frac{1}{4}T$, $\chi(t) = 0$ for $|t| > \frac{1}{2}T$. Actually we regard χ as a function of (x, t) in $\mathscr{T}$, independent of x, the variable in X, and therefore also as an element of $C_c^\infty(\mathbb{R}^{n+1})$. We choose another function $\psi \in C_c^\infty(\Omega)$, $\psi = 1$ on the intersection of supp$(1 - \chi)$ with Ω.

In Ω we write

$$P[(1-\chi)u] = (1-\chi)f - [P, \chi]u$$

and the use the facts that P is elliptic of order m and that $[P, \chi]$ has order $m - 1$ and is supported in the region $\frac{1}{4}T \leqq t \leqq \frac{1}{2}T$, hence in supp$(1 - \chi)$. We obtain at once the *interior estimates*:

$$(5.9) \qquad |||(1-\chi)u|||_{s+m} \le C(|||\psi f|||_s + |||\psi u|||_{s+m-1})$$

valid for any real number s. (C depends of course on s.)

We shall now establish the *estimates at the boundary*, under hypothesis (5.7), or, equivalently, (5.7'). This hypothesis "fixes" the value of the real number s''; we shall avail ourselves of Remark 5.2 to modify the auxiliary number s' according to our needs.

We concentrate on χu and write

$$P(x, t, D_x, \partial_t)(\chi u) = f_\# \quad in\ \Omega \cap \mathcal{T}, \tag{5.10}$$

$$B_j(x, D_x, \partial_t)u|_{t=0} = h_j \quad (1 \leq j \leq \nu).$$

Of course,

$$f_\# = \chi f + [P, \chi]u. \tag{5.11}$$

We define the vector-valued functions $\mathbf{u}, \mathbf{v}, \mathbf{g}, \mathbf{h}_\#$ in the manner described in Section 3, so that (**) holds. We start, not from $u^1 = u$, but from $u^1 = \chi u$.

We begin by applying (1.61), recalling that $\partial_t \mathbf{u} - A^-(t)\mathbf{u} = J\mathbf{v}$ and that all components of $J\mathbf{v}$ are zero, except the last one, equal to $\Lambda^{1-m^-}v^1$:

$$\int_0^T \|\partial_t^k \mathbf{u}\|^2_{s+m-k}\, dt \leq C\Big\{\|\mathbf{u}(0)\|^2_{s+m-1/2} + \tag{5.12}$$

$$\sum_{j \leq \sup(k-1,0)} \int_0^T \|\partial_t^j v^1\|^2_{s+m^+-j}\, dt + R_{s'}(\mathbf{u})\Big\}.$$

We have used, and shall use again, the notation

$$R_{s'}(\mathbf{u}) = \int_0^T \|\mathbf{u}(t)\|^2_{s'}\, dt.$$

In order to estimate $\|\mathbf{u}(0)\|^2_{s+m-1/2}$ we are going to avail ourselves of (5.7), or rather (5.7′) (Remark 5.2). Note that

$$\|\mathbf{u}(0)\|^2_{s'-1/2} = -2\,\mathrm{Re} \int_0^T \{(\mathbf{u}, A^-\mathbf{u})_{s'-1/2} + (\mathbf{u}, \partial_t\mathbf{u} - A^-\mathbf{u})_{s'-1/2}\}\, dt,$$

whence

$$\|\mathbf{u}(0)\|^2_{s'-1/2} \leq C\{R_{s'}(\mathbf{u}) + R_{s'}(v^1)\}. \tag{5.13}$$

From (3.46) we derive

$$\Big\|\sum_{k=1}^{m^-} \mathcal{B}_{jk} u^k(0)\Big\|_\sigma \leq \|h_j\|_\sigma + \|Q_j v(0)\|_\sigma. \tag{5.14}$$

We have used the notation (3.45). We go back to (3.43), keeping in mind that $B_j' = B_j$ under hypothesis (5.8). We see that the total order of Q_j is $m_j - m^-$, and this is also the order of the pseudodifferential operators on X, $Q_{j,k}(t)\Lambda^k$ $(k = 0, \ldots, m^+ - 1)$. We derive that

$$\|Q_j v(0)\|_\sigma \leq C\|\mathbf{v}(0)\|_{\sigma+m_j-m^-}. \tag{5.15}$$

We take $\sigma = s'' + m - m_j - \frac{1}{2}$ in (5.14) and (5.15). We get

$$(5.16)\quad \left\| \sum_{k=1}^{m^-} \mathcal{B}_{jk} u^k(0) \right\|_{s''+m-m_j-1/2} \leq \|h_j\|_{s''+m-m_j-1/2} + C\|\mathbf{v}(0)\|_{s''+m^+-1/2}.$$

In order to estimate the norm of $\mathbf{v}(0)$ we apply the analogue of formula (1.94) for equation (3.36):

$$(5.17)\quad \mathbf{v}(t) = -\int_t^T U_o^+(t, t')[\mathbf{g}(t') - \mathcal{R}\mathbf{u}(t')]\, dt'.$$

We use the fact that all components of $\mathbf{g}$ are zero, except the last one, equal to $\Lambda^{1-m^+} f_{\#}$. We also use (5.11). We write

$$\mathbf{g} = \chi \mathbf{f}_o = \mathbf{g}_1,$$

where all components of $\mathbf{f}_o$ and $\mathbf{g}_1$ are zero, except the last ones, respectively equal to $\Lambda^{1-m^+} f$ and to $\Lambda^{1-m^+}[P, \chi]u$. Thus $\mathbf{g}_1$ vanishes identically for t outside the interval $[\frac{1}{4}T, \frac{1}{2}T]$. But if t stays in this interval, then $U_o^+(0, t)$ is regularizing (in X). Therefore

$$(5.18)\quad \left\| \mathbf{v}(0) + \int_0^T \chi(t) U_o^+(0, t)\mathbf{f}_o(t)\, dt \right\|_{\sigma-1/2}^2 \leq C \sum_{k=0}^{m-1} \int_0^T \|\partial_t^k u\|_{s'}^2\, dt,$$

whatever the real number s' (on which, of course, the constant C depends). We note that

$$\mathbf{w}(t) = -\int_t^T \chi(t') U_o^+(t, t')\mathbf{f}_o(t')\, dt'$$

is the solution of the problem

$$\partial_t \mathbf{w} - A^+ \mathbf{w} = \chi \mathbf{f}_o, \quad 0 \leq t \leq T; \qquad \mathbf{w}(T) = 0.$$

We reason as we did in the derivation of (5.13):

$$\|\mathbf{w}(0)\|_{\sigma-1/2}^2 = -2\,\mathrm{Re} \int_0^T (\mathbf{w}, A^+\mathbf{w})_{\sigma-1/2}\, dt - 2\,\mathrm{Re} \int_0^T (\mathbf{w}, \chi\mathbf{f}_o)_{\sigma-1/2}\, dt$$
$$\leq C \int_0^T (\|\mathbf{w}\|_\sigma^2 + \|\chi\mathbf{f}_o\|_{\sigma-1}^2)\, dt.$$

We combine this with the analogue of (1.61), with $t = T$ in place of $t = 0$ and with $k = 0$:

$$\int_0^T \|\mathbf{w}\|_\sigma^2\, dt \leq C \int_0^T \|\chi\mathbf{f}_o\|_{\sigma-1}^2\, dt,$$

whence

$$\|\mathbf{w}(0)\|^2_{\sigma-1/2} \le C \int_0^T \|\chi f\|^2_{\sigma-m^+}\, dt.$$

We combine this with (5.18) and put $\sigma = s'' + m^+$:

$$\|\mathbf{v}(0)\|^2_{s''+m^+-1/2} \le C\left\{\int_0^T \|\chi f\|^2_{s''}\, dt + \sum_{k=0}^{m-1} R_{s'}(\partial_t^k u)\right\}. \tag{5.19}$$

We combine (5.7′), (5.13), (5.16), and (5.19) and get

$$\|\mathbf{u}(0)\|^2_{s+m-1/2} \le C\left\{\sum_{j=1}^{\nu} \|h_j\|^2_{s''+m-m_j-1/2} + \int_0^T \|\chi f\|^2_{s''}\, dt + \sum_{k=0}^{m-1} R_{s'}(\partial_t^k u) + R_{s'}(v^1)\right\}. \tag{5.20}$$

We must now estimate the sum on the right-hand side of (5.12). But we shall use (5.12) only for $k = 1, \ldots, m^+$. If then $j \le k - 1$ we note that $\partial_t^j v^1 = \Lambda^j v^{j+1}$, and therefore the sum in question does not exceed $\int_0^T \|\mathbf{v}\|^2_{s+m^+}\, dt$. In order to estimate the latter quantity we apply the analogue of (1.61) with $k = 0$ and $t = T$ in the place of $t = 0$:

$$\begin{aligned} \int_0^T \|\mathbf{v}\|^2_{s+m^+}\, dt &\le C\left\{\int_0^T \|\mathbf{g} - \mathscr{R}\mathbf{u}\|^2_{s+m^+-1}\, dt + R_{s'}(\mathbf{v})\right\} \\ &\le C'\left\{\int_0^T \|f_{\#}\|^2_s\, dt + R_{s'}(\mathbf{u}) + R_{s'}(\mathbf{v})\right\} \\ &\le C''\left\{\int_0^T \|\chi f\|^2_s\, dt + |||\psi u|||^2_{s+m-1} + R_{s'}(\mathbf{u}) + R_{s'}(\mathbf{v})\right\}. \end{aligned} \tag{5.21}$$

From the definitions of the u^j and the v^j it follows at once that, given any real number s_1', we can select s' to have

$$R_{s'}(\mathbf{u}) + R_{s'}(\mathbf{v}) \le C \sum_{j=0}^{m-1} R_{s_1'}(\partial_t^j u^1). \tag{5.22}$$

We take (5.22) into account in (5.20) and (5.21) and combine the resulting estimates with (5.12), where we take $k = 0, \ldots, m^+$. Recalling that $u^j = \Lambda^{1-j}\partial_t^{j-1} u^1, j = 1, \ldots, m^-$, and that $u^1 = \chi u$, we obtain

$$\sum_{k=0}^{m-1} \int_0^T \|\chi\, \partial_t^k u\|^2_{s+m-k}\, dt \le C\left\{\sum_{j=1}^{\nu} \|h_j\|^2_{s''+m-m_j-1/2} + \int_0^T \|\chi f\|^2_{s''}\, dt + \sum_{k=0}^{m-1} \int_0^T \|\partial_t^k u\|^2_{s'}\, dt + |||\psi u|||^2_{s+m-1}\right\}. \tag{5.23}$$

We have tacitly assumed $s'' \geq s$ (it always is!). From the expression (3.1) of P in the tubular neighborhood $\mathcal{T}$ we derive

$$\|\chi\, \partial_t^{m+k} u\|_s \leq C\left\{\|\chi\, \partial_t^k f\|_s + \sum_{j=0}^{m-1} \|\chi\, \partial_t^{j+k} u\|_{s+m-j}\right\}. \tag{5.24}$$

We apply this and reason by induction on $k = 0, 1, \ldots$. Starting with (5.23) we get the *boundary estimates*

$$\begin{aligned}\sum_{k=0}^{N} \int_0^T \|\chi\, \partial_t^k u\|_{s+m-k}^2\, dt \leq C\Bigg\{ & \sum_{j=1}^{\nu} \|h_j\|_{s''+m-m_j-1/2}^2 + \int_0^T \|\chi f\|_{s''}^2\, dt \\ & + \sum_{j \leq \sup(N-m,0)} \int_0^T \|\chi\, \partial_t^j f\|_{s-j}^2\, dt \\ & + \sum_{k=0}^{m-1} \int_0^T \|\partial_t^k u\|_{s'}^2\, dt + |||\psi u|||_{s+m-1}^2 \Bigg\}.\end{aligned} \tag{5.25}$$

When s is an integer ≥ 0, the conjunction of the interior estimates (5.9) with the boundary estimates (5.25) takes a relatively simple form:

THEOREM 5.3. *Suppose that* (5.7) *holds for s an integer ≥ 0 and a suitable real number s''. There is a constant $C > 0$ such that, for all $u \in C^\infty(\bar{\Omega}; H)$,*

$$\begin{aligned}\|u\|_{H^{s+m}(\Omega;H)}^2 \leq C\Bigg\{ & \sum_{j=1}^{\nu} \|B_j u|_X\|_{H^{s''+m-m_j-1/2}(X;H)} + \|Pu\|_{H^s(\Omega;H)}^2 \\ & + \int_0^T \|\chi Pu\|_{H^{s''}(X;H)}^2\, dt + \|u\|_{H^{s+m-1}(\Omega;H)}^2 \Bigg\}.\end{aligned} \tag{5.26}$$

From this result we shall now derive the following theorem.

THEOREM 5.4. *If the Calderon operator $\mathcal{B}$ is globally hypoelliptic, the range of the map*

$$u \mapsto (Pu, \{B_j u|_X\}_{j=1,\ldots,\nu}) \tag{5.27}$$

from $C^\infty(\bar{\Omega}; H)$ to

$$C^\infty(\bar{\Omega}; H) \times C^\infty(X; H \otimes \mathbb{C}^\nu), \tag{5.28}$$

is closed.

PROOF. Let $\{u_k\}$ $(k = 1, 2, \ldots)$ be a sequence in $C^\infty(\bar{\Omega}; H)$ whose image under the map (5.27), which we shall denote by Φ, forms a convergent sequence $\{\Phi(u_k)\}$, in the Fréchet space (5.28). Let s be an arbitrary integer ≥ 0 and select s'' such that (5.26) holds. Suppose first that

$$M_k = \|u_k\|^2_{H^{s+m}(\Omega;H)} \leq C < +\infty.$$

Possibly substituting a subsequence for the sequence $\{u_k\}$, we may assume that it converges weakly in $H^{s+m}(\Omega; H)$ to some element u. If s is large enough $\Phi(u)$ makes sense and is the limit of the $\Phi(u_k)$. Thus $\Phi(u)$ belongs to (5.28); by Theorem 5.1, u belongs to $C^\infty(\bar{\Omega}; H)$.

Suppose now that the numbers M_k are not bounded. Again, possibly after substituting a subsequence for the sequence $\{u_k\}$, we may assume that the $M_k \nearrow +\infty$. Also, possibly after replacing each u_k by its orthogonal projection into the orthogonal in $H^{s+m}(\Omega; H)$, W^{s+m}, of the kernel $\mathcal{N}$ of (*) (which is finite dimensional, by Theorem 5.2), we may assume that u_k belongs to W^{s+m} for all k. Set $v_k = M_k^{-1} u_k$. The norm of every v_k in $H^{s+m}(\Omega; H)$ is equal to one; therefore a subsequence, which we can take to be the sequence $\{v_k\}$ itself, converges weakly in that space and by Corollary 2.2, converges strongly in $H^{s+m-1}(\Omega; H)$. On the other hand, the $\Phi(v_k)$ converge to zero in the space (5.28). By virtue of (5.26) we reach the conclusion that the v_k converge strongly in $H^{s+m}(\Omega; H)$, to an element v having norm one and necessarily belonging to W^{s+m}. But $\Phi(v) = 0$, i.e., $v \in \mathcal{N}$, a contradiction. □

It is natural, at this stage, to seek conditions that ensure that the codimension of the range of the map in Theorem 5.4 is finite. In connection with such a property it is natural to introduce the following definition:

DEFINITION 5.2. *We say that the boundary problem* (*) *is Fredholm if the linear map* (5.27), *from* $C^\infty(\bar{\Omega}; H)$ *to the space* (5.28), *is Fredholm.*

On the subject of Fredholm operators we refer to Section 2 of Chapter II. Note that the target space (5.28), as well as the source space, $C^\infty(\bar{\Omega}; H)$, are Fréchet spaces. If we continue (momentarily) to denote by Φ the mapping in Definition 5.2, then its index,

$$\text{(5.29)} \qquad \operatorname{Ind} \Phi = \dim \operatorname{Ker} \Phi - \dim \operatorname{Coker} \Phi,$$

can be taken by definition as the *index of the boundary problem* (*).

We are going to use once more the operators $B_j^\#$ defined in (3.43). Under the current hypothesis (5.8),

$$\text{(5.30)} \qquad B_j^\# = B_j - Q_j M^-,$$

and the degree of $B_j^{\#}$, as a polynomial in ∂_t, does not exceed $m^- - 1$. We introduce the "reduced" boundary problem:

$$(\#) \qquad Pu = f \quad in\ \Omega, \qquad B_j^{\#}u|_X = h_j, \quad j = 1, \ldots, \nu.$$

LEMMA 5.1. *The boundary problem* $(*)$ *is Fredholm if and only if the boundary problem* $(\#)$ *is.*

PROOF. Denote by $\Phi^{\#}$ the linear map

$$(5.31) \qquad u \mapsto (Pu, \{B_j^{\#}u|_X\}_{j=1,\ldots,\nu})$$

from $C^\infty(\bar{\Omega}; H)$, which we denote momentarily by $\mathscr{E}$, to (5.28), which we denote by $\mathscr{F}$. Let $\mathscr{E}_o \subset \mathscr{E}$, $\mathscr{F}_1 \subset \mathscr{F}$ be closed linear subspaces such that

$$\mathscr{E} = \mathscr{E}_o \oplus \operatorname{Ker} \Phi^{\#}, \qquad \mathscr{F} = \operatorname{Im} \Phi^{\#} \oplus \mathscr{F}_1,$$

and call $G^{\#}$ the continuous linear map $\mathscr{F} \to \mathscr{E}$ that is equal to the inverse of $\Phi^{\#}\colon \mathscr{E}_o \to \operatorname{Im}\Phi^{\#}$ on $\operatorname{Im}\Phi^{\#}$, and to zero on $\mathscr{F}_1$.

Next we construct a continuous linear map $\tilde{\Gamma}\colon \mathscr{E} \times \mathscr{E} \to C^\infty(X; H)^\nu$. Let us define the vector-valued function $\mathbf{u}$ starting from u, as indicated in Section 3, and the vector-valued function $\mathbf{v}$ so that equations (3.32) and (3.36) are verified when $\mathbf{g}$ is the m^+-vector with all components zero except the last one, equal to $\Lambda^{1-m^+}Pu$. Let us denote by $\mathbf{v}_*$ the solution of equation (3.36), where the last component of $\mathbf{g}$ is now equal to $\Lambda^{1-m^+}f$ and where

$$(5.32) \qquad \mathbf{v}_* = \mathbf{v} \qquad when\ t = T.$$

Call v_* the first component of $\mathbf{v}_*$ and set

$$\tilde{\Gamma}_j(u, f) = Q_j v_*|_X, \qquad j = 1, \ldots, \nu.$$

Of course $\tilde{\Gamma}(u, f)$ is the ν-vector with components $\tilde{\Gamma}_j(u, f)$. In fact we contend that

$$(5.33) \qquad u \mapsto \tilde{\Gamma}(u, 0) \text{ is a compact operator from } C^\infty(\bar{\Omega}; H) \text{ to } C^\infty(X; H)^\nu.$$

Indeed, when $f \equiv 0$,

$$\mathbf{v}_*(0) = U_o^+(0, T)\mathbf{v}(T) + \int_0^T U_o^+(0, t')\mathscr{R}\mathbf{u}(t')\, dt \qquad \text{(cf. (4.4))}.$$

Then, also, $\mathbf{u} \mapsto \mathbf{v}(T)$ is a continuous linear map from $\mathscr{E}$ to $C^\infty(X; H \otimes \mathbb{C}^{m^+})$. Then our claim follows at once from the fact that $U_o^+(0, T)$ and $\mathscr{R}$ are regularizing operators on X; therefore, for any real number s,

$$\|\mathbf{v}_*(0)\|_s^2 \leq C_s\left(\|\mathbf{v}(T)\|_0^2 + \int_0^T \|\mathbf{u}(t)\|_0^2\,dt\right).$$

Recall that $Q_j v_*|_X$ is the jth component of $\mathscr{Q}\mathbf{v}_*(0)$; see (3.45).

Another observation is that

$$\textit{when } f = Pu,\ v_* = v.$$

This follows at once from (5.32) and from the uniqueness in the backward Cauchy problem for equation (3.36) (using fully the fact that X is compact!).

We now define the following map $\mathscr{F} \to C^\infty(X; H)^\nu$:

$$\Gamma^\#(f, \mathbf{h}) = \tilde{\Gamma}(G^\#(f, \mathbf{h}), f).$$

In fact, let us write

$$\Gamma^\#(f, \mathbf{h}) = \Gamma^* f + K\mathbf{h},$$

where

$$\Gamma^*(f) = \tilde{\Gamma}(G^\#(f, 0), f), \qquad K\mathbf{h} = \tilde{\Gamma}(G^\#(0, \mathbf{h}), 0).$$

(5.34) *The linear operator K on $C^\infty(X; H)^\nu$ is compact.*

Indeed it is the composition of $\mathbf{h} \mapsto G^\#(0, \mathbf{h})$ followed up by $u \mapsto \tilde{\Gamma}(u, 0)$, and the latter is compact by (5.33). We derive from (5.34) that

(5.35) *the linear operator on $\mathscr{F}$,*

(5.36) $$(f, \mathbf{h}) \mapsto (f, \mathbf{h} + \Gamma^\#(f, \mathbf{h}))$$

is Fredholm.

Indeed, (5.36) is the composition of $(f, \mathbf{h}) \mapsto (f, \mathbf{h} + K\mathbf{h})$ followed up by $(f, \mathbf{h}) \mapsto (f, \mathbf{h} + \Gamma^* f)$. The latter is an automorphism of $\mathscr{F}$, and the former is of the kind $I_{\mathscr{F}}$ + compact operator, which is Fredholm by Lemma 2.1 of Chapter II.

Last, restrict the map (5.36) to $\operatorname{Im}\Phi^\#$. Then, with the preceding notation, $u = G^\#(f, \mathbf{h})$, in particular $f = Pu$, and we may take advantage of (5.35). But then

$$\tilde{\Gamma}_j(u, f) = Q_j M^- u|_X,$$

and since $h_j = B_j^\# u|_X$, we have

$$h_j + \tilde{\Gamma}_j(u, f) = B_j u|_X.$$

In other words, $(f, \mathbf{h} + \Gamma^\#(f, \mathbf{h})) = \Phi G^\#(f, \mathbf{h})$.

Thus by (5.35) $\Phi G^{\#}$ is a Fredholm operator on $\mathscr{F}$. The restriction of Φ to $\mathscr{E}_o$ is equal to $(\Phi G^{\#})\Phi^{\#}$, the compose of two Fredholm maps; hence it is Fredholm. This implies at once that dim Coker $\Phi < +\infty$. On the other hand, Ker Φ consists of the sums $u + v$ with $u \in \mathscr{E}_o$ and $v \in \operatorname{Ker} \Phi^{\#}$ such that $\Phi(u) = -\Phi(v)$; hence u belongs to the preimage of $\Phi(\operatorname{Ker} \Phi^{\#})$ under the Fredholm map $\Phi|_{\mathscr{E}_o}$. Since $\Phi(\operatorname{Ker} \Phi^{\#})$ is finite dimensional, so is this preimage and so is Ker Φ.

In order to prove that ($\#$) is Fredholm when this is true of ($*$), it suffices to exchange the maps Φ and $\Phi^{\#}$ (thus $G^{\#}$ and $\Gamma^{\#}$ are replaced by G and Γ respectively) in the preceding argument and replace the map (5.36) by the map $(f, \mathbf{h}) \mapsto (f, \mathbf{h} - \Gamma(f, \mathbf{h}))$. We leave the details to the reader. □

Our next auxiliary result will be a Green formula.

Let P^* denote the adjoint of the operator P, and let $P_j^* = P_j(x, t, D_x)^*$ be the adjoint of P_j. According to (3.1) we have

$$P^* = I(-\partial_t)^m + \sum_{j=1}^{m} (-\partial_t)^{m-j} P_j^*. \tag{5.37}$$

Let u, F be two arbitrary elements of $C^\infty(\bar{\Omega}; H)$; we shall use the cutoff function χ (see (5.9), (5.10)). We have first

$$(P[(1-\chi)u], F)_{L^2(\Omega; H)} = ((1-\chi)u, P^*F)_{L^2(\Omega; H)}. \tag{5.38}$$

On the other hand, recall that $(\ ,\ )_0$ stands for the inner product in $L^2(X; H)$; integration by parts with respect to t gives, in a straightforward manner,

$$\int_0^T (P(\chi u), F)_0\, dt = \int_0^T (\chi u, P^*F)_0\, dt - \sum_{k=1}^{m} (\partial_t^{k-1} u|_X, C_{m-k}F|_X)_0, \tag{5.39}$$

where

$$C_k F|_X = C_k(x, D_x, \partial_t)F|_{t=0} = \sum_{j=0}^{k} (-\partial_t)^j [P_{k-j}(x, t, D_x)^* F]|_{t=0}, \tag{5.40}$$

with the agreement that $P_0 = I$.

Combining (5.38) and (5.39) yields the *Green formula* that we were seeking:

$$(Pu, F)_{L^2(\Omega; H)} = (u, P^*F)_{L^2(\Omega; H)} - \sum_{k=1}^{m} (\partial_t^{k-1} u|_X, C_{m-k}F|_X)_{L^2(X; H)}. \tag{5.41}$$

REMARK 5.3. According to (5.40) we have, for $k = 0, \ldots, m-1$,

$$C_k = I(-\partial_t)^k + \sum_{j=0}^{k-1} C_{k,j}(x, D_x)\partial_t^j.$$

This shows that the operators $C_0, \ldots, C_{m-1}$ make up a *basis* of the (left) module of polynomials in ∂_t of degree $\leq m-1$ with coefficients in the ring $\Psi(X; L(H))$ of pseudodifferential operators on X, valued in $L(H)$.

In connection with the Green formula (5.41), we shall need the following:

LEMMA 5.2. *Let q be an integer ≥ 0 and $F \in H^{-q}(\Omega; H)$ be such that*

$$\langle Pw, \bar{F}\rangle = 0 \qquad \textit{for every } w \in C_c^\infty(\Omega; H). \tag{5.42}$$

Then $P^*F = 0$ in Ω and, as a consequence (i) $F \in C^\infty(\Omega; H)$; (ii) *in the tubular neighborhood $\mathcal{T}$ of X, F is a C^∞ function of $t \in [0, T]$ valued in $\mathcal{D}'(X; H)$.*

PROOF. The only part of Lemma 5.2 that requires a proof is (ii), and this is a variant of Remark 1.2. We have $P^*[\chi(t)F] = F_1 \in C_c^\infty(\Omega \cap \mathcal{T}; H)$. By availing ourselves of (5.37) and denoting by $(-\partial_t)^{-1}$ integration with respect to t from T to $t < T$, we see that

$$(-\partial_t)^{m-k}[\chi(t)F] = -\sum_{j=1}^{m} (-\partial_t)^{m-j-k}P_j^*[\chi(t)F] + (-\partial_t)^{-k}F_1. \tag{5.43}$$

By (2.16) we know that

$$\chi(t)F = \sum_{r=0}^{q} \partial_t^r f_r, \qquad f_r \in L^2(0, T; H^{r-q}(X; H)),$$

and we may take the f_r to be supported in the region $t \leq 3T/4$ (actually even in the region $t \leq T/2$). It suffices then to reason by induction on $k = m+q$, $m+q-1, \ldots, 0, -1, \ldots$, using (5.43). □

We are going to consider $F \in H^{-q}(\Omega; H)$, $S_i \in \mathcal{D}'(X; H)$, $i = 1, \ldots, \nu$, $T_j \in \mathcal{D}'(X; H)$, $j = 1, \ldots, r$, such that, for every $u \in C^\infty(\bar{\Omega}; H)$, we have

$$\langle F, \overline{Pu}\rangle + \sum_{i=1}^{\nu} \langle S_i, \overline{B_i u|_X}\rangle + \sum_{j=1}^{r} \langle T_j, \overline{\partial_t^{j-1} Pu|_X}\rangle = 0. \tag{5.44}$$

(In all this $\langle v, \bar{w}\rangle$ simply stands for the antiduality bracket between H-valued functions and distributions, either in $\bar{\Omega}$ or in X.) By taking $u \in C_c^\infty(\Omega; H)$ we see that (5.42) holds. On the other hand, by virtue of (3.40) and recalling that we are reasoning under hypothesis (5.8),

$$(5.45) \qquad \langle S_i, \overline{B_i u}|_X\rangle = \sum_{k=1}^{m} \langle B_{i,k-1}(x, D_x)^* S_i, \partial_t^{k-1}\bar{u}|_X\rangle.$$

Let us write

$$\partial_t^{j-1} P = I\partial_t^{m+j-1} + \sum_{k=1}^{m+j-1} P_{j,k}(x, t, D_x)\partial_t^{k-1}.$$

We have

$$(5.46) \quad \langle T_j, \partial_t^{j-1}\overline{Pu}|_X\rangle = \langle T_j, \partial_t^{m+j-1}\bar{u}|_X\rangle + \sum_{k=1}^{m+j-1} \langle P_{j,k}^* T_j, \partial_t^{k-1}\bar{u}|_X\rangle.$$

We combine (5.41), (5.45), and (5.46) and use the fact that $P^*F = 0$, thus getting

$$(5.47) \quad \sum_{k=1}^{m} \left\langle \sum_{i=1}^{\nu} B_{i,k-1}^* S_i + \sum_{j=1}^{r} P_{j,k}^* T_j - C_{m-k}F|_X, \partial_t^{k-1}\bar{u}|_X\right\rangle +$$

$$\sum_{k=m+1,k}^{m+r} \left\{\langle T_{k-m}, \partial_t^{k-1}\bar{u}|_X\rangle + \sum_{j=\sup(2,k-m-1)}^{r} \langle P_{j,k}^* T_j, \partial_t^{k-1}\bar{u}|_X\rangle\right\} = 0.$$

But we can take

$$u(x, t) = \chi(t) \sum_{k=0}^{M} \frac{t^k}{k!} u_k(x),$$

with arbitrary $M < +\infty$ and $u_k \in C^\infty(X; H)$. Taking first $u(x, t) = \chi(t)u_k(x)t^k/k!$, with $k = m + r, m + r - 1, \ldots, m + 1$, we reach at once the conclusion that

$$(5.48) \qquad T_j = 0, \qquad j = 1, \ldots, r,$$

and then that we must have

$$(5.49) \qquad C_{m-k}F|_X = \sum_{j=1}^{\nu} B_{j,k-1}^* S_j, \qquad k = 1, \ldots, m.$$

We now prove an intermediate result:

THEOREM 5.5. *Suppose that the Calderon operator $\mathscr{B}$ is globally hypoelliptic in X and defines a Fredholm operator $C^\infty(X; H \otimes \mathbb{C}^{m^-}) \to C^\infty(X; H \otimes \mathbb{C}^\nu)$.*

Then the boundary problem (*) is globally hypoelliptic and Fredholm.

PROOF. By virtue of Lemma 5.1 we may suppose that $B_j = B_j^{\#}$ for every $j = 1, \ldots, \nu$. For simplicity we shall also assume that the degree m_j of B_j is zero for all j.

By Theorems 5.1, 5.2, and 5.4 it suffices to show that in the dual of the Fréchet space (5.28), the orthogonal of the range of the map (5.27) is finite dimensional. Let us use the fact that $C^\infty(\bar{\Omega})$ is equal to the intersection of all the spaces $H^m(\Omega)$; this is a direct consequence of Corollary 2.1. Hence its dual can be identified with the "union" (actually, with the inductive limit) of the duals $(H^m(\Omega))'$. If then we use the isomorphism (2.14), we see that the orthogonal in question consists of the multiplets $(F, \{S_i\}, \{T_j\})$ exactly like those in (5.44). Equation (5.48) tells us that all the T_j must be zero. Since (5.42) holds, we have

$$P^*F = 0 \quad \textit{in } \Omega, \tag{5.50}$$

and the conclusions in Lemma 5.2 hold. And so do the equations (5.49).

But since the degree of every B_j as a polynomial in ∂_t does not exceed $m^- - 1$, we have

$$C_{m-k}F|_X = 0, \qquad k = m^- + 1, \ldots, m. \tag{5.51}$$

By Remark 5.3 and reasoning by descending induction on $m, m - 1, \ldots, m - m^+ + 1$, we see that (5.51) is equivalent to

$$\partial_t^k F|_X = 0, \qquad k = 0, \ldots, m^+ - 1. \tag{5.52}$$

But (5.50)–(5.52) constitute the homogeneous Dirichlet problem for the operator P^*; the corresponding Calderon operator is the identity of $\mathcal{D}'(X; H \otimes \mathbb{C}^{m^+})$, obviously hypoelliptic. By Theorem 5.2 the set of F's satisfying (5.50)–(5.51) form a finite-dimensional linear subspace of $C^\infty(\bar{\Omega}; H)$.

Consider now the equations (5.49) where k stays $\leq m^-$. Multiply both sides by Λ^{k-1}. By virtue of (3.49) this reads

$$\Lambda^{k-1} C_{m-k}F|_X = \sum_{j=1}^{\nu} \mathcal{B}_{j,k}^* S_j, \qquad k = 1, \ldots, m^-. \tag{5.53}$$

Let $\mathbf{S}$ denote the ν-vector with components S_j. Equation (5.53) can be rewritten

$$\mathcal{B}^*\mathbf{S} = \Psi F, \tag{5.54}$$

where Ψ is the obvious linear map $C^\infty(\bar{\Omega}; H) \to C^\infty(X; H \otimes \mathbb{C}^{m^-})$ defined by the left-hand sides in (5.53).

Thus $\mathscr{B}^*\mathbf{S}$ belongs to the image of a finite-dimensional space (of C^∞ functions on X). Since we hypothesize that $\mathscr{B}$ is Fredholm (between the spaces in Theorem 5.5) the kernal of its adjoint, $\mathscr{B}^*$ (acting between the dual of those spaces), is finite dimensional. The set of distributions $\mathbf{S}$ satisfying (5.54) is therefore finite dimensional. □

COROLLARY 5.1. *If the Calderon operator $\mathscr{B}$ and its adjoint $\mathscr{B}^*$ are both globally hypoelliptic, then the boundary problem* $(*)$ *is globally hypoelliptic and Fredholm.*

It suffices to combine Theorem 5.5 with Proposition 5.1.

It is now easy to prove the main result in the present context:

THEOREM 5.6. *The boundary problem* $(*)$ *is Fredholm if and only if the Calderon operator* $\mathscr{B}: C^\infty(X; H \otimes \mathbb{C}^{m^-}) \to C^\infty(X; H \otimes \mathbb{C}^\nu)$ *is Fredholm.*

PROOF. Let us denote by Φ^o the linear map

$$u \mapsto (Pu, \{\Lambda^{1-k}\partial_t^{k-1}u|_X\}_{k-1,\dots,m^-}) \tag{5.55}$$

of $C^\infty(\bar{\Omega}; H)$ into

$$C^\infty(\bar{\Omega}; H) \times C^\infty(X; H \otimes \mathbb{C}^{m^-}). \tag{5.56}$$

By Theorem 5.5, $\dot{\Phi}^o$ is Fredholm (and globally hypoelliptic), since the corresponding Calderon operator is the identity map of $C^\infty(X; H \otimes \mathbb{C}^{m^-})$.

In proving Theorem 5.6 we may replace the problem $(*)$ by the problem $(\#)$, thanks to Lemma 5.1. The corresponding map $\Phi^\#$ is the composition of Φ^o followed up by the operator $I \otimes \mathscr{B}$ acting from (5.56) to (5.28). Since Φ^o is Fredholm, in order for $\Phi^\#$ also to be Fredholm it is necessary and sufficient for the operator $I \otimes \mathscr{B}$ to be Fredholm, and the statement follows. □

Theorem 5.6 truly constitutes an improvement on Theorem 5.5; the operator $\mathscr{B}$ can be self-adjoint and Fredholm (when acting from $C^\infty(X; H \otimes \mathbb{C}^{m^-})$ to itself, assuming now $m^- = \nu$) and yet not be globally hypoelliptic, as shown by the following example.

EXAMPLE 5.1. Let X be the unit circle, $\phi(x)$ a C^∞ function on X everywhere ≥ 0, vanishing only at $x = x_o$, and there to the first order: $\phi(x_o) = 0$, $d\phi(x_o) \neq 0$. Take $m^- = \nu = 1$, and let $\mathscr{B}$ be the operator of multiplication by ϕ. Then $C^\infty(X)$ is the direct sum of the space of C^∞

functions vanishing at x_o and the constant functions; the null space of $\mathscr{B}$ in $C^\infty(X)$ is zero. Thus $\mathscr{B}$ is Fredholm; clearly $\mathscr{B}$ is self-adjoint. On the other hand $\mathscr{B}$ is not globally hypoelliptic, since it annihilates the Dirac measure at x_o.

6. Coercive Boundary Problems

What we call *coercive* boundary problems are often called boundary value problems of the Lopatinski–Shapiro kind. They are the foremost examples of boundary problems (∗) (Section 3). In this section P will be the same elliptic operator as in Section 3 but, for the sake of simplicity, we shall take it to be scalar, i.e., $\dim H = 1$.

Coercivity is characterized by two conditions:

(6.1) *The number ν of boundary conditions is equal to m^-, the number of roots of the polynomial in z, $\sigma(P)(x, t, \xi, z)$, with real part <0.*

(6.2) *The principal symbols $\sigma(B_j)(x, \xi, z)$ of the boundary operators $B_j(x, D_x, \partial_t)$ $(j = 1, \ldots, \nu)$ are linearly independent modulo $\sigma(M^{-0})(x, 0, \xi, z)$ (see (3.7)), for all (x, ξ) in $T^*X\backslash 0$.*

The principal symbols $\sigma(B_j)(x, \xi, z)$ are positive-homogeneous functions of (ξ, z) of respective degrees m_j, with no *a priori* relation to their degrees d_j as polynomials in z. Actually we may multiply each $B_j(x, D_x, \partial_t)$ by Λ^{-m_j} and assume henceforth that their homogeneity degree with respect to (ξ, z) is equal to zero, whatever j. This is what we shall do.

Since, by (6.1), $\nu = \deg_z \sigma(M^{-0})$, we see that the $\sigma(B_j)$ form a *linear basis* of the space of polynomials in the variable z modulo $\sigma(M^{-0}(0))$. We may effect the division of $\sigma(B_j)(x, \xi, z)$ by $\sigma(M^{-0})(x, 0, \xi, z)$, as polynomials in z:

$$\sigma(B_j)(x, \xi, z) = Q_j''(x, \xi, z)\sigma(M^{-0})(x, 0, \xi, z) + b_j(x, \xi, z), \tag{6.3}$$

with $\deg_z b_j \leq m^- - 1$. The polynomials b_j form a basis of the vector space of polynomials in z of degree $<m^-$.

Now, if we go back to the division formulas (3.41) and (3.43), we see that, for each j, $b_j(x, \xi, z)$ is the principal symbol of $B_j^\#(0)$, that is, according to (3.47),

$$b_j(x, \xi, z) = \sum_{k=0}^{m^- - 1} \sigma(B_{j,k}^\#)(x, 0, \xi)z^k. \tag{6.4}$$

Because of the homogeneity of degree zero of b_j with respect to (ξ, z), we see that $\sigma(B^{\#}_{j,k})(x, 0, \xi)$ is positive-homogeneous of degree $-k$ with respect to ξ; therefore, if we set, as in (2.49), $\mathscr{B}_{j,k} = B^{\#}_{j,k-1}(0)\Lambda^{k-1}$, we obtain that $\sigma(\mathscr{B}_{j,k})(x, \xi)$ is positive-homogeneous of degree zero with respect to ξ. It is then clear that the coercivity condition (6.2) is equivalent to the following property:

(6.5) *The vectors* $\beta_j(x, \xi) = \big((\sigma(\mathscr{B}_{j,k})(x, \xi)\big)_{k=1,\ldots,\nu}$ *are linearly independent, for all* $(x, \xi) \in T^*X\backslash 0$.

In other words, condition (6.2) is equivalent to the following statement:

(6.6) *The* $m^- \times m^-$ *matrix* $\sigma(\mathscr{B})(x, \xi) = \big(\sigma(\mathscr{B}_{j,k})(x, \xi)\big)_{j=k,\ldots,m^-}$ *is invertible, for all* (x, ξ) *in* $T^*X\backslash 0$.

Since we can retrace our steps and clearly go back from (6.6) to (6.2), we may state

THEOREM 6.1. *The boundary problem* $(*)$ *is coercive if and only if* $\nu = m^-$ *and the Calderon operator* $\mathscr{B}$ *is elliptic, of order zero under our present agreement that all* B_j *have total order zero.*

By combining Theorems 4.1 and 6.1, we obtain

COROLLARY 6.1. *If the boundary problem* $(*)$ *is coercive, it is hypoelliptic.*

Henceforth we assume that X is the boundary of a bounded open subset Ω of $\mathbb{R}^{n+1}$, as we have in Sections 2 and 5, and we proceed to apply the results of Section 5. We shall make the assumption (5.8), i.e., that the degrees of the B_j as polynomials in ∂_t do not exceed $m - 1$. Of course, under the coercivity hypothesis, problem $(*)$ is globally hypoelliptic and its null space is finite dimensional (Theorem 5.2). The crucial estimate (5.7) is valid here with $s'' = s$, and Theorem 5.3 yields the classical "coercive estimates":

THEOREM 6.2. *Suppose that the boundary problem* $(*)$ *is coercive and let* s *be any integer* ≥ 0. *For a suitable* $C > 0$ *and all* $u \in C^\infty(\bar{\Omega})$,

$$\|u\|_{H^{s+m}(\Omega)} \leq C\Big(\|Pu\|_{H^s(\Omega)} + \sum_{j=1}^{m^-} \|B_j u|_X\|_{H^{s+m-m_j-1/2}(X)} + \|u\|_{H^{s+m-1}(\Omega)}\Big). \tag{6.7}$$

We have kept the orders m_j (not presupposing them to be zero), because this is the way that the estimates are customarily written.

Of course Theorem 5.4 applies, and so does Theorem 5.5:

THEOREM 6.3. *If the boundary problem* (*) *is coercive, it is Fredholm* (*Definition* 5.2).

Actually, because of the great precision in the estimates (6.7), a stronger result is now valid:

THEOREM 6.4. *Suppose that the boundary problem* (*) *is coercive. Whatever the integer* $s \geq 0$, *the map* (5.27), *from* $H^{s+m}(\Omega)$ *to the Hilbert space*

$$H^s(\Omega) \times \prod_{j=1}^{m^-} H^{s+m-m_j-1/2}(X) \tag{6.8}$$

is Fredholm.

The proof is similar to that of Theorem 5.5; the only difference is that we must take the multiplets $(F, \{S_i\}, \{T_j\})$ in the dual of the space (6.8), not merely in the dual of the space (5.28).

EXAMPLE 6.1. The simplest example of coercive boundary conditions is given by the boundary operators

$$B_j = (\partial/\partial t)^j, \qquad j = 0, \ldots, m^- - 1. \tag{6.9}$$

Such boundary conditions are universal in the sense that (6.2) holds regardless of the choice of M^-, provided, of course, that its degree with respect to ∂_t is m^-. The corresponding boundary problem is called the *Dirichlet problem.*

7. The Oblique Derivative Problem. Boundary Problems with Simple Real Characteristics

7.1. An Example: The Oblique Derivative Boundary Problem

Let Ω be an open subset of $\mathbb{R}^{n+1}$ whose boundary is a C^∞ hypersurface X (of dimension n). We assume that Ω lies on one side of X. We select a C^∞ function r in some open neighborhood U of X such that $dr \neq 0$ at every point of U, and $X = \{y \in U; r(y) = 0\}$, $\Omega \cap U = \{y \in U; r(y) < 0\}$. We use momentarily coordinates $y_1, \ldots, y_{n+1}$ in $\mathbb{R}^{n+1}$. We denote by Δ the Laplace operator in $\mathbb{R}^{n+1}$. We suppose that we are given a C^∞ real vector field $L(x, \partial_x)$ and a real-valued C^∞ function $c(x)$ on X, where the variable point is denoted by x, and we consider the following boundary problem:

$$\Delta u = f \qquad \text{in } \Omega, \tag{7.1}$$

$$c(x)\frac{\partial u}{\partial r} + L(x, \partial_x)u = g \qquad \text{in } X. \tag{7.2}$$

We wish to avail ourselves of the results of Section 4 to test the hypoellipticity of problem (7.1)–(7.2) (Definition 4.1).

We reason locally, in the neighborhood of a point of X which we take to be the origin in $\mathbb{R}^{n+1}$. We assume that the coordinates y_j have been labeled so that $(\partial r/\partial y_j)(0) = 0$ if $j \le n$ (and therefore $(\partial r/\partial y_{n+1})(0) \neq 0$). We make the change of coordinates $x_j = y_j$ $(j = 1, \dots, n)$, $r = r(y)$ near the origin, and we get

$$\Delta = R^2\frac{\partial^2}{\partial r^2} + 2M_0\frac{\partial}{\partial r} + h\frac{\partial}{\partial r} + \Delta_x,$$

where we have used the notation

$$R^2 = |dr|^2, \qquad h = \Delta r, \qquad M_0 = \sum_{j=1}^{n} \frac{\partial r}{\partial y_j}\frac{\partial}{\partial x_j}.$$

We seek a pseudodifferential operator $A = A(r, x, D_x)$ on X (near 0) depending smoothly on r (in a neighborhood of $r = 0$) such that

$$\Delta = \left(R\frac{\partial}{\partial r} + A_1\right)\left(R\frac{\partial}{\partial r} - A\right),$$

with

$$A_1 = RAR^{-1} + R^{-1}(2M_0 + h) - \frac{\partial R}{\partial r}.$$

The operator A must satisfy the equation

$$A^2 + R^{-1}(2M_0 + h)A - [A, R]R^{-1}A + R\frac{\partial A}{\partial r} + \Delta_x = 0. \tag{7.3}$$

In all this the functions R, h, etc., stand for the multiplication operators that they naturally define. We seek A in the form of a classical pseudodifferential operator on X; thus we solve (7.3) modulo regularizing operators in X which depend in C^∞ fashion on r. Observe that the principal symbol $\sigma(A)$

must satisfy the quadratic equation

$$\sigma(A)^2 + 2R^{-1}\sigma(M_0)\sigma(A) - |\xi|^2 = 0.$$

We denote by $\xi_1, \ldots, \xi_n$ the coordinates in the cotangent space to X defined by the coordinates $x_1, \ldots, x_n$ in X. We require the real (or the self-adjoint) part of A to be *elliptic positive*, which demands

$$\sigma(A) = \{|\xi|^2 + \sigma(M_0)^2/R^2\}^{1/2} - \sigma(M_0)/R.$$

Note that M_0 is a real vector field on X; therefore its symbol is purely imaginary:

$$\sigma\,(M_0) = \sqrt{-1}\sum_{j=1}^{n} \xi_j \frac{\partial r}{\partial y_j}.$$

In view of this we see that

$$\text{(7.4)} \qquad \operatorname{Re}\sigma(A) = R^{-1}\left\{R^2|\xi|^2 - \left(\sum_{j=1}^{n} \xi_j \frac{\partial r}{\partial y_j}\right)^2\right\}^{1/2} \geq R^{-1}\left|\frac{\partial r}{\partial y_{n+1}}\right||\xi|.$$

By our choice in the labeling of the coordinates y_j, we know that

$$\text{(7.5)} \qquad \frac{\partial r}{\partial y_{n+1}} \neq 0$$

in the neighborhood of the origin. Observe that

$$\sigma(A_1) = \sigma(A).$$

The homogeneous terms of degree $0, -1, \ldots,$ in the total symbols of A and of A_1 are then derived by induction, in pretty much the same manner in which the analogous terms in the symbols of M^+ and of M^- were obtained in Section 3. We write then $t = -r$ and

$$\text{(7.6)} \quad A^-(t) = -R(x, r)^{-1}A(r, x, D_x), \qquad A^+(t) = R(x, r)^{-1}A_1(r, x, D_x).$$

The equation (7.1) thus uncouples into

$$\text{(7.7)} \qquad \frac{\partial u}{\partial t} - A^-(t)u = v,$$

$$\text{(7.8)} \qquad \frac{\partial v}{\partial t} - A^+(t)v = f.$$

We have already neglected any term involving regularizing operators such as $\mathscr{R}\mathbf{u}$ in the right-hand side of (3.36).

Next we take (7.7) into account in (7.2). Taking the Calderon operator to be

$$\mathscr{B} = L(x, \partial_x) - c(x)A^-(0), \tag{7.9}$$

we rewrite (7.2) in the fashion

$$\mathscr{B}u(0) = g + c(x)v(0) \qquad \textit{in } X \textit{ (near the origin)}. \tag{7.10}$$

In (7.10), $u(0)$ and $v(0)$ are distributions in X.

The principal symbol of $\mathscr{B}$ is

$$\sigma(\mathscr{B})(x, \xi) = c(x)R^{-1}(x, 0)\sigma[A(0)] - \sqrt{-1}\, L(x, \xi), \tag{7.11}$$

where $A(0)$ stands for the operator A when $r = 0$. Since the vector field L is real, the characteristic set of $\mathscr{B}$ is defined by the equations

$$c(x) = 0, \qquad L(x, \xi) = 0. \tag{7.12}$$

First, this shows that if $c(0) \neq 0$ the boundary problem (7.1)–(7.2) is coercive (a well-known fact). We shall henceforth assume that

$$c(0) = 0. \tag{7.13}$$

Second, (7.11) shows that $d_\xi\sigma(\mathscr{B})$ vanishes if and only if $c(x) = 0$ and $d_\xi L(x, \xi) = 0$. We shall make the following hypothesis:

(7.14) *At no point of X do the function $c(x)$ and the vector field $L(x, \partial_x)$ vanish at the same time.*

That is, $c(x) + |d_\xi L(x, \xi)| \neq 0$.

If hypothesis (7.14) is valid and if $n = 1$, then *the oblique derivative problem is automatically coercive*, since the second equation in (7.12) cannot be satisfied unless $\xi = 0$. Problem (7.1)–(7.2) is of a new type only if $n \geq 2$. In this case observe that a restatement of (7.14) is that

$$d_\xi\sigma(\mathscr{B})(x, \xi) \neq 0, \qquad \forall (x, \xi) \in T^*X\backslash 0. \tag{7.15}$$

Property (7.15) is often expressed by saying that $\mathscr{B}$ *has simple real characteristics.* For later reference we note that

$$\operatorname{Re}\{-\sqrt{-1}\sigma(\mathscr{B})\} = L(x, \xi) - c(x)R(x, 0)^{-2} \sum_{j=1}^{n} \xi_j \frac{\partial r}{\partial y_j}(x, 0), \tag{7.16}$$

$$\operatorname{Im}\{-\sqrt{-1}\sigma(\mathscr{B})\} = -c(x)R(x, 0)^{-1}\left\{|\xi|^2 - R(x, 0)^{-2}\left(\sum_{j=1}^{n} \xi_j \frac{\partial r}{\partial y_j}(x, 0)\right)^2\right\}^{1/2}. \tag{7.17}$$

Since $\partial r/\partial y_j = 0$ at the origin for $j \le n$, we see that at points of the form $x = 0$, $\xi \ne 0$, we have

$$\text{(7.18)} \qquad \operatorname{Re}\{-\sqrt{-1}\sigma(\mathscr{B})\} = L(0, \xi), \qquad \operatorname{Im}\{-\sqrt{-1}\sigma(\mathscr{B})\} = 0,$$

and by our hypothesis (7.14), $L(0, \xi)$ is not identically zero.

7.2. Boundary Problems with Simple Real Characteristics

We return to the general setup of Section 3. We assume the Hilbert space H to be finite dimensional. We also make the following assumption:

(7.19) *The number* ν *of boundary conditions* (3.39) *is equal to the number* m^- *of characteristic roots with real part* <0 (*see* (3.6)).

Because of this hypothesis we may and shall regard the Calderon operator $\mathscr{B}$ (see (3.49)) as a pseudodifferential operator in X with values in the space of complex $r \times r$ matrices, for some integer $r \ge 1$; these are matrices with scalar entries. For convenience we shall assume that the order of $\mathscr{B}$ is equal to one.

Definition 7.1. *We say that the boundary problem* (*) *of Section 3 has simple real characteristics if* (7.19) *holds, and if moreover,*

$$\text{(7.20)} \qquad \forall (x, \xi) \in T^*X\backslash 0, \qquad \det \sigma(\mathscr{B})(x, \xi) = 0 \Rightarrow d_\xi \det \sigma(\mathscr{B})(x, \xi) \ne 0.$$

When (7.20) is verified Char $\mathscr{B}$ is a subset of a smooth submanifold of $T^*X\backslash 0$ of codimension one, of course conic. When $\sigma(\mathscr{B})$ is real, it is such a manifold; otherwise, in general it is a proper subset of such a manifold.

A first advantage of property (7.20) is that the study of $\mathscr{B}$ can be reduced to that of a certain scalar pseudodifferential operator (also with simple real characteristics) at least microlocally, as we now show.

Lemma 7.1. *Suppose that* (7.20) *holds. Then given any point* (x_o, ξ^o) *in* $T^*X\backslash 0$ *there is a conic open neighborhood* Γ *of* (x_o, ξ^o) *and an elliptic pseudodifferential operator* $\mathscr{Q}$ *of order zero in* Γ, *valued in the space of complex* $r \times r$ *matrices, such that in* Γ,

$$\text{(7.21)} \qquad \mathscr{Q}\mathscr{B} \sim \begin{pmatrix} |D_x| & 0 & 0 & \cdots & 0 & C_1 \\ 0 & |D_x| & 0 & & 0 & C_2 \\ \vdots & & & & & \\ 0 & 0 & 0 & & |D_x| & C_{r-1} \\ 0 & 0 & 0 & & 0 & C_r \end{pmatrix}.$$

PROOF. Let us call $\mathcal{B}_{jk}(j, k = 1, \ldots, r)$ the entries of $\mathcal{B}$, b^0_{jk} their principal symbols. Let us also use the notation $b^0_j = (b^0_{j1}, \ldots, b^0_{jr})$. If $\det \sigma(\mathcal{B}) \neq 0$ at (x_0, ξ^o), we take $\mathcal{Q} = |D_x|\mathcal{B}^{-1}$, in the conic neighborhood suitably chosen. Otherwise we know that, for some $i = 1, \ldots, n$,

$$\frac{\partial}{\partial \xi_i}(b^0_1 \wedge \cdots \wedge b^0_r) \neq 0 \qquad \textit{at } (x_o, \xi^o).$$

Let us choose the labeling of the coordinates x_i and that of the columns and rows in $\mathcal{B}$ so as to have

$$b^0_1 \wedge \cdots \wedge b^0_{r-1} \wedge \frac{\partial b^0_r}{\partial \xi_n} \neq 0 \qquad \textit{at } (x_o, \xi^o). \tag{7.22}$$

Let then $\mathcal{B}'$ denote the matrix-valued pseudodifferential operator in Γ whose entries are $\mathcal{B}'_{jk} = \mathcal{B}_{jk}$ if $j \leq r-1$, $\mathcal{B}'_{rk} = [\mathcal{B}_{rk}, x_n]|D_x|$ $(k = 1, \ldots, r)$. Condition (7.22) tells us that $\mathcal{B}'$ is elliptic. We define $\mathcal{Q}'$ to be the inverse (in Γ) of $\mathcal{B}'$. Then if $\mathcal{Q}'_{kl}$ $(k, l = 1, \ldots, r)$ are the entries of $\mathcal{Q}'$, we have $\sum_k B'_{jk}\mathcal{Q}'_{kl} = \delta_{jl}$ (Kronecker's index). If in $\mathcal{B}_{jk}$ we interpret j as the index defining the columns and k the index defining the rows, we see that

$$\mathcal{Q}'\mathcal{B} = \begin{pmatrix} I & 0 & 0 & \ldots & 0 & C'_1 \\ 0 & I & 0 & & 0 & C'_2 \\ \vdots & & & & & \\ 0 & 0 & 0 & & I & C'_{r-1} \\ 0 & 0 & 0 & & 0 & C'_r \end{pmatrix}.$$

Then $\mathcal{Q} = |D_x|\mathcal{Q}'$ satisfies the requirements of Lemma 4.1. □

COROLLARY 7.1. *The operator $\mathcal{B}$ is hypoelliptic* (i.e., *preserves the wave-front sets*) *in Γ if and only if this is true of C_r.*

The proof is evident. Thus, at least microlocally, we may restrict our attention to scalar pseudodifferential operators.

7.3. Hypoelliptic Pseudodifferential Operators with Simple Real Characteristics

In the remainder of this section we shall completely forget about the boundary problems that our interest originated from and deal instead with a scalar pseudodifferential operator P of order m (an arbitrary real number) in a conic open subset Γ of $T^*X\backslash 0$. P is a section of the sheaf of pseudodifferential operators in X, and it acts on microdistributions in Γ, in the manner described in Section 6 of Chapter I. We shall assume that P is classical and denote by $p(x, \xi)$ its principal symbol. When we specify that P has simple real characteristics (which will not always be the case), this means that

$$(7.23) \qquad \forall (x, \xi) \in \Gamma, \qquad p(x, \xi) = 0 \;\; \textit{implies} \;\; d_\xi p(x, \xi) \neq 0.$$

Let $a(x, \xi)$ be a C^∞ complex function in an open subset $\mathcal{O}$ of $T^*X\backslash 0$. By a *bicharacteristic* of a we mean a C^1 curve $]0, 1[\ni t \mapsto \gamma(t) \in \mathcal{O}$ having the following properties:

(i) *For every* $t \in]0, 1[$, *the tangent vector* $\dot\gamma(t)$ *at the point* $\gamma(t)$ *is* $\neq 0$ *and proportional to the Hamiltonian field of* a,

$$H_a = \sum_{j=1}^{n} \frac{\partial a}{\partial \xi_j}\frac{\partial}{\partial x_j} - \frac{\partial a}{\partial x_j}\frac{\partial}{\partial \xi_j}$$

at the same point;

(ii) a *vanishes identically on the curve* γ.

When a is *real*, then through every point where a vanishes and provided that $da \neq 0$, there passes a unique bicharacteristic of a. When a is nonreal, this is not generally so. Note that any bicharacteristic of a comes equipped with a natural orientation, that defined by the vector field H_a.

DEFINITION 7.2. *We say that the operator P satisfies condition* (Ψ') *at the point* (x_0, ξ^0) *if there exists a conic open neighborhood* Γ^0 *of this point and a complex number z such that the following is true*:

$$(7.24) \qquad \forall (x, \xi) \in \Gamma^0, \qquad zp(x, \xi) = 0 \;\; \textit{implies} \;\; d[\mathrm{Re}(zp)(x, \xi)] \neq 0;$$

(7.25) *along any bicharacteristic of* $\mathrm{Re}(zp)$ *in* Γ^0, *if* $\mathrm{Im}(zp) > 0$ *at some point, then* $\mathrm{Im}(zp) \geq 0$ *at any later point.*

DEFINITION 7.3. *We say that P satisfies the strict condition* (Ψ') *at* (x_0, ξ^0) *if the conic set* Γ^0 *and the complex number z in Definition 7.2 can be chosen so that in addition to (7.24) and (7.25), the following property also holds*:

(7.26) $\mathrm{Im}(zp)$ *does not vanish identically in any nonempty open arc of bicharacteristic of* $\mathrm{Re}(zp)$ *in* Γ^0.

EXAMPLE 7.1. Consider the differential operator $D_1 + ix_1D_2$ in $\mathbb{R}^2(D_j = -\sqrt{-1}(\partial/\partial x_j))$. Its (principal) symbol is $\xi_1 + ix_1\xi_2$. Its characteristic set is defined by the equations

$$\xi_1 = 0, \qquad x_1 = 0. \tag{7.27}$$

It is a line bundle over the x_2-axis from which the zero section has been excised. The bicharacteristics of the real part of the principal symbol are the x_1 (half-)lines. The imaginary part vanishes at $x_1 = 0$ and changes sign at such a point from $-\text{sgn}\, \xi_2$ to $+\text{sgn}\, \xi_2$. Thus the operator satisfies the strict (Ψ') condition when $\xi_2 > 0$ and does not satisfy (Ψ') for $\xi_2 < 0$. The operator $D_1 + ix_1D_2$ is called the *Mizohata operator.*

EXAMPLE 7.2. The operator $D_1 + ix_2D_2$ in $\mathbb{R}^2$ satisfies (Ψ') everywhere, but not the strict (Ψ') condition when $x_2 = 0$.

REMARK 7.1. It can be shown (see Nirenberg and Treves [1]; Treves [4]) that if (7.24) and (7.25) holds for some $z \in \mathbb{C}$, then (7.25) holds for any other complex number for which (7.24) holds. The same is true if we consider the conjunction of (7.25) and (7.26) instead of (7.25) alone.

Condition (Ψ') simplifies somewhat when one deals with pseudodifferential operators that are *antipodal,* that is, such that either $p(x, \xi) = p(x, -\xi)$ for all (x, ξ) in Γ where $(x, -\xi)$ also belongs to Γ, or else $p(x, \xi) = -p(x, -\xi)$ for such (x, ξ).

Indeed let $a(x, \xi)$ be a real-valued C^∞ function in an open subset $\mathcal{O}$ of $T^*X\backslash 0$ which is symmetric with respect to the zero section. Suppose that $a(x, -\xi) = a(x, \xi)$. Then the Hamiltonian field H_a in a neighborhood of $(x, -\xi)$ is oriented in the opposite way to the same field in a neighborhood of (x, ξ), in the following sense: if $b(x, \xi)$ is another C^∞ function in $\mathcal{O}$ such that $b(x, -\xi) = b(x, \xi)$, we have $H_ab(x, -\xi) = -H_ab(x, \xi)$. If instead $a(x, -\xi) = -a(x, \xi)$, it is oriented in the same way. Now, if $b(x, -\xi) = -b(x, \xi)$, we have $H_ab(x, -\xi) = -H_ab(x, \xi)$. In both cases, if b changes sign along a bicharacteristic of a through (x, ξ) [assuming that $a(x, \xi) = 0$] from, say, $-$ to $+$, it changes sign from $+$ to $-$ along the bicharacteristic of a through $(x, -\xi)$. And in order that all changes of sign of b occur from $-$ to $+$, both at (x, ξ) and at $(x, -\xi)$, it is necessary that no change of sign of b occur at all.

If we apply this with P antipodal and $a = \text{Re}\, p$, $b = \text{Im}\, p$, we see that condition (Ψ') reduces to the condition (P) thus defined:

DEFINITION 7.4. *We say that the pseudodifferential operator P satisfies condition (P) at (x_o, ξ^o) if the conic open neighborhood Γ^o of (x_o, ξ^o) and the* *number* $z \in \mathbb{C}$ *can be chosen so that* (7.24) *holds and if the following property also holds*:

(7.28) $\quad$ $\operatorname{Im}(zp)$ *does not change sign along any bicharacteristic of* $\operatorname{Re}(zp)$ *in* Γ^o.

REMARK 7.2. All *differential* operators are antipodal. The Calderon operator $\mathscr{B}$ in the oblique derivative problem (Section 7.1) is not antipodal, unless $c(x)$ vanishes identically (which corresponds rather to a "tangential derivative" boundary problem).

The study of examples and of certain particular cases leads naturally to the following

CONJECTURE. *Suppose that the pseudodifferential operator* P *in* Γ *satisfies the following condition*:

$$\forall (x, \xi) \in \Gamma, \qquad p(x, \xi) = 0 \Rightarrow dp(x, \xi) \neq 0. \tag{7.29}$$

Then, in order that P *be hypoelliptic in* Γ, *it is necessary and sufficient that* P *satisfy the strict condition* (Ψ') *at every point of* Γ.

Let us point out that (7.29) is a weaker requirement than (7.23). In Chapter VIII, Section 7 we shall encounter the following property:

DEFINITION 7.5. *We say that* P *is of principal type in* Γ *if, given any point* (x, ξ) *of* Char P *in* Γ, *there is a complex number* z *such that* $d[\operatorname{Re}(zp)]$ *and the differential form* $\xi_1\, dx_1 + \cdots + \xi_n\, dx_n$ *are linearly independent at that point.*

Let us outline briefly where we stand in our knowledge of the validity of the preceding conjecture.

The necessity of condition (Ψ') has been proved by R. Moyer [1] under the principal-type hypothesis. For antipodal operators of principal type the result of Moyer combined with those of Treves [4] yields the necessity of the strict condition (Ψ'). Still in the antipodal principal-type case the sufficiency of the strict condition (Ψ') is proved in Treves [4]; for a more up-to-date exposition see Hörmander [18, 15].

The conjecture has been completely proved when, in addition to (7.24)–(7.25), one also requires that $\operatorname{Im}(zp)$ have only zeros of finite order along the bicharacteristics of $\operatorname{Re}(zp)$. This corresponds to the subelliptic case, discussed in the next section.

EXAMPLE 7.3. Let $\mathscr{B}$ denote the Calderon operator in the oblique derivative problem of Section 7.1. Let us set $a(x, \xi) = \mathrm{Re}\{-\sqrt{-1}\,\sigma(\mathscr{B})\}$, $b(x, \xi) = \mathrm{Im}\{-\sqrt{-1}\,\sigma(\mathscr{B})\}$. We observe that $a(x, \partial_x)$ is a real vector field near the origin in X, not vanishing there, in view of (7.13), (7.14), and the fact that $\partial r/\partial y_j = 0$ at the origin if $j \leq n$. Furthermore we may write $b(x, \xi) = -c(x)Q(x, \xi)$, with $Q(x, \xi) \geq \kappa|\xi|$ for $x \sim 0$, $\xi \in \mathbb{R}_n$ (κ is some number >0). Consequently, near the origin in X, the sign of b along the bicharacteristics of a is equal to the sign of $-c(x)$ along the integral curves of the vector field $a(x, \partial_x)$. Indeed these integral curves, sometimes called the bicharacteristic curves or the characteristics of a, are the projection into the base of the bicharacteristics of a. (In this connection let us point out that we must assume $n \geq 2$; otherwise there are no zeros of $a(x, \xi)$ unless $\xi = 0$; when $n = 1$, under hypothesis (7.14) the problem (7.1)–(7.2) is coercive, as already indicated.) Thus, in this case, property (Ψ') means that *along any integral curves of* $a(x, \partial_x)$, *if* $c(x) < 0$ *then* $c(x') \leq 0$ *at every point* $x' > x$, and the strict property (Ψ') requires in addition that *c not vanish identically on any nonempty open arc of integral curve of* $a(x, \partial_x)$. All this is near the origin.

7.4. Subelliptic Pseudodifferential Operators

An important subclass of hypoelliptic pseudodifferential operators are the *subelliptic* ones, of which we give now the microlocal definition:

DEFINITION 7.6. *The pseudodifferential operator P of order m in* Γ *is said to be subelliptic in* Γ *if, given any conic open set* Γ_1 *with conically compact closure contained in* Γ, *there is a number* $\delta_1 > 0$ *such that for all (micro)distributions u in* Γ,

(7.30) $\quad Pu \in H^0_{\mathrm{loc}}((\Gamma_1))$ *implies* $u \in H^{m-1+\delta_1}_{\mathrm{loc}}((\Gamma_1))$
(*see Chapter I, Definition* 6.5).

In Definition 7.6 one always has $\delta_1 \leq 1$: *P is elliptic of order m in* Γ *if and only if one can take* $\delta_1 = 1$ *for all* Γ_1.

PROPOSITION 7.1. *Suppose that P is subelliptic in* Γ. *Then, given any conic open set* Γ_1 *with conically compact closure contained in* Γ, *there is* $\delta_1 > 0$ *such that, for all real numbers s and all microdistributions u in* Γ, *we have*

(7.31) $$Pu \in H^s_{\mathrm{loc}}((\Gamma_1)) \Rightarrow u \in H^{s+m-1+\delta_1}_{\mathrm{loc}}((\Gamma_1)).$$

PROOF. Let δ be any number such that $0 < \delta \leq \delta_1$. Let us denote by Z_δ the set of integers (≥ 0 or < 0) j such that for all microdistributions u in Γ, the properties

$$(7.32)_j \qquad Pu \in H^{j\delta}_{\text{loc}}((\Gamma_1)), \qquad u \in H^{j\delta+m-1}_{\text{loc}}((\Gamma_1)),$$

imply

$$(7.33)_j \qquad u \in H^{j\delta+m-1+\delta_1}_{\text{loc}}((\Gamma_1)),$$

By the hypothesis in Proposition 7.1, zero belongs to Z_δ. Let j be an integer such that all integers r such that $|r| < |j|$ belong to Z_δ. Let Q be any elliptic pseudodifferential operator in Γ of order $j\delta/|j|$. Suppose that $(7.32)_j$ holds. Then

$$(7.34) \qquad QPu \in H^{(j-j/|j|)\delta}_{\text{loc}}((\Gamma_1)),$$

and since the order of $[P, Q]$ is equal to $m - 1 + j\delta/|j|$, we also have

$$(7.35) \qquad [P, Q]u \in H^{(j-j/|j|)\delta}_{\text{loc}}((\Gamma_1)).$$

The conjunction of (7.34) and (7.35) means that

$$PQu \in H^{(j-j/|j|)\delta}_{\text{loc}}((\Gamma_1)).$$

Since we also have $Qu \in H^{(j-j/|j|)\delta+m-1}_{\text{loc}}((\Gamma_1))$ and since $j - j/|j| \in Z_\delta$, we reach the conclusion that

$$(7.36) \qquad Qu \in H^{(j-j/|j|)\delta+m-1+\delta_1}((\Gamma_1)).$$

But since Q is elliptic of order $j\delta/|j|$, this implies $(7.33)_j$ and thus j also belongs to Z_δ, which must be the set of all integers, $\mathbb{Z}$.

Let us now suppose that the first relation $(7.32)_j$ is true, $Pu \in H^{j\delta}_{\text{loc}}((\Gamma_1))$, but that the second one it not necessarily true. There certainly is an integer $j' \leq j - 2$ such that $(7.33)_{j'}$ holds. But then we also have $(7.33)_{j'+1}$, and by induction we must have $(7.33)_{j-1}$, which means that we have $(7.32)_j$ and thus $(7.33)_j$. If then s is any real number, we select the integer j such that $\delta = s/j$ satisfies $0 < \delta \leq \delta_1$. □

COROLLARY 7.2. *Suppose that P is subelliptic in Γ. Then given any (micro-)distribution u in Γ, we have*

$$(7.37) \qquad \mathrm{WF}(Pu) = \mathrm{WF}(u).$$

The "global" version of Definition 7.6 is obvious; a pseudodifferential operator P of order m in X is called *subelliptic* if, given any relatively compact open subset X_1 of X, there is a number $\delta_1 > 0$ such that for all distributions u in X, and whatever the real number s,

$$Pu \in H^s_{\text{loc}}(X_1) \Rightarrow u \in H^{s+m-1+\delta_1}_{\text{loc}}(X_1). \tag{7.38}$$

If P is not properly supported, we may replace it by an equivalent properly supported operator.

To return to the microlocal version, if (x_o, ξ^o) is a point of Γ, we shall say that the pseudodifferential operator P in Γ is *subelliptic with loss of* λ *derivatives at* (x_o, ξ^o) if there is a conic open neighborhood $\Gamma_1 \subset \Gamma$ of (x_o, ξ^o) such that (7.30) holds with $1 - \delta_1 = \lambda$ and if there is no other open neighborhood of the same point, contained in Γ_1, such that the analogous property (7.30) holds with $1 - \delta_1 < \lambda$. (Thus we always have, here, $\lambda < 1$.)

The following result of Yu. V. Egorov is the main theorem in [2]; for a proof see Hörmander [20].

THEOREM 7.1. *The pseudodifferential operator P in Γ is subelliptic at (x_o, ξ^o) with loss of λ derivatives if and only if there is a conic open neighborhood $\Gamma^o \subset \Gamma$ of (x_o, ξ^o) and a complex number z such that the following is true*:

To every $(x, \xi) \in \Gamma^o$ there is an integer $k(x, \xi) \leq \lambda/(1-\lambda)$ such that

$$H^j_{\text{Re}(zp)}[\text{Im}(zp)](x, \xi) \tag{7.39}$$

is zero if $j < k(x, \xi)$, nonzero if $j = k(x, \xi)$, and positive if $j = k(x, \xi)$ and if $k(x, \xi)$ is odd. Furthermore,

$$k(x_o, \xi^o) = \lambda/(1-\lambda). \tag{7.40}$$

If we compare the necessary and sufficient condition for subellipticity in Theorem 7.1 with that in Definition 7.2, we see easily that the condition in Theorem 7.1 states that P must satisfy condition (Ψ') in the neighborhood Γ^o. Moreover, it also states that when the complex number z is chosen so that (7.24) holds, the zeros of $\text{Im}(zp)$ at points along the bicharacteristics of $\text{Re}(zp)$ in Γ^o must be finite. In addition it links the order of such zeros to the derivative loss λ.

EXAMPLE 7.4. Let a, b denote respectively the real and the imaginary part of the principal symbol p of P. Suppose we assume that

$$\forall (x, \xi) \in \Gamma, \qquad p(x, \xi) = 0 \Rightarrow \{a, b\}(x, \xi) \neq 0, \tag{7.41}$$

where $\{\ ,\ \}$ stands for the Poisson bracket (see (4.33), Chapter I). Then P is of course elliptic near every point that does not lie on its characteristic set. At a point $(x_o, \xi^0) \in \Gamma$ such that $p(x_o, \xi^o) = 0$, P is not subelliptic if $\{a, b\}(x_o, \xi^o) = H_a b(x_o, \xi^o) < 0$ and it is subelliptic with loss of one-half derivative (that is, $\lambda = 1/2$) if $\{a, b\}(x_o, \xi^o) > 0$. This is exactly the situation of the Mizohata operator $D_1 + ix_1D_2$ in $\mathbb{R}^2$ (Example 7.1). Indeed $a = \xi_1$, $b = x_1\xi_2$, $\{a, b\} = 2\xi_2$. The characteristic points are of the form $(0, x_2, 0, \xi_2)$ with $\xi_2 \neq 0$. Thus the subellipticity of the operator holds at those points where $\xi_2 > 0$ (with the integer $k(x, \xi)$ being equal, there, to one) and does not hold at the points where $\xi_2 < 0$.

EXAMPLE 7.5. If we return once again to the Calderon operator $\mathcal{B}$ in the oblique derivative problem (Example 7.3), we see that it is subelliptic if and only if the zeros of $c(x)$ along the integral curves of $a(x, \partial_x)$ all have finite order and if c changes sign from $+$ to $-$ whenever it changes sign at one of these zeros.

8. Example of a Boundary Problem with Double Characteristics: The $\bar{\partial}$-Neumann Problem in Subdomains of $\mathbb{C}^N$

As a last example of a noncoercive boundary problem for an elliptic equation we now describe the $\bar{\partial}$-Neumann problem in an open subset Ω of $\mathbb{C}^N$. We assume Ω to be a bounded set and have a smooth, i.e., C^∞, boundary, denoted by X; we also assume that Ω lies on only one side of X. Our primary aim is to obtain enough information about the Calderon operator $\mathcal{B}$ (see (3.49)) to be able to apply Theorem 4.1. At that point we shall refer to recent results about pseudodifferential operators with double characteristics: when combined with the information about $\mathcal{B}$, they will show that under suitable hypotheses, the $\bar{\partial}$-Neumann problem is hypoelliptic in the sense of Definition 4.2.

We do not motivate, here, the importance of the study of $\bar{\partial}$-Neumann problem. Its interest lies in the applications to holomorphic functions of several complex variables. About those applications the reader might consult Folland and Kohn [1].

8.1. Description of the $\bar{\partial}$-Neumann Problem

First our notation: The real coordinates in $\mathbb{R}^{2N}$ will be $x_1, \ldots, x_N, y_1, \ldots, y_N$, which define the coordinates $z_j = x_j + \sqrt{-1}\, y_j$ ($j = 1, \ldots, N$) in $\mathbb{C}^N$. We write $z = (z_1, \ldots, z_N)$, $x = \operatorname{Re} z$, $y = \operatorname{Im} z$. We identify $\mathbb{R}^N$ to the subspace $y_1 = \cdots = y_N = 0$. Of course $\bar{z} = x - \sqrt{-1}\, y$.

We shall deal with differential forms of a special kind on $\mathbb{R}^{2N}$, or on some subset of $\mathbb{R}^{2N}$. About differential forms see Chapter I, Example 7.2, and Chapter VII, Section 2.1. The forms we shall be interested in are of type $(0, q)$, that is, of the kind

$$\alpha = \sum_{|J|=q} \alpha_J \, d\bar{z}_J.$$

where $J = (j_1, \ldots, j_q)$ is a multi-index of length q such that $1 \le j_1 < \cdots < j_q \le N$; if $q = 0$, α is a function or a distribution. Generally speaking the coefficients α_J are complex-valued functions or distributions in a subset $\mathfrak{M}$ of $\mathbb{R}^{2N}$, which for us will be either Ω or its closure $\bar{\Omega}$. If these coefficients are C^m functions (for $0 \le m \le +\infty$) we write

$$\alpha \in C^m(\mathfrak{M}; \Lambda^{0,q});$$

if these coefficients are L^p functions, distributions, etc., we replace C^m in this notation by L^p, $\mathcal{D}'$, etc.

By definition the *Cauchy–Riemann* operator $\bar{\partial}$ is the linear operator $C^{m+1}(\mathfrak{M}; \Lambda^{0,q}) \to C^m(\mathfrak{M}; \Lambda^{0,q+1})$ $(q \le N-1)$ defined by

$$\bar{\partial}\alpha = \sum_{j=1}^{N} \sum_{|J|=q} \frac{\partial \alpha_J}{\partial \bar{z}_j} \, d\bar{z}_j \Lambda \, d\bar{z}_J. \tag{8.1}$$

(All this is taken up again, from a slightly different viewpoint, in Section 5 of Chapter IX.)

We may use the canonical basis $dx_I \wedge dy_J$, where I, J are ordered multi-indices of various lengths, in the exterior algebra over the dual of $\mathbb{R}^{2N}$ to extend the hermitian product and norm of $L^2(\mathbb{R}^{2N})$ to differential forms on $\mathbb{R}^{2N}$. If

$$f = \sum_{I,J} f_{I,J} \, dx_I \wedge dy_J, \qquad g = \sum_{I,J} g_{I,J} \, dx_I \wedge dy_J,$$

their L^2 inner product is $(f, g)_0 = \sum_{I,J} \int f_{I,J} \bar{g}_{I,J} \, dx \, dy$. (We assume that the coefficients $f_{I,J}$, $g_{I,J}$ are square integrable.) With this definition, if α is a form of type $(0, q)$ as before, and $\beta = \sum_{|J'|=q'} \beta_{J'} \, d\bar{z}_{J'}$ is a form of type $(0, q')$, both with square-integrable coefficients, we see that $(\alpha, \beta)_0 = 0$ if $q \ne q'$, whereas, if $q = q'$,

$$(\alpha, \beta)_0 = 2^q \sum_{|J|=q} \int \alpha_J \bar{\beta}_J \, dx \, dy. \tag{8.2}$$

In all this, $dx\,dy = dx_1 \cdots dx_N \, dy_1 \cdots dy_N$ is the Lebesgue measure in $\mathbb{R}^{2N}$. We may then define the formal adjoint of $\bar{\partial}$ for the L^2 inner product (8.2). It is a linear operator $\vartheta: C^{m+1}(\mathfrak{M}; \Lambda^{0,q}) \to C^m(\mathfrak{M}; \Lambda^{0,q-1})$ $(q > 0)$;

(8.3) $$\vartheta\alpha = -2 \sum_{|J'|=q-1} \sum_{\substack{k=1,\ldots,N \\ k \notin J'}} \varepsilon_J^{kJ'} \frac{\partial \alpha_J}{\partial z_k} d\bar{z}_{J'},$$

where $\alpha = \sum_{|J|=q} \alpha_J \, d\bar{z}_J$. In the summation at the right in (8.3), J is obtained by adjunction of $k \notin J'$ to J' and reordering; $\varepsilon_J^{kJ'} = +1$ or -1 according to the parity of the permutation that brings the set $\{k\} \cup J'$ to its ordered form, J; this notation will systematically be used from now on. The differential operator $\square = \bar{\partial}\vartheta + \vartheta\bar{\partial}$ is called the *complex Laplacian* in $\mathbb{C}^N$. An easy computation shows that with α as before,

(8.4) $$\square\alpha = -\tfrac{1}{2} \sum_{|J|=q} \Delta\alpha_J \, d\bar{z}_J,$$

where Δ is the ordinary Laplace operator in $\mathbb{R}^{2N}$,

$$\Delta = 4 \sum_{j=1}^{N} \frac{\partial^2}{\partial z_j \, \partial \bar{z}_j}.$$

We may identify the cotangent bundle over $\mathbb{R}^{2N}$ to $\mathbb{C}^N \times \mathbb{C}_N$ and thus view the principal symbol $\sigma(\vartheta)$ of ϑ as a C^∞ function of the variable point (z, ζ) in that space, valued in the space of linear mappings

$$\Lambda^{0,q}\mathbb{C}^N \to \Lambda^{0,q-1}\mathbb{C}^N \qquad (q > 0).$$

We use the notation $\Lambda^{0,q}\mathbb{C}^N$ to indicate the obvious subspace of the *complex* exterior algebra over $\mathbb{R}^{2N}$. Therefore, if we substitute for ζ a C^∞ section of the cotangent bundle (over some set) we obtain a C^∞ function valued in the preceding space of linear mappings; such a function can be identified to a (generally nonsquare) matrix with C^∞ entries, by using the canonical basis in $\mathbb{C}^N$.

Let us then select a C^∞ real-valued function r in $\mathbb{C}^N$ such that X is defined by the equation $r = 0$, with $dr \neq 0$ in a neighborhood of X and $r < 0$ in Ω. We thus get the matrix-valued function of the point z varying in X:

$$\sigma(\vartheta)(z, dr(z)) = (\sigma_{J'}^J)_{|J|=q, |J'|=q-1}.$$

By virtue of (8.3) we obtain

(8.5) $$\sigma_{J'}^J = -2i\varepsilon_{J'}^{kJ} \frac{\partial r}{\partial z_k} \quad \text{if } J' = J\backslash\{k\}, \qquad \sigma_{J'}^J = 0 \quad \text{if } J' \not\subset J.$$

DEFINITION 8.1. *The $\bar{\partial}$-Neumann problem in Ω is the problem of finding, given a form $f \in C^\infty(\Omega; \Lambda^{0,q})$, another form $u \in C^2(\bar{\Omega}; \Lambda^{0,q})$ verifying*

(8.6) $$\square u = f \qquad \text{in } \Omega,$$

(8.7) $$\sigma(\vartheta)(z, dr)u = 0 \qquad on\ \partial\Omega,$$

(8.8) $$\sigma(\vartheta)(z, dr)\bar{\partial}u = 0 \qquad on\ \partial\Omega.$$

One could vary the requirements on the regularity of f and look, for instance, at the case where $f \in L^2(\Omega; \Lambda^{0,q})$. The requirement that u be C^2 in $\bar{\Omega}$ is simply meant to ensure that equations (8.6)–(8.8) have an obvious meaning; but there also the requirement could have been substantially weakened. At any rate our purpose is to show that when the solution u exists and certain hypotheses are satisfied , then u is a C^∞ function up to any piece of the boundary $X = \partial\Omega$ where this is true of f.

On the reasons for posing boundary conditions such as (8.7)–(8.8), we refer the reader to Chapter I of Folland and Kohn [1].

By virtue of (8.5) we may rewrite (8.7) as follows when $q > 0$:

(8.9) *When $r = 0$ we have, for all multi-indices J', $|J'| = q - 1$,*

$$\sum_{j \notin J'} \varepsilon_J^{jJ'} \frac{\partial r}{\partial z_j} u_J = 0,$$

where J is the multi-index obtained by adjoining j to J' and reordering; u_J is the corresponding coefficient of the form u.

Equation (8.8) can be rewritten as follows, provided that q is >0;

(8.10) *When $r = 0$, for all multi-indices J, $|J| = q$, we have*

$$\sum_{k \notin J, j \in J^*} \varepsilon_{J^*}^{kJ} \varepsilon_{J^*}^{jK} \frac{\partial r}{\partial z_k} \frac{\partial u_K}{\partial \bar{z}_j} = 0,$$

where J^* is derived from J by adjoining k and reordering, while $K = J^* \backslash \{j\}$.

When $q = 0$ (in which case u is a scalar function), (8.8) reads

(8.11) $$\bar{L}u = 0 \qquad when\ r = 0,$$

with

(8.12) $$L = \sum_{k=1}^{N} \frac{\partial r}{\partial \bar{z}_k} \frac{\partial}{\partial z_k}.$$

REMARK 8.1. If $q = N$, condition (8.8) is, of course, void. Furthermore, to every J' of length $N - 1$ there is a unique $j \notin J'$, and thus (8.9) reads

$$\frac{\partial r}{\partial z_j} u_{(1,\ldots,N)} = 0 \ when\ r = 0, \quad j = 1, \ldots, N.$$

Since $dr \neq 0$ when $r = 0$, this is equivalent to

$$u_{(1,\dots,N)} = 0 \qquad \textit{on } \partial\Omega. \tag{8.13}$$

Since equation (8.6) according to (8.4) reads

$$\Delta u_{(1,\dots,N)} = -2f \qquad \textit{in } \Omega, \tag{8.14}$$

we see that when $q = N$, *the* $\bar{\partial}$*-Neumann problem reduces to the Dirichlet problem.*

At this stage we localize the analysis near an arbitrary point of the boundary of Ω, which we shall take to be the origin in $\mathbb{C}^N$, and choose the coordinates in $\mathbb{C}^N$ so as to have, near the origin,

$$r(z) = \operatorname{Re} z_N + 2 \operatorname{Re} \sum_{j,k=1}^{N-1} \frac{\partial^2 r(0)}{\partial z_j\, \partial z_k} z_j z_k + 2\mathscr{L}_0(r; z) + 0(|z|^3), \tag{8.15}$$

where

$$\mathscr{L}_0(r; w) = \sum_{j,k=1}^{N-1} \frac{\partial^2 r(0)}{\partial z_j\, \partial \bar{z}_k} w_j \bar{w}_k \qquad (w \in \mathbb{C}^N) \tag{8.16}$$

is called the *Levi form* of $\partial\Omega$ (or of Ω) at the point 0.

By availing ourselves of the expression (8.15) we are going to rewrite in a more practically useful fashion the boundary conditions (8.9), (8.10). We shall skip all computations, which are simple manipulations of indices. We regard the form $u = \sum_{|J|=q} u_J\, d\bar{z}_J$ as a vector, with components u_J. Actually we replace it by one of its linear transformations:

$$u^{\#} = u - S(z)u, \tag{8.17}$$

defined as follows: When $q = 0$, $S(z) \equiv 0$. When $q > 0$,

$$\text{If } N \in J,\ u_J^{\#} = u_J - (-1)^q \left(\frac{\partial r}{\partial z_N}\right)^{-1} \sum_{\substack{k=1\\ k \notin J}}^{N-1} \varepsilon_J^{kJ'} \frac{\partial r}{\partial z_k} u_K, \tag{8.18}$$

where $J' = J\backslash\{N\}$ *and* K *is obtained by adjoining* k *to* J'.

$$\text{If } N \notin J,\ u_J^{\#} = u_J - 4|dr|^{-2} \sum_{j \in J, k \notin J'} \varepsilon_J^{jJ'} \varepsilon_K^{kJ'} \frac{\partial r}{\partial \bar{z}_j} \frac{\partial r}{\partial z_k} u_K, \tag{8.19}$$

with $J' = J\backslash\{j\} = K\backslash\{k\}$.

In fact, let us use the direct-sum decomposition

$$u^{\#} = u' + u'', \tag{8.20}$$

where $u'_J = 0$ if $N \in J$ and $u''_J = 0$ if $N \notin J$.

PROPOSITION 8.1. *Condition* (8.9) (*which presumes* $q > 0$) *is equivalent to*

(8.21) $$u'' = 0 \qquad \text{when } r = 0.$$

In our reformulation of (8.10) we shall use the differential operator L defined by (8.12). We shall also use the matrix γ, acting on vectors of the kind u' (and transforming them into like vectors), defined by

(8.22) $$(\gamma u')_J = \sum_{j\in J, k\notin J'} \varepsilon_J^{jJ'} \varepsilon_K^{kJ'} \left(\frac{\partial^2 r}{\partial \bar{z}_j\, \partial z_k} - \frac{\partial^2 r}{\partial \bar{z}_j\, \partial z_N} \frac{\partial r}{\partial z_k} \Big/ \frac{\partial r}{\partial z_N}\right) u'_K,$$

$$\text{where } N \notin J,\ J' = J\backslash\{j\} = K\backslash\{k\}.$$

Noting that $u^{\#} = u'$ when $q = 0$ (there is no u'' then), we state

PROPOSITION 8.2. *Condition* (8.10) (*which presumes* $q > 0$) *is equivalent to*

(8.23) $$(\bar{L} + \gamma)u' = 0 \qquad \text{when } r = 0.$$

We may also take (8.23) as expressing the boundary condition (8.11), which presumes $q = 0$, if we agree that $\gamma = 0$ when $q = 0$.

One may say that $\bar{\partial}$-*Neumann boundary conditions* (8.7)–(8.8) are equivalent to the conjunction of *Dirichlet conditions* on the component u'' of $u^{\#}$ with "*Neumann-like*" *boundary conditions* on the component u'.

Last, we wish to rewrite equation (8.6) in terms of $u^{\#}$. We look closely at (8.18)–(8.19) and take into account the fact, following from (8.15), that

(8.24) $$\frac{\partial r}{\partial z_j} = O(|z|) \quad \text{if } j < N, \qquad 2\frac{\partial r}{\partial z_N} = 1 + O(|z|^2), \quad z \sim 0.$$

(We have already used the fact that $\partial r/\partial z_N \neq 0$ in a neighborhood of 0.) According to this we see that $S(0) = 0$ and that, consequently, $I - S(z)$ is invertible in a suitable neighborhood of the origin. We then set $u = (I - S(z))^{-1}u^{\#}$ in (8.6). This equation becomes

(8.25) $$\Delta u^{\#} + Tu^{\#} = f^{\#},$$

where

(8.26) $$T = -2(I - S(z))[\Delta, (I - S(z))^{-1}].$$

(8.27) $$f^{\#} = -2(I - S(z))f.$$

It is now convenient to switch coordinates, from Re z_j, Im z_j ($j = 1, \ldots, N$) to

$$(8.28) \qquad x_j = \operatorname{Re} z_j, \qquad y_j = \operatorname{Im} z_j \ (j = 1, \ldots, N-1), \quad r, \quad y_N = \operatorname{Im} z_N.$$

Note that in these coordinates (near the origin) the boundary $X = \partial\Omega$ can be identified to a piece of the hyperplane of the coordinates x_j, y_k ($j < N$, $k \leq N$). Let us set (cf. p. 191)

$$(8.29) \quad 2\frac{\partial r}{\partial \bar{z}_j} = p_j + iq_j \quad (p_j, q_j \text{ real}; j = 1, \ldots, N); \quad R = |dr|; \quad h = \Delta r.$$

By virtue of (8.24) we have

$$(8.30) \qquad p_j = O(|z|), \quad q_j = O(|z|) \qquad \text{if } j < N;$$
$$p_N = 1 + O(|z|^2), \qquad q_N = O(|z|^2), \qquad R = 1 + O(|z|^2).$$

Let us introduce the additional notation:

$$(8.31) \qquad \begin{aligned} M_0 &= \sum_{j=1}^{N-1} \left(p_j\frac{\partial}{\partial x_j} + q_j\frac{\partial}{\partial y_j}\right) + q_N\frac{\partial}{\partial y_N}, \\ M_1 &= \sum_{j=1}^{N-1} \left(q_j\frac{\partial}{\partial x_j} - p_j\frac{\partial}{\partial y_j}\right) - p_N\frac{\partial}{\partial y_N}. \end{aligned}$$

An easy computation shows that

$$(8.32) \quad \Delta = R^2\frac{\partial^2}{\partial r^2} + 2M_0\frac{\partial}{\partial r} + h\frac{\partial}{\partial r} + \Delta', \qquad 4L = R^2\frac{\partial}{\partial r} + M_0 + iM_1,$$

where Δ' is the Laplace operator on X:

$$\Delta' = \sum_{j=1}^{N-1} \left(\frac{\partial^2}{\partial x_j^2} + \frac{\partial^2}{\partial y_j^2}\right) + \frac{\partial^2}{\partial y_N^2}.$$

From (8.26) and (8.32) we derive

$$(8.33) \qquad T = h_1\frac{\partial}{\partial r} + T',$$

where h_1 is a C^∞ matrix-valued function near 0 and T' is a first-order differential operator, with C^∞ matrix-valued coefficients, whose principal part is tangential to X. We factorize directly equation (8.25):

$$(8.34) \qquad I\Delta + T = \left(IR\frac{\partial}{\partial r} + A_1\right)\left(IR\frac{\partial}{\partial r} - A\right).$$

Here as in the sequel we completely disregard the error terms coming from regularizing operators. They can be handled exactly as in Section 4, and they have no effect on the reasonings or on their conclusions. We get

$$(8.35)\qquad A_1 = RAR^{-1} + \left(2R^{-1}M_0 + R^{-1}h - \frac{\partial R}{\partial r}\right)I + R^{-1}h_1,$$

$$(8.36)\qquad A^2 + \frac{2}{R}M_0A + \left(\frac{h}{R} + \frac{h_1}{R} - \frac{\partial R}{\partial r} - [A, R]R^{-1}\right)A + R\left[\frac{\partial}{\partial r}, A\right] + I\Delta' + T' = 0.\dagger$$

Equation (8.36) shows that A is a pseudodifferential operator of order one on X, depending smoothly on r; this implies that $[\partial/\partial r, A] = \partial A/\partial r$. We may and shall take the principal symbol of A to be

$$(8.37)\qquad \sigma(A) = -\sigma(M_0)/R + \{\sigma(M_0)^2/R^2 - \sigma(\Delta')\}^{1/2}.$$

It will be shown later that $\operatorname{Re}\sigma(A) > 0$ in the complement of the zero section in T^*X. Since M_0 is a real vector field on X, its principal symbol $\sigma(M_0)$ is purely imaginary; we shall see that $\sigma(-\Delta') + [\sigma(M_0)/R]^2 > 0$. The square root in (8.37) is the positive one. From (8.35) and (8.37) we get

$$(8.38)\qquad \sigma(A_1) = \sigma(M_0)/R + \{\sigma(M_0)^2/R^2 - \sigma(\Delta')\}^{1/2}.$$

We decompose the equation (8.25) into the system

$$(8.39)\qquad R\frac{\partial u^\#}{\partial r} - Au^\# = v^\#,$$

$$(8.40)\qquad R\frac{\partial v^\#}{\partial r} + A_1v^\# = f^\#,$$

to which we must adjoin the boundary conditions (8.21), (8.23), which we rewrite here:

$$(8.41)\qquad R\frac{\partial u'}{\partial r} + \frac{1}{R}(M_0 - iM_1 + 4\gamma)u' = 0,\ u'' = 0 \quad \textit{when } r = 0.$$

Let us then define the pseudodifferential operator A_0 on X, with values in the space of matrices that transform vectors of the kind u' into like ones, as follows:

$$(8.42)\qquad A_0u'(0) = (Au^\#)'|_{r=0}.$$

† Notice the similarity between these equations and (7.7)–(7.8). The treatment here is much like that in the oblique derivative problem (Section 7.1).

If we extract $R(\partial u'/\partial r)$ from (8.39) and put it into (8.41), we may rewrite the latter as

$$\mathscr{B}'u'(0) = -v'(0), \qquad u''(0) = 0, \tag{8.43}$$

where $u'(0)$, $u''(0)$, $v'(0)$ are the values of u', u'', v' at $r = 0$ (these values are functions in X) and where

$$\mathscr{B}' = A_0 + R^{-1}[I(M_0 - iM_1) + 4\gamma]|_{r=0}. \tag{8.44}$$

The local representation of the $\bar{\partial}$-Neumann problem (8.21), (8.23), (8.25) provided by (8.39), (8.40), (8.43) is the analogue of the decomposition of (*) into (**) in Section 3. It should indeed be noted that the equations (8.39), (8.40) must be satisfied in the portion $r < 0$ of a neighborhood of the origin, and that the role of the variable t in Section 3 is played here by $-r$.

REMARK 8.2. When $q = 0$, all the matrices $S(z)$, γ, T, T', h_1 vanish identically, and the pseudodifferential operators A, A_0, $\mathscr{B}'$ are scalar, as they should be.

8.2. The Principal Symbol of the Calderon Operator† $\mathscr{B}'$

We shall use the notation $\zeta_j = \xi_j + i\eta_j$ $(j = 1, \ldots, N-1)$, $\zeta_N = i\eta_N$. Then we have

$$\sigma(-\Delta') = |\zeta|^2, \qquad \sigma(M_0 - iM_1) = 2i\zeta \cdot \partial r, \tag{8.45}$$

We derive at once from (8.37), (8.44), and (8.45):

$$\sigma(\mathscr{B}') = \{|\zeta|^2 - (\operatorname{Re}\zeta \cdot \partial r/|\partial r|)^2\}^{1/2} - \operatorname{Im}\zeta \cdot \partial r/|\partial r|. \tag{8.46}$$

In all this ∂r is the "vector" with components $\partial r/\partial z_j$ $(j = 1, \ldots, N)$; we observe that $R|_{r=0} = 2|\partial r||_{r=0}$. By virtue of (8.46) we see that $\sigma(\mathscr{B}') \geq 0$. Let us multiply $\sigma(\mathscr{B}')$ by

$$B_0 = \{|\zeta|^2 - (\operatorname{Re}\zeta \cdot \partial r/|\partial r|)^2\}^{1/2} + \operatorname{Im}\zeta \cdot \partial r/|\partial r|, \tag{8.47}$$

and set

$$F = |\partial r|^2 B_0 \sigma(\mathscr{B}') = |\zeta|^2|\partial r|^2 - |\zeta \cdot \partial r|^2. \tag{8.48}$$

† It is evocative but inaccurate to call $\mathscr{B}'$ the Calderon operator in the $\bar{\partial}$-Neumann problem: according to (8.43) the Calderon operator $\mathscr{B}$ is the direct sum of $\mathscr{B}'$, acting on vectors u' and of the identity acting on vectors u''. The latter corresponds to the Dirichlet problem for the part u'' of the solution.

The *characteristic set* of $\mathscr{B}'$, i.e., the zero set of $\sigma(\mathscr{B}')$, is contained in the set

$$(8.49)\qquad |\zeta \cdot \partial r| = |\zeta||\partial r|, \qquad \operatorname{Im} \zeta \cdot \partial r \geq 0.$$

The first one of these of these conditions requires $\zeta = c\bar{\partial} r$ for some complex function c. By the fact that $\zeta_N = i\eta_N$ this in turn requires $c = i\eta_N/(\partial r/\partial \bar{z}_N)$ (cf. (8.24)), and therefore (8.49) implies

$$(8.50)\qquad \xi_j + i\eta_j - i\eta_N \frac{\partial r}{\partial \bar{z}_j} \Big/ \frac{\partial r}{\partial \bar{z}_N} = 0, \qquad j = 1, \ldots, N-1.$$

We also observe that the second condition (8.49) can be rewritten $\operatorname{Im} c \geq 0$ which, for z small, is equivalent to $\eta_N \geq 0$. But if $\eta_N = 0$, (8.50) implies $\xi_j = \eta_j = 0$ for all $j = 1, \ldots, N-1$; hence if we restrict the concept of characteristic set to the complement of the zero section in the cotangent bundle (here over a portion of X), we see that we must complement (8.50) with

$$(8.51)\qquad \eta_N > 0.$$

In the region $\eta_N < 0$, $\mathscr{B}'$ is elliptic. Note also that $B_0 > 0$ in the region (8.51). Recalling that $\dim T^*X = 2(2N-1)$, we state

PROPOSITION 8.3. Char $\mathscr{B}'$ *is a C^∞ submanifold of dimension $2N$ of $T^*X\backslash 0$. If z_o is sufficiently close to the origin (in X), the intersection of* Char $\mathscr{B}'$ *with the fiber $T^*_{z_o}X$ is a single ray $\xi_j = \rho\xi^o_j$, $\eta_j = \rho\eta^o_j$ $(j = 1, \ldots, N-1)$, $\eta_N = \rho > 0$.*

When $N = 1$ there are no equations (8.50); Char $\mathscr{B}'$ is defined simply by the inequality (8.51). In the remainder of this section we suppose $N \geq 2$.

Let us introduce the following complex vector fields:

$$(8.52)\qquad Z_j = \frac{\partial}{\partial z_j} - \left(\frac{\partial r}{\partial z_j} \Big/ \frac{\partial r}{\partial z_N}\right)\frac{\partial}{\partial z_N}, \qquad j = 1, \ldots, N-1.$$

The complex conjugates $\bar{Z}_1, \ldots, \bar{Z}_{N-1}$ define what is called the *induced Cauchy–Riemann operator* on X; for more information on this important topic we refer to Chapter 5 of Folland and Kohn [1] and to Section 5 of Chapter IX.

It is seen at once that equation (8.50) can be rewritten:

$$(8.53)\qquad \sigma(Z_j) = 0, \qquad j = 1, \ldots, N-1.$$

On the other hand, let us denote by w the $(N-1)$-vector with components $w_j = (\partial r/\partial z_j)/(\partial r/\partial z_N)$, and by $\sigma(Z)$ the one with components $\sigma(Z_j)$. Then

$$(8.54)\qquad (2|\partial r|)^{-2}F = |\sigma(Z)|^2 - |w \cdot \overline{\sigma(Z)}|^2/(1 + |w|^2).$$

Thus in the region $\eta_N > 0$, we have (cf. (8.47))

$$\sigma(\mathcal{B}') = 4B_0^{-1} \sum_{j=1}^{N-1} |\sigma(Z_j)|^2 - \sum_{j,k=1}^{N-1} c_{jk}\sigma(Z_j)\overline{\sigma(Z_k)}, \tag{8.55}$$

where (c_{jk}) is a self-adjoint positive semidefinite $(N-1)\times(N-1)$ matrix depending smoothly on the variable point in a suitable neighborhood of the origin, in the base X. Since, by (8.24), $w_j(0) = 0$ for all $j < N$, we have

$$c_{jk}(0) = 0, \qquad \forall j, k = 1, \ldots, N-1. \tag{8.56}$$

From all this we derive

PROPOSITION 8.4. *The principal symbol* $\sigma(\mathcal{B}')$ *vanishes exactly of order two on the (smooth) manifold* Char $\mathcal{B}'$.

REMARK 8.3. Let us show that $\text{Re}\,\sigma(A) > 0$ on the portion of $T^*X\backslash 0$ which lies over a sufficiently small neighborhood of the origin in X (and for small enough values of $-r$). Using the notation of (8.45), we derive from (8.37),

$$\sigma(A) = \{|\zeta|^2 - (\text{Re}\,\zeta\cdot\partial r/|\partial r|^2\}^{1/2} + i\,\text{Re}\,\zeta\cdot\partial r/|\partial r|. \tag{8.57}$$

Thus $\text{Re}\,\sigma(A) \geq 0$ near 0 and $\text{Re}\,\sigma(A) = 0$ only if $\text{Re}\,\zeta\cdot\partial r = |\zeta||\partial r|$ and therefore $\text{Im}\,\zeta\cdot\partial r = 0$. But we have just seen that this conjunction implies $\zeta = 0$. □

8.3. The Subprincipal Symbol of the Calderon Operator $\mathcal{B}'$

We continue to represent the variable point in the cotangent bundle over X by $(x_1, \ldots, x_{N-1}, y_1, \ldots, y_N, \xi_1, \ldots, \xi_{N-1}, \eta_1, \ldots, \eta_N)$. Since we are dealing here with classical pseudodifferential operators we can consider their total symbol which is a formal series of symbols that are positive-homogeneous of integral degrees with respect to the fiber variable (ξ, η). Thus the total symbol of A is $a_0 + a_1 + \cdots$, with $\deg a_j = 1 - j$, and the one of $\mathcal{B}'$ is $b_0 + b_1 + \cdots$ with $\deg b_j = 1 - j$ also $(j = 0, 1, \ldots)$. By definition, the *subprincipal symbol* of $\mathcal{B}'$ is the quantity

$$\sigma_1(\mathcal{B}') = b_1 - \frac{1}{2i}\left\{\sum_{j=1}^{N-1} \frac{\partial^2 b_0}{\partial x_j\,\partial\xi_j} + \sum_{k=1}^{N} \frac{\partial^2 b_0}{\partial y_k\,\partial\eta_k}\right\}. \tag{8.58}$$

The subprincipal symbol is easily shown to be invariant under a change of coordinates, provided that it is restricted to the set of zeros where the principal symbol and all its first derivatives vanish. By Proposition 8.4 the latter is exactly what happens in the case of $\mathscr{B}'$, and therefore the restriction of $\sigma_1(\mathscr{B}')$ to Char $\mathscr{B}'$ is invariant.

The computation of $\sigma_1(\mathscr{B}')$ is routine: it is based on the information provided by (8.30), (8.36) and, of course, the definition of $\mathscr{B}'$, (8.44). Let us here content ourselves with giving its value at a point $\omega_o \in$ Char $\mathscr{B}'$ which lies directly above the origin (i.e., in the cotangent space to X at 0). By Proposition 8.3 such a point is completely determined by requiring that $\eta_N = 1$; since $\sigma_1(\mathscr{B}')$ is positive-homogeneous of degree zero, it does not really matter what value of $\eta_N > 0$ we choose. Recalling that $h = \Delta r$, one easily finds

$$\sigma_1(\mathscr{B}') = 4\gamma(0) - \tfrac{1}{2}\Delta r(0) I \qquad at\ \omega_o. \tag{8.59}$$

We recall that γ is the matrix defined in (8.22). Its generic entry is a scalar γ_J^K where J, K are multi-indices of length q such that $N \notin J, N \notin K$. We can compute γ_J^K by using (8.22). But the computation of this and of $\sigma_1(\mathscr{B}')$ is made easier if we assume that the Levi form (8.16) has been diagonalized at the origin:

$$\frac{\partial^2 r(0)}{\partial z_j\, \partial \bar{z}_k} = \lambda_j \delta_{jk} \qquad (j, k = 1, \ldots, N-1), \tag{8.60}$$

which is always possible by a linear change of coordinates in $\mathbb{C}^N$. Then

$$\gamma_J^K(0) = \sum_{j \in J} \lambda_j \quad if\ J = K, \qquad \gamma_J^K(0) = 0 \quad if\ J \neq K. \tag{8.61}$$

Also $-\frac{1}{2}\Delta r(0) = -2\sum_{j=1,\ldots,N-1} \lambda_j$.

PROPOSITION 8.5. *Suppose that* (8.60) *holds. The restriction of* $\sigma_1(\mathscr{B}')$ *to the intersection of* Char $\mathscr{B}'$ *with the cotangent space to X at the origin is a diagonal matrix with diagonal entries equal to*

$$2\Big(\sum_{j \in J} \lambda_j - \sum_{j \notin J} \lambda_j\Big) \tag{8.62}$$

where J ranges over the collection of multi-indices with length q which do not contain N.

8.4. Hypoellipticity with Loss of One Derivative. Condition $Z(q)$

Propositions 8.3–8.5 enable us to use recent results of various authors (see mainly Boutet de Monvel [3] and Hörmander [16]) to obtain necessary and sufficient conditions for $\mathscr{B}'$ to be *hypoelliptic with loss of one derivative,* which means that for any open subset $\mathcal{O}$ of the boundary X, any distribution u' in $\mathcal{O}$, and any real number s,

$$\mathcal{B}'u' \in H^s_{\mathrm{loc}}(\mathcal{O}; V') \quad \textit{implies } u' \in H^s_{\mathrm{loc}}(\mathcal{O}; V'). \tag{8.63}$$

We have denoted by V' the space of vectors of the kind u'. Property (8.63) indeed evidences the loss of one order of smoothness, since $\mathcal{B}'$ is of order *one*: if $\mathcal{B}'$ were elliptic, which is taken to be the case of no derivative loss, $\mathcal{B}'u'$ in H^s_{loc} would imply u' in H^{s+1}_{loc}. Property (8.63) is equivalent to local estimates (valid provided that the open set $\mathcal{O}$ is small enough):

$$\|u'\|_s \leq \mathrm{const}\|\mathcal{B}'u'\|_s, \qquad u' \in C^\infty_c(\mathcal{O}; V'). \tag{8.64}$$

Among other things (8.64) implies that the analogue of the basic estimate (5.7) holds here with $s'' = s + 1$; we recall that here the number m of Section 5 is equal to 2, and that "some" orders m_j are equal to zero, others to one.

We proceed now to describe under which conditions property (8.63) holds.

We continue to use the notation of Section 8.3. In particular ω_o is the point in Char $\mathcal{B}'$ defined by $z = 0$, $\eta_N = 1$. We shall denote by E_o the *tangent* space to T^*X at ω_o, and by $E^{\mathbf{C}}_o$ its complexification.

The Taylor expansion of $\sigma(\mathcal{B}')$ about ω_o begins with the quadratic form

$$\tfrac{1}{2}\sum_{j=1}^{N-1} |\sigma_j|^2 \tag{8.65}$$

where, for each $j < N$, σ_j is the linear part of $(2i)\overline{\sigma(Z_j)}$ at ω_o (cf. (8.52)):

$$\sigma_j = \zeta_j - 2i\lambda_j z_j - \sum_{k=1}^{N-1} \frac{\partial^2 r(0)}{\partial \bar{z}_j\, \partial \bar{z}_k} \bar{z}_k. \tag{8.66}$$

We denote by $Q(\theta, \theta')$ the bilinear form on E_o defined by the quadratic form (8.65); we extend it bilinearly (not sesquilinearly!) to $E^{\mathbf{C}}_o$. We see that

$$Q(\theta, \bar{\theta}) \geq 0, \qquad \forall \theta \in E^{\mathbf{C}}_o. \tag{8.67}$$

We introduce now the fundamental symplectic form on E_o,

$$\Sigma = \sum_{j=1}^{N-1} d\xi_j \wedge dx_j + \sum_{j=1}^{N} d\eta_j \wedge dy_j, \tag{8.68}$$

and extend it as a bilinear form to $E^{\mathbf{C}}_o$. Since Σ is nondegenerate there is an endomorphism of $E^{\mathbf{C}}_o$, Φ, such that

$$Q(\alpha, \beta) = i\Sigma(\alpha, \Phi\beta), \qquad \alpha, \beta \in E^{\mathbf{C}}_o. \tag{8.69}$$

The following is not difficult to prove:

PROPOSITION 8.6. *The eigenvalues of the endomorphism* Φ *are the real numbers* $2\lambda_j, -2\lambda_j$ $(j = 1, \dots, N-1)$ (where, we recall, the λ_j are the eigenvalues of the Levi form of X at the origin).

We come now to the results in Boutet de Monvel [3] and Hörmander [16]. They tell us that if the open set $\mathcal{O}$ contains the origin and if (8.64) holds, then we must have the following:

(8.70) *Let* χ_j $(j = 1, \dots, r)$ *be the positive eigenvalues of* Φ. *Then, whatever the eigenvalue* μ *of* $\sigma_1(\mathcal{B}')$, *the vector* θ *in* $E_o^{\mathbf{C}}$ *such that* $\Phi^d\theta = 0$ *for some* $d > 0$, *the r-tuple* $m_1, \dots, m_r$ *of nonnegative integers,*

$$\mu + Q(\theta, \bar{\theta}) + \sum_{j=1}^{r} (2m_j + 1)\chi_j \neq 0.$$

According to Proposition 8.6 we may take $\chi_j = 2|\lambda_j|, j = 1, \dots, r$ (possibly after changing the indices of the λ_j); of course $r < N$, and $\lambda_j = 0$ if $r < j \leq N-1$. From Proposition 8.5 we derive the following:

(8.71) *For all eigenvalues* μ *of* $\sigma_1(\mathcal{B}')$, $\mu + \sum_{j=1}^{r} \chi_j \geq 0$.

If then we also take into account (8.67) we see that condition (8.70) is equivalent to the following one.

(8.72) *Whatever the eigenvalue* μ *of* $\sigma_1(\mathcal{B}')$, $\mu + \sum_{j=1}^{r} \chi_j > 0$.

It is obvious that condition (8.72), if it holds at ω_o, will also hold in a full neighborhood of ω_o (in Char $\mathcal{B}'$). Since (8.67) also holds in a neighborhood of ω_o, we derive that condition (8.70) will hold in a full neighborhood of ω_o as soon as (8.72) holds at ω_o. The results in Boutet de Monvel [3] then tell us that the estimates (8.64) hold if the set $\mathcal{O}$ is small enough.

Now, according to Proposition 8.5, (8.72) can be restated by saying that

(8.73) $$\sum_{j \in J} \lambda_j - \sum_{j \notin J} \lambda_j + \sum_{j=1}^{N-1} |\lambda_j| > 0$$

whatever the multi-index J, of length q, such that $N \notin J$.

Property (8.73) can, in turn, be rephrased as follows:

$Z(q)_0$ *The Levi form of X at the origin has at least $N - q$ eigenvalues which are strictly positive or at least $q + 1$ which are strictly negative.*

Thus, according to the main result in Boutet de Monvel [3] and Hörmander [16], we may state

THEOREM 8.1. *In order that $\mathscr{B}'$ be hypoelliptic with loss of one derivative in some open neighborhood of the origin in X, it is necessary and sufficient that condition $Z(q)_0$ hold.*

REMARK 8.4. When $q = N$, condition $Z(q)$ is trivially satisfied. We know that in this case the $\bar{\partial}$-Neumann problem reduces to the Dirichlet problem.

REMARK 8.5. The open set Ω (or its boundary X) is said to be *strongly pseudoconvex* at the point $z = 0$ if the Levi form at that point, (8.16), is positive definite, i.e., every one of its eigenvalues λ_j is >0. In this case it is clear that $Z(q)_0$ holds provided that $q \geq 1$. For $q = 0$ it does not; the operator $\mathscr{B}'$ is *not* hypoelliptic (with any regularity loss!) when $q = 0$ and Ω is strongly pseudoconvex. The $\bar{\partial}$-Neumann problem in this case is not even Fredholm.

It is not difficult to check that, when condition $Z(q)_0$ holds, the adjoint $\mathscr{B}'^*$ of $\mathscr{B}'$ is also hypoelliptic with loss of one derivative in the neighborhood of the origin in X. Indeed, as a simple and straightforward calculation shows, property (8.70) does not distinguish between $\mathscr{B}'$ and its adjoint. We may therefore apply Theorem 5.5, or rather Corollary 5.1:

THEOREM 8.2. *Suppose that condition $Z(q)$ is satisfied at every point of X. Then the $\bar{\partial}$-Neumann problem is hypoelliptic and Fredholm.*

We are tacitly using the fact that the "true" Calderon operator $\mathscr{B}$ of the $\bar{\partial}$-Neumann problem is the direct sum of $\mathscr{B}'$ acting on the vectors of the kind u' and the identity on vectors of the kind u''.

By refining the derivation of estimates of the kind (5.7), starting from the estimates of the kind (8.64), one can derive the so-called $\frac{1}{2}$-subelliptic estimates for the $\bar{\partial}$-Neumann problem. On this we refer to Folland and Kohn [1].

IV

Pseudodifferential Operators of Type (ρ, δ)

In order to generalize the parametrix construction of Section 1, Chapter I to certain classes of nonelliptic (yet hypoelliptic) equations, one must enlarge the pool of amplitudes that one is willing to use. Such a construction is carried out in Section 1 of this chapter. In an embryonic form it was first, I believe, attempted in Treves [1]. The classes of symbols best suited for such an extension were introduced and described by Hörmander in [3], and given the name $S_{\rho,\delta}$ classes. Roughly speaking the properties of pseudodifferential operators of type $(1, 0)$, our "standard" pseudodifferential operators (Chapter I), generalize well to type (ρ, δ) as long as we keep $0 \leq \delta < \frac{1}{2} < \rho$, as shown in Section 2. Those properties, and particularly the symbolic calculus, are based on the fact that the asymptotic expansion, constructed from an amplitude $a(x, y, \xi)$ and leading to the symbol

$$\sum_{\alpha} \frac{1}{\alpha!} \partial_\xi^\alpha D_y^\alpha a(x, y, \xi)|_{y=x},$$

consists of terms that get better as $|\alpha| \nearrow +\infty$. This basic property remains true if the deterioration resulting from differentiation in x, which is measured by δ, is more than matched by the improvement resulting from differentiation in ξ, measured by ρ, i.e., when $\delta < \rho$. This ceases to be the case when $\rho = \delta$, in particular when $\rho = \delta = \frac{1}{2}$. Yet, even in this case, the pseudodifferential operator Op a retains an important property of the better types, specifically the continuity property, say in the L^2 sense if we assume that $a(x, y, \xi)$ has degree zero. This is the celebrated *Calderon–Vaillancourt theorem* (Theorem 3.1; see Calderon and Vaillancourt [1, 2]).

Following an argument of Beals–Fefferman [1], we apply the Calderon–Vaillancourt theorem to prove the so-called *sharp Gårding inequality*, first stated and proved by Hörmander in [48]. In a nutshell, this inequality tells us that a self-adjoint pseudodifferential operator of order one (and of type (1, 0)), whose symbol is equal to a positive symbol (of degree one) plus a symbol of degree zero, is bounded from below—if we regard it as an unbounded linear operator on L^2. A refinement of this result, due to A. Melin, computes the best constant in the lower bound, accordingly called *Melin's constant.* We do not give Melin's result here and refer the reader to his article [1]. A further strengthening of the Gårding inequality is given in Fefferman and Phong [1].

1. Parametrices of Hypoelliptic Linear Partial Differential Equations

The construction of parametrices for elliptic linear PDEs with constant coefficients outlined in Section 1 of Chapter I can easily be extended to a much wider class of PDEs with constant coefficients. In order for formula (1.7) of Chapter I to make sense, the only thing that is required is that $|P(\xi)|$ be bounded away from zero outside some bounded set. The operator R defined in (1.9) of Chapter I is then automatically regularizing; formula (1.13) of Chapter I means that the operator K is a parametrix of $P(D)$. The question is then whether we can interpret it as some kind of pseudodifferential operator. We have seen that this is the case when the polynomial $P(\xi)$ is elliptic. Note that the distribution kernel associated with the operator K is equal to $K(x-y)$, where

$$K(x) = (2\pi)^{-n} \int e^{ix\cdot\xi}\chi(\xi)P(\xi)^{-1}\,d\xi. \tag{1.1}$$

If K is to be any kind of pseudodifferential operator, it must be pseudolocal. But if this is so, the differential operator $P(D)$ has to be hypoelliptic. Linear partial differential operators with constant coefficients which are hypoelliptic have been characterized by Hörmander (see [2]). There are several equivalent characterizations; the best suited to our purposes here is the following.

THEOREM 1.1. *In order that the linear partial differential operator with constant coefficients in $\mathbb{R}^n$, $P(D)$, be hypoelliptic, it is necessary and sufficient that there be a number $\rho > 0$ and two constants C, $R \geq 0$, such that the following is true*:

(1.2) *Given any n-tuple α, and any $\xi \in \mathbb{R}_n$, $|\xi| \geq R$,*

$$|P^{(\alpha)}(\xi)/P(\xi)| \leq C|\xi|^{-|\alpha|\rho}.$$

In particular we may select α with length $|\alpha| = m = \deg P$ such that $P^{(\alpha)} = \text{constant} \neq 0$. We derive therefore from (1.2) that, if $|\xi| \geq R$,

$$|P(\xi)^{-1}| \leq C|\xi|^{-m\rho}. \tag{1.3}$$

More generally, for all $\alpha \in \mathbb{Z}_+^n$, $\xi \in \mathbb{R}_n$, $|\xi| \geq R$,

$$|D_\xi^\alpha P(\xi)^{-1}| \leq C_\alpha |\xi|^{-(m+|\alpha|)\rho}. \tag{1.4}$$

We assume that the cutoff function χ used in (1.1) has its support in the region $|\xi| \geq R$ of (1.2) and is equal to one in the complement of some larger ball $|\xi| \geq R' > R$. We derive that the Fourier transform of $x^\alpha K$, which is equal to $(-D_\xi)^\alpha[\chi(\xi)/P(\xi)]$, is bounded by $C'_\alpha|\xi|^{-(m+|\alpha|)\rho}$. By letting $|\alpha|$ increase as much as needed, we may obtain that $x^\alpha K$ is as regular as we wish. In particular given any integer $M \geq 0$, there is another integer j such that $|x|^{2j}K(x) \in C^M(\mathbb{R}^n)$. This implies at once that K is C^∞ in the complement of the origin. As a consequence we obtain that the operator K is pseudolocal and that the differential operator $P(D)$ is hypoelliptic, thus proving the *sufficiency* of property (1.2). The proof of the *necessity* of (1.2) is based on the Seidenberg–Tarski theorem; see Hörmander [2].

EXAMPLE 1.1. The archetypical example of a hypoelliptic differential operator that is not elliptic is the *heat* operator (here in $\mathbb{R}^n$, with $n \geq 2$):

$$P(D) = \frac{\partial}{\partial x^n} - \sum_{j=1}^{n-1} \left(\frac{\partial}{\partial x^j}\right)^2.$$

The reader can easily check that (1.2) is satisfied with $\rho = \frac{1}{2}$.

REMARK 1.1. It follows from (1.3) that the numer ρ in (1.2) can be taken equal to one if and only if P is elliptic.

It was recognized quite early after Hörmander had proved Theorem 1.1 that the preceding method could be extended to some special classes of hypoelliptic linear partial differential operators $P(x, D)$ in an open subset Ω of $\mathbb{R}^n$ with variable (C^∞) coefficients (see Hörmander [1]). Exactly as in the elliptic case it is natural to try a formula of the kind (1.22) of Chapter I; the symbol $k(x, \xi)$ is then taken to be a formal series of the kind (1.28), Chapter I, except that it is now convenient to modify the equations used to determine successively the terms k_j. The reason for this is that, now, contrary to what happens when one deals with an elliptic operator $P(x, D)$, the essential information and means of control are not necessarily contained in the principal symbol $P_m(x, \xi)$. Thus we replace the expansion (1.27) of Chapter I by

(1.5) $$P(x, D_x + \xi) = P(x, \xi) + \sum_{\alpha \neq 0} \frac{1}{\alpha!} \partial_\xi^\alpha P(x, \xi) D_x^\alpha.$$

Here, as in Chapter I, the problem is always that of inverting the symbol $P(x, \xi)$ in the sense of composition of symbols, which means that we seek k so as to have (cf. Chapter I, (4.24))

(1.6) $$P(x, \xi) \odot k(x, \xi) = \sum_\alpha \frac{1}{\alpha!} \partial_\xi^\alpha P(x, \xi) D_x^\alpha k(x, \xi) = 1.$$

The idea is then to take (cf. Chapter I, (1.29), (1.30))

(1.7) $$k_0 = 1/P,$$

(1.8) $$k_j = -\sum_{1 \leq |\gamma| \leq j} \frac{1}{\gamma!} \partial_\xi^\gamma P(x, \xi) D_x^\gamma k_{j-|\gamma|}(x, \xi)/P(x, \xi).$$

What we wish for is clear: each term k_j should define a pseudolocal operator K_j, and there should be some "improvement" as j increases.

We reason in analogy with the constant coefficient case (cf. (1.3), (1.4)). We assume that $|P(x, \xi)|$ is bounded away from zero outside a set $|\xi| \geq R$, but it is natural here to let $R = R(x)$ depend (continuously) on x. On the other hand, we must assume that each differentiation with respect to ξ increases the decay at infinity of $1/P(x, \xi)$, by a fixed amount. Once we do this we may recognize that differentiation with respect to x should be allowed to decrease that decay—in other words to deteriorate our symbols—provided that it does not do so with excess. We shall therefore assume that there exist two real numbers ρ, δ such that the following holds true: Given any compact subset $\mathscr{K}$ of Ω and any pair $\alpha, \beta \in \mathbb{Z}_+^n$ there is $C = C(\mathscr{K}, \alpha, \beta) > 0$ such that

(1.9) $$|D_\xi^\alpha D_x^\beta \{1/P(x, \xi)\}| \leq C|P(x, \xi)|^{-1}|\xi|^{-|\alpha|\rho + |\beta|\delta}$$

for all x in $\mathscr{K}$, ξ in $\mathbb{R}_n$, $|\xi| \geq R(x)$.

By induction on $|\alpha + \beta|$ one can prove that (1.9) is equivalent to

(1.10) $$|D_\xi^\alpha D_x^\beta P(x, \xi)| \leq C'|P(x, \xi)|\,|\xi|^{-|\alpha|\rho + |\beta|\delta},$$

for the same x, ξ. Suppose then that the k_j are determined by (1.7), (1.8). What are the implications, concerning the k_j, to be derived from (1.9)? By the Leibniz formula we derive from (1.8)

(1.11) $$|D_\xi^\alpha D_x^\beta k_j| \leq \text{const} \sum |D_\xi^{\alpha'} D_x^{\beta'} k_o|\,|D_\xi^{\bar{\alpha}+\gamma} D_x^{\bar{\beta}} P|\,|D_\xi^{\tilde{\alpha}} D_x^{\tilde{\beta}+\gamma} k_{j-|\gamma|}|.$$

The summation with respect to γ is the same as in (1.8); the summation with respect to $\bar{\alpha}$, $\bar{\bar{\alpha}}$, a' is submitted to the constraint $\bar{\alpha} + \bar{\bar{\alpha}} + \alpha' = \alpha$, and that with respect to $\bar{\beta}$, $\bar{\bar{\beta}}$, β' to $\bar{\beta} + \bar{\bar{\beta}} + \beta' = \beta$. If we combine this with (1.9) and (1.10) we get, for $j = 1, 2, \ldots,$

$$|D_\xi^\alpha D_x^\beta k_j| \leq C'' \sum |\xi|^{-|\alpha'+\bar{\alpha}+\gamma|\rho+|\beta'+\bar{\beta}|\delta} |D_\xi^{\bar{\bar{\alpha}}} D_x^{\bar{\bar{\beta}}+\gamma} k_{j-|\gamma|}|. \tag{1.12}$$

By induction on j we derive from this (for x, ξ as in (1.9)):

$$|D_\xi^\alpha D_x^\beta k_j| \leq C^{\#} |P|^{-1} |\xi|^{-j(\rho-\delta)-|\alpha|\rho+|\beta|\delta}. \tag{1.13}$$

Thus, if any improvement is to occur from one term k_j to the next, we must have

$$\delta < \rho. \tag{1.14}$$

Under this hypothesis we can select cutoff functions $\chi_j(x, \xi) \in C^\infty(\Omega \times \mathbb{R}_n)$ with $\chi_j(x, \xi) = 0$ for $|\xi| < R_j(x)$, $\chi_j(x, \xi) = 1$ for $|\xi| > R_j'(x) > R_j(x) > R(x)$, such that

$$\sum_{j=0}^{+\infty} \chi_j(x, \xi) k_j(x, \xi) \tag{1.15}$$

converges in $C^\infty(\Omega \times \mathbb{R}_n)$ toward a function $k(x, \xi)$ satisfying inequalities of the kind

$$|D_\xi^\alpha D_x^\beta k(x, \xi)| \leq C |P(x, \xi)|^{-1} |\xi|^{-|\alpha|\rho+|\beta|\delta} \tag{1.16}$$

$$\textit{for } x \textit{ in } \Omega,\ \xi \textit{ in } \mathbb{R}_n,\ |\xi| \geq R(x),$$

where $C = C_{\alpha,\beta}(x)$ is a strictly positive continuous function in Ω.

We then ask the standard questions: Does the symbol k define a linear operator K which is very regular (cf. Chapter I, Section 2)? Is $P(x, D)K - I$ regularizing?

Consider the associated kernel

$$K(x, y) = (2\pi)^{-n} \int e^{i(x-y)\cdot\xi} k(x, \xi)\, d\xi. \tag{1.17}$$

Let us look at the kernel (1.17) near a point (x_o, y_o) such that $x_o \neq y_o$. Given any integer $N \geq 0$ we may write

$$K(x, y) = (2\pi)^{-n} \int e^{i(x-y)\cdot\xi} |x - y|^{-2N} \Delta_\xi^N k(x, \xi)\, d\xi,$$

and we use the fact that

$$|\Delta_\xi^N k(x, \xi)| \leq C |\xi|^{-2N\rho}.$$

We have tacitly used the hypothesis that $|P(x\ \xi)|$ is bounded away from zero for $|\xi| \geq R(x)$. If we suppose that

$$\rho > 0, \tag{1.18}$$

then it follows at once that $K(x, y)$ is a C^∞ function off the diagonal in $\Omega \times \Omega$.

Let us now show that $K(x, y)$ *is separately regular with respect to x and to y* (see Chapter I, Section 2). Whatever $\beta \in \mathbb{Z}_+^n$ and the integer $N \geq 0$,

$$D_x^\beta K(x, y) = (1 - \Delta_y)^N \left\{ (2\pi)^{-n} \int e^{i(x-y)\cdot\xi}(1 + |\xi|^2)^{-N} k_\beta(x, \xi)\, d\xi \right\},$$

where we have used the notation

$$k_\beta(x, \xi) = \sum_{\beta' \leq \beta} \binom{\beta}{\beta'} \xi^{\beta-\beta'} D_x^{\beta'} k(x, \xi).$$

It follows at once from (1.16) that, given any β, we may select N such that

$$(2\pi)^{-n} \int e^{i(x-y)\cdot\xi}(1 + |\xi|^2)^{-N} k_\beta(x, \xi)\, d\xi$$

is a continuous function of (x, y). We derive from this fact that the associated linear operator K maps continuously $C_c^\infty(\Omega)$ into $C^\infty(\Omega)$.

On the other hand, we have

$$\begin{aligned} D_y^\beta K(x, y) &= (2\pi)^{-n} \int e^{i(x-y)\cdot\xi} \xi^\beta k(x, \xi)\, d\xi \\ &= (2\pi)^{-n} \int (1 - \Delta_x)^N (e^{i(x-y)\cdot\xi})(1 + |\xi|^2)^{-N} \xi^\beta k(x, \xi)\, d\xi \\ &= \sum_{|\gamma+\gamma'| \leq 2N} c_{\gamma,\gamma'}^N D_x^\gamma K_{\gamma'}(x, y), \end{aligned}$$

where

$$K_{\gamma'}(x, y) = (2\pi)^{-n} \int e^{i(x-y)\cdot\xi}(1 + |\xi|^2)^{-N} \xi^\beta D_x^{\gamma'} k(x, \xi)\, d\xi.$$

From (1.16) we derive

$$(1 + |\xi|^2)^{-N} |\xi^\beta D_x^{\gamma'} k(x, \xi)| \leq C|\xi|^{-2N(1-\delta)+|\beta|}.$$

Thus, if we assume

$$\delta < 1, \tag{1.19}$$

we reach the conclusion that, whatever $\beta \in \mathbb{Z}_+^n$, we can find an integer N

large enough that $D_y^\beta K(x, y)$ can be expressed as a finite sum of derivatives of order $\leq 2N$ of continuous functions. This implies that the operator K extends as a continuous linear map $\mathscr{E}'(\Omega) \to \mathscr{D}'(\Omega)$.

Finally, in order to prove that $P(x, D)K - I$ is regularizing, one must study the "remainder" series

$$r = (1 - \chi_0) + \sum_\alpha \sum_{j=0}^{+\infty} \frac{1}{\alpha!} \partial_\xi^\alpha P[D_x^\alpha(\chi_j k_j) - \chi_{j+|\alpha|} D_x k_j]. \tag{1.20}$$

It is a good exercise for the reader to prove that the cutoff functions $\chi_j(x, \xi)$, together with the positive functions $R_j(x)$, $R_j'(x)$ which determine the choice of the χ_j, can be selected in such a way that $r(x, \xi)$ will decay fast at infinity (that is, as $|\xi| \to +\infty$) in an obvious sense, implying that the kernel

$$R(x, y) = (2\pi)^{-n} \int e^{i(x-y)\cdot\xi} r(x, \xi)\, d\xi \tag{1.21}$$

is a C^∞ function in $\Omega \times \Omega$. The reader should take advantage of the fact that in the summation at the right in (1.20), $|\alpha|$ does not exceed the degree of the polynomial $P(x, \xi)$ with respect to ξ.

The inequalities (1.9) suggest what is to be done when looking for classes of (nonstandard) pseudodifferential operators that include hypoelliptic differential operators of the kind of $P(x, D)$, and their parametrices such as (1.15). In principle one tries to duplicate all the aspects of the theory of standard pseudodifferential operators; but extending each of the basic "principles" of that theory leads to conditions concerning ρ and δ, as we shall now see.

2. Amplitudes and Pseudodifferential Operators of Type (ρ, δ)

As usual, Ω is an open subset in $\mathbb{R}^n$. As in Section 1, ρ and δ are real numbers.

DEFINITION 2.1. *We shall denote by $S^m_{\rho,\delta}(\Omega, \Omega)$ the space of C^∞ functions in $\Omega \times \Omega \times \mathbb{R}_n$, $a(x, y, \xi)$, having the following property:*

(2.1) *Given any compact subset $\mathscr{K}$ of $\Omega \times \Omega$ and any triplet of n-tuples α, β, γ, there is a constant $C_{\alpha,\beta,\gamma}(\mathscr{K}) > 0$ such that*

$$|D_\xi^\alpha D_x^\beta D_y^\gamma a(x, y, \xi)| \leq C_{\alpha,\beta,\gamma}(\mathscr{K})(1 + |\xi|)^{m-|\alpha|\rho+|\beta+\gamma|\delta}, \tag{2.2}$$

$$\forall (x, y) \in \mathscr{K},\ \xi \in \mathbb{R}_n.$$

We refer to the elements of $S^m_{\rho,\delta}(\Omega, \Omega)$ as *amplitudes* of degree m and of type (ρ, δ) in $\Omega \times \Omega$. The space $S^m_{\rho,\delta}(\Omega, \Omega)$ is naturally equipped with a Fréchet space topology, derived from the inequalities (2.2). If $m \leq m'$, $\rho \geq \rho'$, and $\delta \leq \delta'$, $S^m_{\rho,\delta}(\Omega, \Omega)$ is continuously embedded in $S^{m'}_{\rho',\delta'}(\Omega, \Omega)$.

We shall be of course interested in the distribution kernels

$$A_o(x, y) = (2\pi)^{-n} \int e^{i(x-y)\cdot\xi} a(x, y, \xi)\, d\xi \tag{2.3}$$

and in the associated linear operator, which we shall denote by Op a.

THEOREM 2.1. *Suppose* $0 < \rho$ *and* $\delta < 1$. *Then if* $a \in S^m_{\rho,\delta}(\Omega, \Omega)$, *the kernel* $A_o(x, y)$ *is very regular.*

As we recall, to say that $A_o(x, y)$ is very regular is to say that it is *separately regular*, i.e., it is a C^∞ function of each one of its arguments, x or y, with values in the space of distributions in Ω with respect to the other one, and that it is a C^∞ function of (x, y) off the diagonal in $\Omega \times \Omega$.

The proof of Theorem 2.1 is a routine extension of the proof, described in Section 1, of the same property for the kernel (1.17).

DEFINITION 2.2. *Assume* $\rho > 0$, $\delta < 1$. *We denote by* $\Psi^m_{\rho,\delta}(\Omega)$ *the space of continuous linear operators* $A : \mathscr{E}'(\Omega) \to \mathscr{D}'(\Omega)$ *such that there is an amplitude* $a \in S^m_{\rho,\delta}(\Omega, \Omega)$ *such that* $A = \text{Op}\, a$.

From there, notation and terminology are extended exactly as in the case of standard pseudodifferential operators. The latter correspond to the case $\rho = 1$, $\delta = 0$.

The intersection of all the spaces $S^m_{\rho,\delta}(\Omega, \Omega)$ (resp., $\Psi^m_{\rho,\delta}(\Omega)$) as m varies over $\mathbb{R}$ is identical to that of all the spaces $S^m(\Omega, \Omega)$ (resp., $\Psi^m(\Omega)$); it is $S^{-\infty}(\Omega, \Omega)$ (resp., $\Psi^{-\infty}(\Omega)$). Their union is sometimes denoted by $S_{\rho,\delta}(\Omega, \Omega)$ (resp., $\Psi_{\rho,\delta}(\Omega)$). Henceforth we shall always assume $\rho > 0$, $\delta < 1$. Because of the symmetry between x and y in the definition of the amplitudes, we have (cf. Chapter I, Theorem 3.1, Remarks 3.1, 3.2)

THEOREM 2.2. *If* $A \in \Psi^m_{\rho,\delta}(\Omega)$, *the transpose* ${}^tA$ *of* A *and the adjoint* A^* *of* A *also belong to* $\Psi^m_{\rho,\delta}(\Omega)$. *If* $a(x, y, \xi) \in S^m_{\rho,\delta}(\Omega, \Omega)$ *is such that* $A = \text{Op}\, a$, *then* ${}^tA = \text{Op}\, a'$ *and* $A^* = \text{Op}\, a^*$, *where* ${}^ta(x, y, \xi) = a(y, x, -\xi)$ *and* $a^*(x, y, \xi) = \overline{a(y, x, \xi)}$.

Properly supported pseudodifferential operators of type (ρ, δ) are defined exactly as in the case of standard pseudodifferential operators

(Definition 3.1). Then if $A \in \Psi^m_{\rho,\delta}(\Omega)$ and $B \in \Psi^{m'}_{\rho',\delta'}(\Omega)$, and if B is properly supported, we may form their compose $A \circ B$. Suppose $A = \operatorname{Op} a$, $B = \operatorname{Op} b$; then $A \circ B = \operatorname{Op} k$ with $k(x, y, \xi)$ given by (3.11) of Chapter I. Refinement of the proof of Theorem 3.2 of Chapter I (by using the reasoning in the last part of the proof of Theorem 3.2 of the present chapter) will enable the reader to establish the following

THEOREM 2.3. *If* $\delta'' = \sup(\delta, \delta') \le \rho'' = \inf(\rho, \rho')$, *given an operator* $A \in \Psi^m_{\rho,\delta}(\Omega)$ *and a properly supported operator* $B \in \Psi^{m'}_{\rho',\delta'}(\Omega)$, *we have* $A \circ B \in \Psi^{m+m'}_{\rho'',\delta''}(\Omega)$.

As we now see, the "invariance under diffeomorphism" will require more restrictions on the choice of ρ and δ. We deal with a diffeomorphism ϕ of Ω onto another open subset of $\mathbb{R}^n$, Ω'; we define the transfer A^ϕ of A exactly as in the standard case (see Chapter I, (3.14)). We know that if $A = \operatorname{Op} a$, then $A^\phi = \operatorname{Op} a^\phi$, with a^ϕ given by (3.20) of Chapter I. We must take a close look at the question whether (3.20) of Chapter I defines an element of $S^m_{\rho,\delta}(\Omega, \Omega)$. Actually we could have started with an amplitude $a(x, x', \xi)$ supported in an arbitrary neighborhood of the diagonal, such as the neighborhood $\mathscr{W}_0$ of Lemma 3.1 of Chapter I. In this case we may content ourselves with looking at

$$a^\phi(y, y', \eta) = |\det \mathscr{J}(y)| a(\overset{-1}{\phi}(y), \overset{-1}{\phi}(y'), {}^t\mathscr{J}_o^{-1}(y, y')\eta), \tag{2.4}$$

where $\mathscr{J}(y)$ is the Jacobian matrix of $\overset{-1}{\phi}$ at the point y, and $\mathscr{J}_o$ is the matrix defined in (3.22), Chapter I.

We assume that $a(x, x', \xi) \in S^m_{\rho,\delta}(\Omega, \Omega)$. A number $|\alpha|$ of differentiations of (2.4) with respect to η has an effect similar to that of the same number of differentiations of a with respect to ξ; it "improves" the decay by $|\alpha|\rho$ degrees. But differentiation with respect to y or y' has a more complicated effect since, by the chain rule, it involves differentiations of a with respect to ξ and brings out powers of η. Thus if we differentiate a^ϕ once with respect to y, say, we increase the degree of growth (as $|\eta| \to +\infty$) by $\sup(\delta, 1 - \rho)$. Thus we may state the following theorem. The proof is left to the reader.

THEOREM 2.4. *Suppose that* $1 - \delta \le \rho$ *and* $\delta < 1$. *Then if* $A \in \Psi^m_{\rho,\delta}(\Omega)$, *we have* $A^\phi \in \Psi^m_{\rho,\delta}(\Omega')$.

We come now to the development of a *symbolic calculus* for the operators belonging to $\Psi^m_{\rho,\delta}(\Omega)$. Let us deal with an operator $\operatorname{Op} a$ with

$a \in S^m_{\rho,\delta}(\Omega, \Omega)$. Formally, the symbol of $\operatorname{Op} a$ and of all the operators equivalent to $\operatorname{Op} a$ must be given by

$$\sum_{\alpha \in \mathbf{Z}^n_+} \frac{1}{\alpha!} \partial^\alpha_\xi D^\alpha_y a(x, x, \xi). \tag{2.5}$$

But of course we want the degree with respect to ξ of the sucessive terms $\partial^\alpha_\xi D^\alpha_y a(x, x, \xi)$ to decrease as $|\alpha|$ increases. When x remains in a compact subset of Ω, we have, by virtue of (2.2),

$$|D^{\alpha'}_\xi D^{\beta'}_x \partial^\alpha_\xi D^\alpha_y a(x, x, \xi)| \leq \text{const}(1 + |\xi|)^{m-|\alpha+\alpha'|\rho+|\beta'+\alpha|\delta}. \tag{2.6}$$

In other words, $\partial^\alpha_\xi D^\alpha_y a(x, x, \xi) \in S^{m-|\alpha|(\rho-\delta)}_{\rho,\delta}(\Omega)$. (The single Ω indicates that there is no more variable y; $S^m_{\rho,\delta}(\Omega)$ is the space of *symbols* of degree m and type (ρ, δ) in Ω.) Thus to have a symbolic calculus, paralleling that for standard pseudodifferential operators, we must have (1.14), that is, $\delta < \rho$. Then all the results that are valid in the case $\rho = 1$, $\delta = 0$ extend; in particular, Theorem 4.1 of Chapter I extends:

THEOREM 2.5. *Suppose $\delta < \rho$. The mapping $a(x, \xi) \mapsto \operatorname{Op} a$ of $S^m_{\rho,\delta}(\Omega)$ into $\Psi^m_{\rho,\delta}(\Omega)$ induces a bijection of $S^m_{\rho,\delta}(\Omega)/S^{-\infty}_{\rho,\delta}(\Omega)$ onto $\Psi^m_{\rho,\delta}(\Omega)/\Psi^{-\infty}_{\rho,\delta}(\Omega)$.*

The proof of Theorem 2.5 calls for one remark: establishing the injectivity of the map $a \mapsto \operatorname{Op} a$ requires an extension of Theorem 3.1, Chapter VI, to operators whose symbol is of type (ρ, δ). The extension is quite straightforward, and we shall not embark on it (see Hörmander [4]).

Theorem 2.5 gives us the right to talk of the symbol (or of the symbol class) of any equivalence class of pseudodifferential operators of type (ρ, δ).

We may avail ourselves of the results about symbols to prove a partial result about the L^2 continuity of operators belonging to $\Psi^0_{\rho,\delta}(\Omega)$:

THEOREM 2.6. *Suppose that $\rho > 0$, $\delta < 1$, and $\delta < \rho$. Every operator $A \in \Psi^0_{\rho,\delta}(\Omega)$ defines a continuous linear map $L^2_c(\Omega) \to L^2_{\text{loc}}(\Omega)$.*

PROOF. We shall prove that, for every integer $j = 0, 1, \ldots$, any operator $K \in \Psi^{-j(\rho-\delta)}_{\rho,\delta}(\Omega)$ defines a continuous linear map $L^2_c(\Omega) \to L^2_{\text{loc}}(\Omega)$. It is clearly true if j is sufficiently large. We reason therefore by descending induction on j. We may and shall assume that K is properly supported, and denote by $k(x, \xi)$ a representative of its symbol class. We select a positive C^∞ function $C(x)$ in Ω such that

$$\sup_{\xi \in \mathbf{R}_n} |k(x, \xi)|^2 < C(x)(1 + |\xi|^2)^{-j(\rho-\delta)}, \qquad \forall x \in \Omega. \tag{2.7}$$

Let B denote a properly supported element of $\Psi_{\rho,\delta}^{-j(\rho-\delta)}(\Omega)$ with symbol

$$\{C(x)(1+|\xi|^2)^{-j(\rho-\delta)} - |k(x,\xi)|^2\}^{1/2}. \tag{2.8}$$

If this is so, $R = (1-\Delta)^{-j(\rho-\delta)/2}C(x)(1-\Delta)^{-j(\rho-\delta)/2} - K^*K - B^*B$ belongs to $\Psi_{\rho,\delta}^{-(j+1)(\rho-\delta)}(\Omega)$. By the induction on j, the operator R maps continuously $L_c^2(\Omega)$ into $L_{\text{loc}}^2(\Omega)$. This means that for a suitable choice of the positive continuous function C_1 in Ω,

$$\int_\Omega \overline{u(x)}\,[(1-\Delta)^{-j(\rho-\delta)/2}C(x)(1-\Delta)^{-j(\rho-\delta)/2}u(x)]\,dx \geq$$

$$\int_\Omega |Ku|^2\,dx + \int_\Omega |Bu|^2\,dx - \int_\Omega C_1(x)|u|^2\,dx,$$

for every $u \in C_c^\infty(\Omega)$. From this we conclude at once that K is continuous. □

COROLLARY 2.1. *Let ρ and δ be as in Theorem 2.6. Let m, s be any two real numbers. Any operator $A \in \Psi_{\rho,\delta}^m(\Omega)$ defines a continuous linear map $H_c^s(\Omega) \to H_{\text{loc}}^{s-m}(\Omega)$.*

PROOF. Indeed we can write $A \sim (1-\Delta)^{m/2}A_0$ with $A_0 \in \Psi_{\rho,\delta}^0(\Omega)$ properly supported. □

We may summarize all the preceding results by the following remark:

REMARK 2.1. All the definitions and theorems in the theory of standard pseudodifferential operators extend in a straightforward manner to operators in the classes $\Psi_{\rho,\delta}^m$ provided that we have

$$1-\rho \leq \delta < \rho, \qquad \delta < 1. \tag{2.9}$$

Notice that (2.9) demands

$$1/2 < \rho. \tag{2.10}$$

We conclude this section by reinterpreting the results of Section 1 within the framework of the spaces $S_{\rho,\delta}^m$ and $\Psi_{\rho,\delta}^m$. For this we return to (1.9), (1.10), and (1.13). We make the further assumption that there is a real number m' such that

$$|P(x,\xi)^{-1}| \leq C|\xi|^{m'} \qquad \text{if } x \in \mathscr{K},\ |\xi| \geq R(x), \tag{2.11}$$

where $\mathscr{K}$ is an arbitrary compact subset of Ω and $C > 0$ depends on $\mathscr{K}$. We assume that $\delta < \rho$, $\rho > 0$, $\delta < 1$. Let $k(x,\xi)$ be the sum of the series (1.15).

The estimates (1.16) show that $K = \mathrm{Op}\, k$ belongs to $\Psi^{m'}_{\rho,\delta}(\Omega)$. And $P(x, D)K - I$ is regularizing, in other words, K is a parametrix of $P(x, D)$.

REMARK 2.2. Unless we assume that all conditions (2.9) are verified, the properties (1.9) and (2.11) of $P(x, \xi)$ are not invariant under diffeomorphism, as shown in the following example.

EXAMPLE 2.1. Make the following change of variables:

$$y^j = x^j, \quad j = 1, \ldots, n-1, \qquad y^n = x^n + (x^1)^2/2. \tag{2.12}$$

The *heat operator* in Example 1.1 becomes

$$-\left(\frac{\partial}{\partial y^1}\right)^2 - 2y^1\left(\frac{\partial}{\partial y^1}\right)\left(\frac{\partial}{\partial y^n}\right) - \left(y^1\frac{\partial}{\partial y^n}\right)^2 - \sum_{j=2}^{n-1}\left(\frac{\partial}{\partial y^j}\right)^2 \tag{2.13}$$

whose (total) symbol is $(\eta_1 - y^1\eta_n)^2 + \eta_2^2 + \cdots + \eta_n^2$, whose zero set is "big"; it is the set

$$\eta_1 = y^1\eta_n, \qquad \eta_2 = \cdots = \eta_n = 0. \tag{2.14}$$

Certainly this symbol does not satisfy (2.11).

REMARK 2.3. If we suppose that (1.9) and (2.11) hold, with ρ, δ as in (2.9), then it follows from Theorem 2.4 that a diffeomorphism does not affect the essential properties of the parametrix K and that therefore (2.11) remains valid after a change of coordinates. (This follows from the fact that $P(x, \xi) \odot k(x, \xi) \sim 1$.)

For instance, if the condition (1.2) is satisfied with $\rho > 1/2$ the theory applies. Differential operators with constant coefficients such that this is true do indeed exist:

EXAMPLE 2.2. Consider the polynomial in two variables:

$$P(\xi_1, \xi_2) = \xi_1^6 + \xi_1^4\xi_2^4 + \xi_2^6. \tag{2.15}$$

It satisfies (1.2) with $\rho = 3/4$. That ρ cannot be $>3/4$ is seen at once by computing the derivatives $P^{(i,j)}$ with $j = 1, 2, 3, 4$, and putting $\xi_1 = \xi_2 \to +\infty$. Simple estimates show that it can be taken equal to $3/4$.

Variable coefficients can be highly degenerate, as the following example of Hörmander shows.

EXAMPLE 2.3. Take, in $\mathbb{R}^n$,

$$P(x, \xi) = 1 + |x|^{2\nu}|\xi|^{2\mu}. \tag{2.16}$$

Suppose that $\delta = \mu/\nu < 1$. If we then determine the k_j by the equations (1.7), (1.8) it is not difficult to check that $k = \sum_j k_j \in S^0_{1,\delta}(\mathbb{R}^n)$. Since Op k is a parametrix of $P(x, D)$, the latter is hypoelliptic. This is not true when $\mu = \nu$ (see Hörmander [1, 4]).

3. The Calderon–Vaillancourt Theorem and the Sharp Gårding Inequality

The reader will have noticed that some of the results of Section 2 do not apply to the case where $\delta = \rho$ ($\frac{1}{2} \leq \rho < 1$). Mainly what fails to work in this case is the symbolic calculus, which requires $\delta < \rho$ for the successive terms in the formal symbols to show an "improvement." Our proof of the L^2 continuity of the pseudodifferential operators of type (ρ, δ) (Theorem 2.6) relies on it and therefore does not apply. That the statement is nevertheless true is the content of the *Calderon–Vaillancourt theorem* [1], stated and proved as Theorem 3.1. Following ideas of Beals and Fefferman [1] we derive from it the so-called *sharp Gårding inequality* (first proved by Hörmander in [5]). Beals and Fefferman have introduced classes of symbols (denoted by $S^{M,m}_{\Phi,\phi}$) which generalize the classes $S^m_{\rho,\delta}$, and permit an extension of the symbolic calculus, with interesting applications to solvability and hypoellipticity of pseudodifferential equations (see Beals [1, 2] and Beals and Fefferman [3]). Recent work of Hörmander [19] further generalizes such "calculi."

THEOREM 3.1. *Suppose that* $0 < \rho < 1$. *Every operator* $A \in \Psi^0_{\rho,\rho}(\Omega)$ *defines a continuous linear map* $L^2_c(\Omega) \to L^2_{\text{loc}}(\Omega)$.

PROOF. We take $A = \text{Op}\, a$ with $a(x, y, \xi) \in S^0_{\rho,\rho}(\Omega, \Omega)$. Actually we may assume that the (x, y)-projection of the support of $a(x, y, \xi)$ is contained in a compact subset of $\Omega \times \Omega$ (cf. beginning of the proof of Theorem 2.1, Chapter I).

We select a large integer N and write, after integration by parts

$$Au(x) = (2\pi)^{-n} \iint e^{i(x-y)\cdot\xi} b(x, y, \xi)u(y)\, dy\, d\xi, \tag{3.1}$$

where

$$b(x, y, \xi) = [1 + (-\Delta_\xi)^N \phi(\xi)^{2N}]\{[1 + \phi(\xi)^{2N}|x - y|^{2N}]^{-1} a(x, y, \xi)\}, \tag{3.2}$$

$$\phi(\xi) = (1 + |\xi|^2)^{\rho/2}. \tag{3.3}$$

By hypothesis we have

$$|D_\xi^\alpha D_x^\beta a| + |D_\xi^\alpha D_y^\beta a| \le C\phi^{|\beta|-|\alpha|}. \tag{3.4}$$

Since $|D_\xi^\alpha \phi| \le C(1 + |\xi|)^{-|\alpha|}\phi$, we derive

$$|D_\xi^\alpha D_x^\beta b| + |D_\xi^\alpha D_y^\beta b| \le C'\phi(\xi)^{|\beta|-|\alpha|}/[1 + \phi(\xi)^{2N}|x - y|^{2N}]. \tag{3.5}$$

First we apply (3.5) with $\alpha = \beta = 0$. If $2N > n$, we derive right away

$$\iint |b(x, y, \xi)b(x, y', \xi)|\, dx\, dy' \le C_o^2 \phi(\xi)^{-2n}. \tag{3.6}$$

Let us then define

$$b^\#(x, y, \xi) = e^{i(x-y)\cdot\xi} b(x, y, \xi), \tag{3.7}$$

$$B(\xi)u(x) = (2\pi)^{-n} \int b^\#(x, y, \xi)u(y)\, dy. \tag{3.8}$$

We apply the following lemma:

LEMMA 3.1. *Let X, Y be two measure spaces, with respective measures dx, dy, $b(x, y)$ a measurable function in $X \times Y$ such that*

$$\iint |b(x, y)b(x, y')|\, dx\, dy' \tag{3.9}$$

belongs to $L^\infty(Y)$ and has L^∞ norm $\le C^2$. The operator B defined by $Bu(x) = \int b(x, y)u(y)\, dy$ is bounded from $L^2(Y)$ to $L^2(X)$ with norm $\le C$.

PROOF. If $u \in L^2(Y)$, and if we write $b^\#(x, y, y') = b(x, y)b(x, y')$ (symmetric in y, y'), then

$$\begin{aligned}\int |Bu|^2\, dx &= \iiint b(x, y)\bar{b}(x, y')u(y)\bar{u}(y')\, dy\, dy'\, dx \\ &\le \iiint |b^\#(x, y, y')|^{1/2}|u(y)||b^\#(x, y', y)|^{1/2}|u(y')|\, dy\, dy'\, dx \\ &\le \iiint |b^\#(x, y, y')||u(y)|^2\, dy\, dy'\, dx \le C^2 \int |u|^2\, dy.\end{aligned}$$

□

We conclude that the operator $B(\xi)$ in (3.8) is bounded from $L^2(\Omega)$ to itself, with norm $\le C_o\phi(\xi)^{-n}$.

Next we estimate the operator norm (on L^2) of $B(\xi)^*B(\eta)$ and that of $B(\xi)B(\eta)^*$. This is because we are going to take advantage of the following lemma of M. Cotlar.

LEMMA 3.2. *Let Ξ be a measure space with measure $d\xi$, H a Hilbert space, with inner product $(\ ,\ )_H$, $B(\xi)$ a measurable function of ξ with values in the Banach space of bounded linear maps of H into itself, $L(H)$, where the norm is denoted by $\|\ \ \|$. Assume that there is a constant $C > 0$ such that, for all ξ in Ξ,*

$$\text{(3.10)}\qquad \int \|B(\xi)^*B(\eta)\|^{1/2}\,d\eta \le C, \qquad \int \|B(\xi)B(\eta)^*\|^{1/2}\,d\eta \le C.$$

Then given any two elements f, g of H, the function $(B(\xi)f, g)_H$ is integrable over Ξ. If we set

$$\text{(3.11)}\qquad (Bf, g)_H = \int (B(\xi)f, g)_H\,d\xi,$$

then B is a bounded linear operator on H with norm $\le C$.

PROOF. Let X be a measurable subset of Ξ having finite measure and on which $\|B(\xi)\| \le M < +\infty$; let us define an operator B_X like B except that we replace $B(\xi)$ by zero in the complement of X. Given any sequence of bounded linear operators $B_1, \ldots, B_j$ on H, we have

$$\|B_1 \cdots B_j\| \le \|B_1\|\|B_2B_3\| \cdots \|B_{2p}B_{2p+1}\|,$$

$$\|B_1 \cdots B_j\| \le \|B_1B_2\| \cdots \|B_{2p-1}B_{2p}\|\|B_{2p+1}\|,$$

where $j \le 2p+1 \le j+1$ and we agree that $B_{2p+1} = I$ when $j = 2p$, and therefore, after multiplication and root extraction,

$$\text{(3.12)}\qquad \|B_1 \cdots B_j\| \le \|B_1\|^{1/2}\|B_1B_2\|^{1/2}\|B_2B_3\|^{1/2} \cdots \|B_{j-1}B_j\|^{1/2}\|B_j\|^{1/2}.$$

Observe also that $\|B_X\|^2 = \|B_X^*B_X\|$; hence for all $j = 1, 2, \ldots,$

$$\text{(3.13)}\qquad \|B_X\|^{2^j} = \|(B_X^*B_X)^{2^{j-1}}\| =$$

$$\sup \iint_{X\times X} (B(\xi)^*B(\eta)(B_X^*B_X)^{2^{j-1}-1}u, u)\,d\xi\,d\eta =$$

$$\sup \int_{X^{2\nu}} (B(\xi_1)^*B(\eta_1) \cdots B(\xi_\nu)^*B(\eta_\nu)u, u)\,d\xi_1 \cdots d\xi_\nu d\eta_1 \cdots d\eta_\nu,$$

where the *supremum* is taken over the unit sphere of H and $\nu = 2^{j-1}$. By combining (3.12) and (3.13) we obtain

$$(3.14)\quad \|B_X\|^{2^j} \le \int_{X^{2\nu}} \|B(\xi_1)^*\|^{1/2}\|B(\xi_1)^*B(\eta_1)\|^{1/2}\|B(\eta_1)B(\xi_2)^*\|^{1/2} \cdots \|B(\xi_\nu)^*B(\eta_\nu)\|^{1/2}\|B(\eta_\nu)\|^{1/2}\, d\xi_1 \cdots d\xi_\nu\, d\eta_1 \cdots d\eta_\nu.$$

We take advantage of (3.10) and of the fact that $\|B(\xi)\|$ is bounded by M on X. We get

$$\|B_X\|^{2^j} \le M(\text{meas } X)^2 C^{2^j},$$

that is, $\|B_X\| \le [M(\text{meas } X)^2]^{2^{-j}} C$. Letting j go to $+\infty$ yields $\|B_X\| \le C$. Letting X grow to Ξ yields what was sought. □

In the proof of Theorem 3.1, $B(\xi)B(\eta)^*$ is the operator associated with the kernel

$$c(x, y, \xi, \eta) = e^{i(x\cdot\xi - y\cdot\eta)} \int e^{-iz\cdot(\xi-\eta)} b(x, z, \xi)\bar{b}(y, z, \eta)\, dz.$$

Let us introduce the following differential operator with respect to z:

$$L = [\phi(\xi)^2 + \phi(\eta)^2 + |\xi - \eta|^2]^{-1}[\phi(\xi)^2 + \phi(\eta)^2 - \Delta_z].$$

Obviously $L \exp\{-iz\cdot(\xi-\eta)\} = \exp\{-iz\cdot(\xi-\eta)\}$. We derive from (3.5), for any integer $J \ge 0$,

$$(3.15)\quad |L^J[b(x, z, \xi)\bar{b}(y, z, \eta)]| \le C\{1 + [\phi(\xi) + \phi(\eta)]^{-1}|\xi - \eta|\}^{-2J}[1 + \phi(\xi)|x - z|]^{-2N}[1 + \phi(\eta)|y - z|]^{-2N}.$$

Consequently, by integration by parts with respect to z in the integral expressing $c(x, y, \xi, \eta)$, we get

$$(3.16)\quad \left\{\iint |c(x, y, \xi, \eta)c(x, y', \xi, \eta)|\, dx\, dy'\right\}^{1/2} \le C'\{1 + [\phi(\xi) + \phi(\eta)]^{-1}|\xi - \eta|\}^{-2J}\phi(\xi)^{-n}\phi(\eta)^{-n}.$$

We may apply Lemma 3.1: $B(\xi)B(\eta)^*$ is a bounded linear operator on $L^2(\Omega)$ with norm bounded by the right-hand side of (3.16).

We are seeking an estimate of $\int_{\mathbb{R}_n} \|B(\xi)B(\eta)^*\|^{1/2}\, d\eta$. Therefore we must estimate the integral with respect to η of the quantity

$$F(\xi, \eta) = \{1 + [\phi(\xi) + \phi(\eta)]^{-1}|\xi - \eta|\}^{-J}\phi(\xi)^{-n/2}\phi(\eta)^{-n/2}.$$

In the region $|\xi - \eta| \geq |\eta|/2$, where we also have $|\xi| \leq 3|\xi - \eta|$, and

$$1 + [\phi(\xi) + \phi(\eta)]^{-1}|\xi - \eta| \geq c(1 + |\eta|)^{1-\rho}, \qquad c > 0,$$

we have

$$|F(\xi, \eta)| \leq C''(1 + |\eta|)^{-n\rho/2 - J(1-\rho)};$$

consequently, the integral of $F(\xi, \eta)$ over that region is bounded independently of ξ, provided that J is large enough.

Suppose now that $|\xi - \eta| \leq |\eta|/2$. Then there is $c' > 0$ such that $\phi(\xi) \leq c'^{-1}\phi(\eta) \leq c'^{-2}\phi(\xi)$, and therefore

$$|F(\xi, \eta)| \leq C''[1 + |\xi - \eta|/\phi(\xi)]^{-J}\phi(\xi)^{-n}.$$

The integral of F over the latter region is also bounded independently of ξ, provided $J > n$. □

From Theorem 3.1 we derive the following important result, known as the *sharp Gårding inequality*:

THEOREM 3.2. *Let $a(x, \xi) \in S^1_{1,0}(\Omega)$ have the following property*:

(3.17) $$a(x, \xi) \geq 0 \qquad \text{for all } x \text{ in } \Omega,\ \xi \text{ in } \mathbb{R}_n.$$

Given any compact subset $\mathcal{K}$ of Ω there is a constant $C > 0$ such that

(3.18) $$\operatorname{Re} \int_K \bar{u}(\operatorname{Op} a)u\, dx + C\int |u|^2\, dx \geq 0 \qquad \text{for all } u \in C^\infty_c(\mathcal{K}).$$

PROOF. Let Ω' be any relatively compact open subset of Ω and let $g \in C^\infty_c(\Omega)$ be nonnegative and equal to one in Ω'; we set

$$b(x, \xi) = g(x)[1 + a(x, \xi)]^{1/2},$$

and write $B = \operatorname{Op} b$.

We shall apply the following elementary lemma:

LEMMA 3.3. *Let a be as in Theorem 3.2. To every compact subset $\mathcal{K}$ of Ω there is a constant $C > 0$ such that*

$$(1 + |\xi|)^{-1/2}|d_x a(x, \xi)| + (1 + |\xi|)^{1/2}|d_\xi a(x, \xi)| \leq Ca(x, \xi)^{1/2}$$

for all x in $\mathcal{K}$, ξ in $\mathbb{R}_n$.

PROOF. It suffices to reason in sufficiently thin *closed* conic subsets of $\Omega \times \mathbb{R}_n$ and then use a finite covering of $\mathcal{K} \times \mathbb{R}_n$ by such sets. Let x_o be a point close to x in Ω, ξ^o a point in $\mathbb{R}_n$ such that $|\xi - \xi^o| \leq \frac{1}{2}(1 + |\xi|)$. From the fact that $a \in S^1_{1,0}(\Omega)$ and from the classical expression of the remainder in a finite Taylor expansion, here of order two, we derive

$$0 \leq a(x_o, \xi^o) \leq a(x, \xi) + (x - x_o) \cdot a_x(x, \xi) + (\xi - \xi^o) \cdot a_\xi(x, \xi) + M \sum_{|\alpha+\beta|=2} |(x - x_o)^\alpha (\xi - \xi^o)^\beta| (1 + |\xi|)^{1-|\beta|}.$$

Here M is a positive constant that depends solely on the compact set $\mathcal{K}$ and on the symbol a. Let $\varepsilon > 0$ be suitably small. We take $x_0 = x + \varepsilon a_x(x, \xi)(1 + |\xi|)^{-1}$, $\xi^o = \xi + \varepsilon a_\xi(x, \xi)(1 + |\xi|)$; then from the preceding inequality we get

$$(1 + |\xi|)^{-1} |a_x|^2 + (1 + |\xi|) |a_\xi|^2 \leq a + M\varepsilon^2 \sum_{|\alpha+\beta|=2} |a_x^\alpha| |a_\xi^\beta| (1 + |\xi|)^{(|\beta|-|\alpha|)/2}.$$

Then we get the result by taking $M\varepsilon^2$ very small in comparison to one. □

It is then a routine application of Lemma 3.3 and of the chain rule that for a suitable choice of the positive constant $C_{\alpha,\beta}$ in Ω,

$$|D_\xi^\alpha D_x^\beta b| \leq C_{\alpha,\beta} (1 + a)^{1/2 - |\alpha+\beta|/2} (1 + |\xi|)^{(|\beta|-|\alpha|)/2}. \tag{3.19}$$

The compose B^*B is equal to Op c with

$$c(x, \xi) = (2\pi)^{-n} \iint e^{-i(x-z)\cdot(\xi-\eta)} b(z, \xi) b(z, \eta) \, dz \, d\eta. \tag{3.20}$$

Let us write

$$b(z, \eta) = b(z, \xi) + \sum_{|\gamma|=1} (\eta - \xi)^\gamma b_\gamma(z, \xi, \eta), \tag{3.21}$$

$$b_\gamma(z, \xi, \eta) = \int_0^1 \partial_\xi^\gamma b(z, \xi + t(\eta - \xi)) \, dt. \tag{3.22}$$

After integration by parts this implies

$$c(x, \xi) = b(x, \xi)^2 - r(x, \xi) = [1 + a(x, \xi)] g(x)^2 - r(x, \xi), \tag{3.23}$$

with

$$r(x, \xi) = (2\pi)^{-n} \iint e^{-i(x-z)\cdot(\xi-\eta)} p(z, \xi, \eta) \, dz \, d\eta, \tag{3.24}$$

where

$$p(x, \xi, \eta) = \sum_{|\gamma|=1} D_z^\gamma b(z, \xi) b_\gamma(z, \xi, \eta). \tag{3.25}$$

Theorem 3.2 will be proved if we prove that $r \in S^0_{1/2,1/2}(\Omega)$. Indeed, by Theorem 3.1, Op r maps $L^2_c(\Omega)$ continuously into $L^2_{\text{loc}}(\Omega)$. But in Ω', (3.23) tells us that Op $r \sim I + \text{Op}\, a - B^*B$; then the conclusion in Theorem 3.2 follows.

Note that given any pair of n-tuples α, β, we have

$$D^\alpha_\xi D^\beta_x r(x,\xi) = \sum \binom{\alpha}{\alpha'}(2\pi)^{-n} \iint e^{-i(x-z)\cdot(\xi-\eta)} D^{\alpha'}_\xi D^\beta_z D^{\alpha-\alpha'}_\eta p(z,\xi,\eta)\, dz\, d\eta.$$

By availing ourselves of (3.25) and by the Leibniz formula, we see that $D^\alpha_\xi D^\beta_x r$ is a finite linear combination of terms of the form

$$\iint e^{-i(x-z)\cdot(\xi-\eta)} q(z,\xi,\eta)\, dz\, d\eta, \tag{3.26}$$

$$q(z,\xi,\eta) = [D^{\bar\alpha}_\xi D^{\bar\beta+\gamma}_z b(z,\xi)][D^{\bar{\bar\alpha}}_\xi D^{\bar{\bar\beta}}_z D^{\alpha^*}_\eta b_\gamma(z,\xi,\eta)], \tag{3.27}$$

where $\bar\alpha + \bar{\bar\alpha} + \alpha^* = \alpha$, $\bar\beta + \bar{\bar\beta} = \beta$, $|\gamma| = 1$. It follows at once from (3.19) that for any pair of n-tuples α', β', we have

$$|D^{\alpha'}_\eta D^{\beta'}_z q(z,\xi,\eta)| \le C(1+|\xi|)^{(|\beta+\beta'|-|\alpha+\alpha'|)/2}, \tag{3.28}$$

assuming that

$$|\xi - \eta| \le |\xi|/2. \tag{3.29}$$

We may write $q = q_1 + q_2$, with q_j satisfying the same estimates as (3.28) and supp q_1 contained in the region (3.29), supp q_2 contained in the set $|\xi - \eta| \ge |\xi|/4$. We may write

$$K_2(x,\xi) = \iint e^{-i(x-z)\cdot(\xi-\eta)}(1+|\xi-\eta|^2)^{-k}(1-\Delta_z)^k q_2(z,\xi,\eta)\, dz\, d\eta. \tag{3.30}$$

But we have

$$|(1-\Delta_z)^k q_2| \le C(1+|\xi-\eta|)^{k+(|\beta-|\alpha|)/2}$$

and also, of course, $|\eta| \le 5|\xi-\eta|$. It follows at once, by selecting k sufficiently large in (3.30), that $|K_2| \le \text{const}(1+|\xi|)^{(|\beta|-|\alpha|)/2}$.

We call $K_1(x,\xi)$ the integral (3.26) with q replaced by q_1. Let us set (cf. (3.3)) $\phi(\xi) = (1+|\xi|^2)^{1/4}$ and introduce the differential operator

$$L_\# = [1+\phi(\xi)^2|x-z|^2 + \phi(\xi)^{-2}|\xi-\eta|^2]^{-1}[1-\phi(\xi)^2\Delta_\eta - \phi(\xi)^{-2}\Delta_z].$$

By integration by parts we have

$$K_1(x,\xi) = \iint e^{-i(x-z)\cdot(\xi-\eta)} L^N_\# q_1(z,\xi,\eta)\, dz\, d\eta.$$

We derive at once from (3.28) that

$$|L_{\#}^{N}q_1| \le C(1+|\xi|)^{(|\beta|-|\alpha|)/2}/[1+\phi(\xi)^2|x-z|^2+\phi(\xi)^{-2}|\xi-\eta|^2]^{-N}. \tag{3.31}$$

Let us integrate both sides in (3.31) with respect to (z, η) over the whole space $\mathbb{R}^n \times \mathbb{R}_n$, assuming $N > n$. We obtain at once

$$|K_1| \le \int |L_{\#}^{N}q_1|\, dz\, d\eta \le C'(1+|\xi|)^{(|\beta|-|\alpha|)/2}.$$

If we combine this inequality with the analogous one for K_2, we reach the conclusion that (3.26) and consequently also $D_\xi^\alpha D_x^\beta r$ satisfy similar inequalities. This proves that $r \in S^0_{1/2,1/2}(\Omega)$. □

EXAMPLE 3.1. Let $a \in S^1_{1,0}(\Omega)$ be exactly as in Theorem 3.2. Application of Lemma 3.3 and of the chain rule shows that for every compact subset $\mathscr{K}$ of Ω and for every pair of n-tuples α, β, there is a constant $C = C_{\alpha,\beta}(\mathscr{K}) > 0$ such that

$$|D_\xi^\alpha D_x^\beta(e^{-a})| \le Ce^{-a/2}(1+|\xi|)^{(|\beta|-|\alpha|)/2}, \tag{3.32}$$

in particular,

$$e^{-a} \in S^0_{1/2,1/2}(\Omega). \tag{3.33}$$

Thus $\operatorname{Op} e^{-a}$ defines a bounded linear operator $L^2_c(\Omega) \to L^2_{\mathrm{loc}}(\Omega)$, by the Calderon–Vaillancourt theorem (Theorem 3.1).

EXAMPLE 3.2. This example is closely related to Example 3.1. Let X be a *compact* C^∞ manifold, A a classical pseudodifferential operator on X, of order one, whose principal symbol, $a(x, \xi)$, is everywhere ≥ 0.

The abstract theory of evolution equations (see, for instance, Treves [3], Part III) shows that the following initial value problem, where the unknown is an operator-valued function of $t \ge 0$,

$$\partial_t U + AU = 0 \quad in\ X \times \bar{\mathbb{R}}_+, \qquad U(0) = I \quad in\ X, \tag{3.34}$$

has a unique solution. The argument is quite similar to that used in Section 1.3 of Chapter III. It is based on certain *energy inequalities*, which we now derive. Let us suppose that X carries a Riemannian structure and deal with the space $L^2(X)$ defined by such a structure. Let $(\ ,\ )_0$ and $\|\ \|_0$ denote the inner product and the norm in $L^2(X)$ and let $u(t)$ be a smooth function of

$t \geq 0$ valued in $C^\infty(X)$. We have, for each $t \geq 0$,

$$\|u(t)\|_0^2 - \|u(0)\|_0^2 = 2\,\mathrm{Re} \int_0^t \left(\frac{\partial u}{\partial t}(t'), u(t')\right)_0 dt'$$

$$= 2\,\mathrm{Re} \int_0^t \left(\frac{\partial u}{\partial t}(t') + Au(t'), u(t')\right)_0 dt'$$

$$- 2\,\mathrm{Re} \int_0^t (Au(t'), u(t'))\, dt'.$$

At this point one can invoke the sharp Gårding inequality (Theorem 3.2) and derive from the fact that

$$-2\,\mathrm{Re}(Au, u)_0 \leq \mathrm{const}\|u\|_0^2$$

the following inequality:

$$\|u(t)\|_0^2 \leq \|u(0)\|_0^2 + C_o \int_0^t \|u_t - Au\|_0^2\, dt' + C_1 \int_0^t \|u\|_0^2\, dt'. \tag{3.35}$$

Such an inequality plays, to some extent, the role of the Gronwall inequality for ordinary differential equations. In particular, notice that, if

$$u_t = -Au \tag{3.36}$$

then, by the Gronwall inequality,

$$\|u(t)\|_0^2 \leq e^{C_1 t}\|u(0)\|_0^2. \tag{3.37}$$

The latter estimate implies, among other things, the *uniqueness* of the solution of (3.34). In connection with this it should be said that the operator $U(t)$ assigns $u(t)$ to $u(0)$ (under condition (3.36)).

Actually, because the principal symbol of A is ≥ 0 and because of the sharp Gårding inequality, the following facts are true: the anti-self-adjoint part of A is a pseudodifferential operator of order zero on X, the self-adjoint part of A is bounded from below, $A \geq -cI$, in the standard notation of operator theory. Under these conditions one can construct the continuous semigroup e^{-tA}, $t \geq 0$ (see Treves [3], Section 45) and this is what the operator $U(t)$ is.

V

Analytic Pseudodifferential Operators

"Generalized functions," playing the role vis-à-vis analytic functions that distributions play vis-à-vis C^∞ functions, do exist: They are the hyperfunctions of M. Sato. The analogues of pseudodifferential (and even hyperdifferential) operators can be made to act on them, and many results in this book have hyperfunction parallels. On this vast subject we refer the reader to Sato, Kawai, and Kashiwara [1]. In general, that is, in the absence of precise information, such operators transform hyperfunctions, and among them distributions, into hyperfunctions. As a consequence there may still be some justification for a theory of pseudodifferential operators that transform distributions into distributions while preserving their analyticity. This chapter is devoted to the definition and study of standard pseudodifferential operators (in the sense of Chapter I) having this property.

Section 1 introduces the notion of *analytic wave-front set*, due to Hörmander. That it coincides (for distributions) with the *essential singular support*) (or *spectrum*) of Sato, and to another notion due to Bros and Iagolnitzer, is proved in Bony [1]. Section 1 also introduces our main tool in the subsequent section for microlocalization: the cutoff functions of Andersson–Hörmander. (We modify somewhat the original construction of Andersson [1].) It may be worthwhile to point out right away that the operators to which these functions (which are defined in the cotangent bundle) give rise are not standard pseudodifferential operators, contrary to what happens in the C^∞ case. But they serve, effectively, the same purpose. Recently Sjöstrand has presented a theory of analytic pseudodifferential operators that avoids the use of those cutoff functions (see Sjöstrand [5]).

Section 2 defines the amplitudes $a(x, y, \xi)$ used to define the pseudodifferential operators. We have slightly modified the commonly accepted definition, by requiring that $a(x, y, \xi)$ be analytic with respect to (x, y) and that $\overline{\partial}_\xi a$ be of exponential decay as $|\xi| \to +\infty$. This enables us to freely use cutoff functions with respect to ξ, equal to zero in a ball centered at the origin and to one outside a bigger such ball, and it has no effect on the properties of the operators. Actually, the class of operator thus defined is the same as the one in Boutet de Monvel and Kree [1] (see also Boutet de Monvel [2]), as we show in Lemma 2.4.

Section 3 develops the parallel theory, in the analytic framework, of the theory developed in Chapter I in the C^∞ framework.

At this stage, as we deal with "concrete" operators, we continue to be hindered by the standard drawback of analytic theory, that there are no analytic cutoff functions (in the base). For instance, we are forced to prove directly the invariance of analytic wave-front sets under analytic diffeomorphism, unable to use pseudodifferential operators as in the C^∞ case. The way to circumvent these difficulties is well known: introduce *sheaves*. Since it cannot be disguised here, as it can in the C^∞ category, the step is taken in Section 4, and we are now essentially free to do all we might reasonably want and were able to do in the C^∞ case, provided that we resign ourselves to dealing with germs of distributions (modulo microlocally analytic functions) and with germs of pseudodifferential operators (modulo microlocally analytic-regularizing ones). As an example of our newly acquired freedom we prove the classical Holmgren theorem (following Hörmander in [14]).

The closing section of the chapter, Section 5, is devoted to a difficult proof, that of the analyticity up to the boundary in (coercive) boundary problems for elliptic equations. It is intended more for the specialist than for the general reader.

1. Analyticity in the Base and in the Cotangent Bundle

Throughout the chapter the adjective "analytic" will mean "real analytic": a C^∞ function f in Ω is *analytic* if it has the following three equivalent properties:

(i) Every point x_o of Ω has a neighborhood $U(x_o)$ in which the Taylor expansion of f about x_o converges (absolutely, uniformly) to f.

(ii) Given any compact set $\mathscr{K}$ of Ω there are two constants $C_o, C_1 > 0$ such that, for all n-tuples $\alpha \in \mathbb{Z}_+^n$,

$$|\partial^\alpha f(x)| \leq C_o C_1^{|\alpha|} \alpha!, \qquad \forall x \in \mathscr{K}. \tag{1.1}$$

(iii) There is an open neighborhood $\Omega^{\mathbb{C}}$ of Ω in $\mathbb{C}^n$ to which f can be extended as a holomorphic function of $z = x + iy$.

A C^∞ function h in an open subset $\mathcal{O}$ of $\mathbb{C}^n$ is *holomorphic* if it is a solution of the *homogeneous Cauchy–Riemann equations*:

$$\frac{\partial h}{\partial \bar{z}^j} = \frac{1}{2}\left(\frac{\partial h}{\partial x^j} + \sqrt{-1}\,\frac{\partial h}{\partial y^j}\right) = 0, \qquad j = 1, \ldots, n. \tag{1.2}$$

We shall systematically shorten (1.2) into $\bar{\partial} h = 0$ and refer to $\bar{\partial}$ as the Cauchy–Riemann operator in $\mathbb{C}^n$.

A map $f: \Omega \to \mathbb{R}^d$ is analytic if following up with any linear functional on $\mathbb{R}^d$ gives an analytic function in Ω. A *real analytic manifold* X is defined like a C^∞ manifold, except that the coordinate changes must be real analytic: if $(\mathcal{O}, \chi)$ and $(\tilde{\mathcal{O}}, \tilde{\chi})$ are two local charts, $\tilde{\chi} \circ \overset{-1}{\chi}$ must be an analytic map of $\chi(\mathcal{O} \cap \tilde{\mathcal{O}})$ onto $\tilde{\chi}(\mathcal{O} \cap \tilde{\mathcal{O}})$, which are open subsets of one and the same Euclidean space $\mathbb{R}^n$ ($n = \dim X$). An analytic diffeomorphism is simply a diffeomorphism which is an analytic map (for then its inverse is also analytic).

It hardly needs recalling that the fundamental difficulty in dealing with analytic functions is that there are no such functions with nonempty compact support. For us it means mainly that we cannot count on using analytic cutoff functions in order to localize, even less to microlocalize. We are going to show, however, that if one is willing to rely on *sequences* of cutoff functions, rather than on single ones, one can reap many of the profits of localization (and microlocalization).

Let U be an open subset of $\mathbb{R}^n$ with compact closure contained in Ω, and let V be an arbitrary open subset of Ω containing the closure of U. What we would like to have at our disposal is a function f that vanishes outside V, is equal to one in U and has the property (iii). Although this is not possible, a lesser demand can be satisfied: that the estimates (1.1) hold, not for all n-tuples, but only for those whose length $|\alpha|$ does not exceed some preassigned (but arbitrary) integer N, in such a way that the constants depend on N in a manner suited to the use we are going to make of these estimates. More precisely, we have

LEMMA 1.1. *There is a constant $C_* > 0$, depending only on n, such that given any open subset U of $\mathbb{R}^n$, any number $d > 0$, any integer $N > 0$, there is a C^∞ function g_N in $\mathbb{R}^n$, having the following properties.*

$$0 \le g_N \le 1 \textit{ everywhere}, \; g_N = 1 \textit{ in } U, \qquad g_N(x) = 0 \textit{ if } \operatorname{dist}(x, U) > d; \tag{1.3}$$

$$|D^\alpha g_N| \le (C_* N/d)^{|\alpha|} \qquad \textit{for all } \alpha \in \mathbb{Z}_+^n \textit{ such that } |\alpha| \le N. \tag{1.4}$$

PROOF. We select a C^∞ function $\psi \geq 0$ in $\mathbb{R}^n$ with support in the unit ball and such that $\int \psi\, dx = 1$. For any $\varepsilon > 0$ we write $\psi_\varepsilon(x) = \varepsilon^{-n}\psi(x/\varepsilon)$. Let $\tilde{\chi}_U$ denote the function equal to one in the set $\{x; \operatorname{dist}(x, U) < d/2\}$ and to zero everywhere else. Set

$$(1.5)\qquad g_N = \overbrace{\psi_{d/2N} * \cdots * \psi_{d/2N}}^{N} * \tilde{\chi}_U \qquad (*\text{: } convolution).$$

Since the support of a convolution is contained in the vector sum of the supports of the factors (see (0.17)), property (1.3) is verified. Let $D_j = -\sqrt{-1}\,\partial/\partial x^j$ $(j = 1, \ldots, n)$. If $r \leq N$, we have

$$(1.6)\qquad D_{j_1} \cdots D_{j_r} g_N = (D_{j_1}\psi_{d/2N}) * \cdots * (D_{j_r}\psi_{d/2N}) * \overbrace{\psi_{d/2N} * \cdots * \psi_{d/2N}}^{N-r} * \tilde{\chi}_U.$$

By virtue of Hölder's inequalities for convolution (see (0.13)) we have

$$(1.7)\qquad \|D_{j_1} \cdots D_{j_r} g_N\|_{L^\infty} \leq (2C_o N/d)^r, \qquad C_o = \sup_{1 \leq j \leq n} \|D_j \psi\|_{L^1}. \qquad \square$$

REMARK 1.1. Note that $(CN)^{|\alpha|} \leq e^{CN}\alpha!$.

We can now prove

LEMMA 1.2. *In order for $u \in \mathcal{D}'(\Omega)$ to be an analytic function in Ω, it is necessary and sufficient for each point x_o of Ω to have an open neighborhood $U(x_o)$ with the following property:*

(1.8) *Given any integer $N \geq 0$ there is a function $\phi_N \in C_c^\infty(\Omega)$, $\phi_N = 1$ in $U(x_o)$, such that*

$$(1.9)\qquad |(\widehat{\phi_N u})(\xi)| \leq C^{N+1} N!(1 + |\xi|)^{-N},$$

with C independent of N.

PROOF. The condition is necessary. Let $U = U(x_o)$ have compact closure contained in Ω, and let d be a number strictly less than the distance from U to the complement of Ω. We take ϕ_N to be the function g_N of Lemma 1.1. If $|\alpha| \leq N$,

$$(1.10)\qquad \xi^\alpha (\widehat{g_N u})(\xi) = \int e^{-ix\cdot\xi} D^\alpha (g_N u)\, dx = \sum_{\beta \leq \alpha} \binom{\alpha}{\beta} \int e^{-ix\cdot\xi} D^\beta g_N D^{\alpha-\beta} u\, dx.$$

We apply (1.4), in conjunction with Remark 1.1. Noting that the support of

g_N is contained in a compact subset of Ω, independent of N, we have (cf. (1.1)):

(1.11) $$|\xi^\alpha(\widehat{g_N u})| \leq C_o e^{CN} \alpha! \sum_{\beta \leq \alpha} C_1^{|\alpha - \beta|}.$$

By adding up these inequalities for every α such that $|\alpha| \leq N$, we obtain (1.9) (after redefinition of C).

Let us show that the condition is sufficient. We note that if $N > k + n$, then (1.9) implies that $\phi_N u$ is a C^k function, which of course means that u is C^∞, and so is $\phi_N u$. We may write

(1.12) $$\phi_N u(x) = (2\pi)^{-n} \int e^{ix \cdot \xi}(\widehat{\phi_N u})(\xi)\, d\xi,$$

and therefore, if $|\alpha| = N - n - 1$,

$$|D^\alpha(\phi_N u)| \leq (2\pi)^{-n} C^{N+1} N! \int (1 + |\xi|)^{-n-1}\, d\xi \leq B_n C_*^{|\alpha|} \alpha!.$$

Since $\phi_N = 1$ in $U(x_o)$, this implies that u is analytic in that set and therefore in Ω. □

We shall sometimes denote by $\mathscr{A}(\Omega)$ the space of analytic functions in the open set Ω. The *analytic singular support* of a distribution v, which we shall denote by sing supp$_a$ v, is the intersection of all closed sets in the complement of which v is an analytic function. If v and w are distributions in $\mathbb{R}^n$, and if at least one of them is compactly supported, we have

(1.13) $$\text{sing supp}_a(v * w) \subset \text{sing supp}_a\, v + \text{sing supp}_a\, w.$$

The plus sign on the right stands for vector addition. The inclusion (1.13) will be used later. We leave its proof as an exercise to the reader; it is a modified and simplified version of the proof of Theorem 2.1 in Section 2.

Lemma 1.2 makes it obvious what must be done if we are to lift the notion of analyticity from the base, here Ω, to the cotangent bundle over Ω, as we have done in Chapter I for the notions of C^∞ and H^s regularity.

DEFINITION 1.1. *Let $x_o \in \Omega$, $\xi^o \in \mathbb{R}_n \backslash \{0\}$. We shall say that a distribution u in Ω is analytic near (x_o, ξ^o) if there is an open neighborhood U of x_o in Ω, an open cone Γ in $\mathbb{R}_n$ containing ξ^o and a constant $C > 0$ such that, for each integer $N = 0, 1, \ldots,$ one can find a function $\phi_N \in C_c^\infty(\Omega)$, $\phi_N = 1$ in U, $\phi_N = 0$ outside a compact subset $\mathscr{K}$ of Ω independent of N, such that*

(1.14) $$|(\widehat{\phi_N u})(\xi)| \leq C^{N+1} N! (1 + |\xi|)^{-N}, \qquad \forall \xi \in \Gamma.$$

We say that u is analytic in a conic open subset Γ *of* $\Omega \times (\mathbb{R}_n\backslash\{0\})$ *if u is analytic near every point of* Γ. *The complement of the set of points in* $\Omega \times (\mathbb{R}_N\backslash\{0\})$ *near which u is analytic is a closed conic subset of* $\Omega \times (\mathbb{R}_n\backslash\{0\})$ *called the analytic wave-front set of u and denoted by* $\mathrm{WF}_a(u)$.

Definition 1.1 has various disadvantages. For one, it is not clear that the notion of analyticity in a neighborhood of a point (x_o, ξ^o) subsists if we restrict the distribution under study, u, to an open subset Ω' of Ω still containing x_o (but perhaps not containing the compact set $\mathcal{K}$ outside of which the cutoffs ϕ_N vanish). Second, it is not clear either that the base projection of the analytic wave-front set is equal to the analytic singular support, as one should expect it to be. We are going to see that both properties are in fact true. But in order to do this we must refine our analysis in the cotangent bundle and have at our disposal tools that allow us to cut down the supports with respect to the ξ-variables, down to appropriate conic sets, without jeopardizing the analyticity in the base of the distributions being cut down.

Let us define the problem we are confronted with more precisely. We are seeking a function $g(\xi)$ in $\mathbb{R}_n\backslash\{0\}$ which should satisfy the following *desiderata*: g should be C^∞ everywhere; it should be equal to one in some arbitrarily given open cone Γ^o in $\mathbb{R}_n\backslash\{0\}$, to zero outside another such cone Γ'^o containing the closure (in $\mathbb{R}_n\backslash\{0\}$) of Γ^o. Also g should be tempered and, in fact, as nearly positive-homogeneous of degree zero as possible. Last but not least, the operator of convolution with the inverse Fourier transform $G(x)$ of $g(\xi)$ must be analytical pseudolocal. It follows at once from (1.13) that a necessary and sufficient condition in order that the latter be true is that $G(x)$ be analytic in the complement of the origin in $\mathbb{R}^n$. Thus, one of the first questions we must answer is how to make sure, by putting the proper conditions on $g(\xi)$, that $G(x)$ is analytic for $x \neq 0$.

We shall answer this question in part by applying a very particular case of Stokes' theorem. Let F be a C^∞ function of (ξ, η) in the subset of $\mathbb{R}_{2n}$ defined by

(1.15) $$|\eta| < \delta_o|\xi| \qquad (\delta_o > 0).$$

And let η^o denote a certain vector in $\mathbb{R}_n$, $|\eta^o| < \delta_o$. We shall assume that the decay of $F(\xi, t|\xi|\eta^o)$ as $|\xi| \mapsto +\infty$ (with $0 \le t \le 1$) is sufficiently fast that all the integrals considered here are absolutely convergent, and first

(1.16) $$J_0 = \int F(\xi, 0)\, d\xi.$$

Observe that $\xi \mapsto \zeta^t = (\zeta_1^t, \ldots, \zeta_n^t)$, where $\zeta_j^t = \xi_j + \sqrt{-1}\, t|\xi|\eta_j^o$, defines a homeomorphism of $\mathbb{R}_n$ onto an n-dimensional C^0 submanifold S^t of the set (1.15) (actually the mapping is Lipschitz continuous and so is ζ^t; both are C^∞ off the origin). The Jacobian determinant $D\zeta^t/D\xi$ has generic entry

$$\delta_{jk} + \sqrt{-1}\, t\eta_j^o \xi_k/|\xi| \qquad (j, k = 1, \ldots, n),$$

which is L^∞. Then set

$$J_t = \int F(\xi, t|\xi|\eta^o) \frac{D\zeta^t}{D\xi}\, d\xi. \tag{1.17}$$

Of course, we may regard $F(\xi, \eta)$ as a function of $\zeta = \xi + i\eta$, and read (1.17) as

$$J_t = \int_{S^t} F(\zeta)\, d\zeta. \tag{1.18}$$

Let us denote by $\eta^o \cdot \bar{\partial}_\zeta$ the first-order complex vector field $\sum_{j=1}^n \eta_j^o (\partial/\partial\bar{\zeta}_j)$ ($\zeta_j = \xi_j + \sqrt{-1}\, \eta_j$). By Stokes' theorem we have

$$J_o - J_1 = -\sqrt{-1} \int_0^1 \int_{S^t} (\eta^o \cdot \bar{\partial}_\zeta F)(\zeta^t)|\xi| \frac{D\zeta^t}{D\xi}\, d\xi\, dt. \tag{1.19}$$

PROOF. The n-chain S^0–S^1 can be regarded as the oriented boundary of the $(n + 1)$-chain $\bigcup_{0 \le t \le 1} S^t$. It suffices therefore to observe that

$$d(F(\zeta)\, d\zeta_1 \wedge \cdots \wedge d\zeta_n) = \sum_{j=1}^n \frac{\partial F}{\partial \bar{\zeta}_j} d\bar{\zeta}_j \wedge \cdots \wedge d\zeta_n$$

and that $\zeta_j = \xi_j + \sqrt{-1}\, t|\xi|\eta_j^o$, $j = 1, \ldots, n$, on that $(n + 1)$-chain; hence this differential is equal to

$$-\sqrt{-1} \left(\sum_{j=1}^n \frac{\partial F}{\partial \bar{\zeta}_j} \eta_j^o \right) |\xi|\, dt \wedge d\zeta_1 \wedge \cdots \wedge d\zeta_n,$$

whence (1.19). □

We shall need the following application of (1.19):

LEMMA 1.3. *Let $g(\xi, \eta)$ be a C^∞ function in the "sector" (1.15), having the following properties:*

There are constants $\delta > 0$, $\delta < \delta_o$, C, $r \ge 0$ and m real such that

$$|g(\xi, 0)| \le C(1 + |\xi|)^m, \tag{1.20}$$

$$|g(\xi, \eta)| \le C e^{\delta r |\xi|} \qquad \text{if } |\eta| \le \delta|\xi|, \tag{1.21}$$

$$|\bar{\partial}_\zeta g(\xi, \eta)| \le C e^{-c|\xi|}. \tag{1.22}$$

Then the inverse Fourier transform of $g(\xi, 0)$, $G(x)$, *which is a tempered distribution* (*by* (1.20)), *can be extended as a holomorphic function in the set*

$$(1.23)\qquad z = x + iy \in \mathbb{C}^n; \qquad |x| > r, \quad |y| < \delta(|x| - r), \quad |y| < c.$$

Proof. Let k be an integer $>\frac{1}{2}(m + n)$. It suffices to show that the (unique) tempered distribution G_1 such that $G = (1 - \Delta)^k G_1$ can be extended holomorphically to (1.23), for this will then be true of G. Since the Fourier transform of G_1 is $(1 + |\xi|^2)^{-k} g(\xi, 0)$, it is integrable; we may as well assume that (1.20) holds with m replaced by $m - 2k$. Let us then set

$$F(z; \xi, \eta) = g(\xi, \eta)\, e^{iz \cdot \zeta},$$

and let x_o be an arbitrary point in $\mathbb{R}^n$ such that $|x_o| > r$. Take η^o to be the vector $\delta x_o/|x_o|$; then $(\xi, t|\xi|\eta^o)$ belongs to (1.15) for all t, $0 \leq t \leq 1$. We have

$$|F(z; \xi, t|\xi|\eta^o)| = |g(\xi, t|\xi|\eta^o)| \exp(-y \cdot \xi - t\delta|\xi| x \cdot x_o/|x_o|).$$

By (1.21) we obtain

$$(1.24)\qquad |F(z; \xi, |\xi|\eta^o)| \leq C \exp\{-y \cdot \xi - \delta|\xi|(x \cdot x_o/|x_o| - r)\}.$$

Restrict the variation of z to an open subset of (1.23) defined by

$$(1.25)\qquad \frac{x \cdot x_o}{|x_o|} - r > \frac{\delta'}{\delta}(|x_o| - r), \qquad |y| < \delta''(|x_o| - r)$$

for some choice of numbers $0 < \delta'' < \delta' < \delta$. We derive from (1.24)

$$(1.26)\qquad |F(z; \xi, |\xi|\eta^o)| \leq C e^{-(\delta' - \delta'')|\xi|},$$

which has the immediate consequence that, in the notation of (1.17)

$$J_1(z) = \int F(z; \xi, |\xi|\eta^o) \frac{D\zeta^1}{D\xi}\, d\xi$$

is holomorphic in (1.25). Next we use (1.22), to obtain

$$(1.27)\qquad |\bar{\partial}_\zeta F(z; \xi, t|\xi|\eta^o)| \leq C \exp\left(-y \cdot \xi - t\delta|\xi| \frac{x \cdot x_o}{|x_o|} - c|\xi|\right).$$

Therefore, if $|y|$ remains $<c$, the right-hand side in (1.19) defines a holomorphic function of z. The equality (1.19) shows then that $\int e^{ix \cdot \xi} g(\xi, 0)\, d\xi$ can be extended holomorphically to (1.23). □

COROLLARY 1.1. *Let g be as in Lemma 1.3; moreover suppose that to every $\varepsilon > 0$ there is a $\delta > 0$, $\delta < \delta_o$, and $C_\varepsilon > 0$ such that*

$$|g(\xi, \eta)| \le C_\varepsilon\, e^{\delta\varepsilon|\xi|} \qquad \text{if } |\eta| < \delta|\xi|. \tag{1.28}$$

Then $G(x)$ is analytic in $\mathbb{R}^n \backslash \{0\}$.

LEMMA 1.4. *Let Γ, Γ^* be two open cones in $\mathbb{R}_n \backslash \{0\}$ such that the closure of Γ is contained in Γ^*. There is a number $C > 0$ such that the following is true:*

Given any number $R > 0$, there is a C^∞ function g^R in $\mathbb{C}_n \backslash \sqrt{-1}\,\mathbb{R}_n$, where the variable is $\zeta = \xi + \sqrt{-1}\,\eta$, having the following properties:

$$0 \le g^R(\xi) \le 1, \qquad \forall \xi \in \mathbb{R}_n \backslash \{0\}; \tag{1.29}$$

(1.30) *for every ξ in $\mathbb{R}_n \backslash \{0\}$ and every α in $\mathbb{Z}^n_+$,*

$$|D^\alpha g^R(\xi)| \le (C/R)^{|\alpha|} \qquad \text{if } |\xi| \ge R|\alpha|;$$

$$g^R(\zeta) = 1 \quad \text{if } \xi \in \Gamma, \qquad g^R(\zeta) = 0 \quad \text{if } \xi \notin \Gamma^*; \tag{1.31}$$

$$|g^R(\zeta)| \le C \exp(C|\eta|/R); \tag{1.32}$$

$$|\bar\partial g^R(\zeta)| \le C \exp\{(C|\eta| - |\xi|)/R\}. \tag{1.33}$$

PROOF. We begin by constructing a special partition of unity on the real line. Let $\phi_r \in C^\infty(\mathbb{R}^1)$, $0 \le \phi_r \le 1$ everywhere, $\phi_r(t) = 0$ for $t > 1$, $\phi_r(t) = 1$ for $t < 0$, and $|D^l \phi_r| \le (C_* r)^l$ if $l \le r$.

Let us then set $h_0(t) = \phi_1(R^{-1}t - 1)$ and, for $N = 1, 2, \dots$,

$$h_N(t) = 1 \qquad \text{if } \tfrac{1}{2}4^N R \le t \le 4^N R; \tag{1.34}$$

$$h_N(t) = 1 - \phi_{4^{N-1}}(4^{1-N}R^{-1}t - 1) \qquad \text{if } t < \tfrac{1}{2}4^N R; \tag{1.35}$$

$$h_N(t) = \phi_{4^N}(4^{-N}R^{-1}t - 1) \qquad \text{if } t > 4^N R. \tag{1.36}$$

We have:

$$\sum_{N=0}^{+\infty} h_N \equiv 1. \tag{1.37}$$

(1.38) *On* supp h_N, $2^{2N-2} \le t/R \le 2^{2N+1}$, *and any point t of $\mathbb{R}^1$ belongs to the support of at most two functions h_N, $h_{N'}$.*

$$|D_t^l h_N| \le (C_*/R)^l \qquad \text{if } l \le \tfrac{1}{2}t/R. \tag{1.39}$$

Let us denote by θ the variable in the unit sphere S_{n-1} of $\mathbb{R}_n$. We shall denote by $\mathcal{O}$ (resp., $\mathcal{O}^*$) the intersection of Γ (resp., of Γ^*) with S_{n-1}. We shall assume that $\mathcal{O}$ is not dense in S_{n-1}; if it were we would have $\Gamma^* = \mathbb{R}_n\backslash\{0\}$, and we would take $g^R \equiv 1$. Let d be a small number >0, such that the set of points in S_{n-1}, $\mathcal{O}_d$, whose distance to $\mathcal{O}$ is $<d$, has its closure contained in $\mathcal{O}^*$. We avail ourselves once again of Lemma 1.1 and select, for each $N = 0, 1, \ldots$, a C^∞ function ψ_N on S_{n-1} having the following properties:

$$(1.40) \qquad 0 \le \psi_N \le 1 \textit{ everywhere}, \qquad \psi_N = 1 \textit{ in } \mathcal{O}, \qquad \psi_N = 0 \textit{ outside } \mathcal{O}_d;$$

$$(1.41) \qquad |D_\theta^\gamma \psi_N| \le (C_* 4^{N-1}/d)^{|\gamma|} \qquad \textit{if } |\gamma| \le 4^{N-1}.$$

In (1.41) D_θ stands for the system of $\frac{1}{2}n(n-1)$ vector fields tangent to S_{n-1}, $\xi_j D_{\xi_k} = \xi_k D_{\xi_j}$, $1 \le j < k \le n$; $\gamma \in \mathbb{Z}_+^{n(n-1)/2}$.

If we set $f_N(\xi) = \psi_N(\xi/|\xi|)$, we see that

$$|D_\xi^\alpha f_N(\xi)| \le (C_1 4^{N-1}/|\xi|)^{|\alpha|} \qquad \textit{if } |\alpha| \le 4^{N-1},$$

from which we get

$$(1.42) \qquad |D^\alpha f_N(\xi)| \le (C_1/R)^{|\alpha|} \qquad \textit{if } |\alpha| \le 4^{N-1} \le |\xi|/R.$$

At this point we set

$$(1.43) \qquad g^R = \sum_{N=0}^{+\infty} g_N^R, \qquad g_N^R(\xi) = f_N(\xi) h_N(|\xi|).$$

Since $f_N(\xi) = 1$ if $\xi \in \Gamma$, for all N, (1.37) implies that $g^R = 1$ in Γ. Since $f_N(\xi) = 0$ if $\xi/|\xi| \notin \mathcal{O}_d$ for all N, again by virtue of (1.37) we see that the support of g^R is contained in a closed cone Γ_d, itself contained in Γ^*. By virtue of (1.37) and of the fact that $0 \le f_N \le 1$, we have $0 \le g^R \le 1$.

On the other hand, by the Leibniz formula, by (1.39), (1.41), and by noting that when $h_N(|\xi|) \ne 0$ we must have $|\xi| \ge 4^{N-1}R$, we see that

$$(1.44) \qquad |D^\alpha g_N^R| \le (C_2/R)^{|\alpha|} \qquad \textit{if } |\alpha| \le \tfrac{1}{2}|\xi|/R.$$

Finally concerning the definition of g^R in $\mathbb{R}_n\backslash\{0\}$, we see that we have established the following properties:

$$(1.45) \quad 0 \le g^R \le 1 \textit{ everywhere}, \qquad g^R = 1 \textit{ in } \Gamma, \qquad g^R = 0 \textit{ off } \Gamma_d \subset \Gamma^*;$$

$$(1.46) \qquad |D^\alpha g^R| \le (C_3/R)^{|\alpha|} \qquad \textit{if } |\xi| \ge 2R|\alpha|.$$

The constant C_3 in (1.46) depends on d (like the constants C_1, C_2). In establishing (1.46), we have used the fact that any point ξ in $\mathbb{R}_n\backslash\{0\}$ belongs to the supports of at most two functions $g_N^R, g_{N'}^R$.

The next step is to extend g^R to $\mathbb{C}_n\backslash\sqrt{-1}\,\mathbb{R}_n$. We introduce a sequence of C^∞ functions χ_α in $\mathbb{R}_n$ with the following properties:

$$(1.47) \qquad \chi_0 \equiv 1; \qquad \textit{if } |\alpha| \ge 1, \textit{ then } 0 \le \chi_\alpha \le 1 \textit{ everywhere};$$

(1.48) $\chi_\alpha(\xi) = 0 \quad \text{if } |\xi| < 2R|\alpha|, \quad \chi_\alpha(\xi) = 1 \quad \text{if } |\xi| > 3R|\alpha| \quad (|\alpha| \geq 1);$

$$\sup_\alpha |d\chi_\alpha| < +\infty. \tag{1.49}$$

Then for $\zeta = \xi + \sqrt{-1}\,\eta$, $\xi \neq 0$, we set

$$g^R(\zeta) = \sum_\alpha \frac{1}{\alpha!}(-\eta)^\alpha D_\xi^\alpha g^R(\xi)\chi_\alpha(\xi). \tag{1.50}$$

From (1.46) we deduce (recalling that $g^R(\xi) \leq 1$)

$$|g^R(\zeta)| \leq C_3 \exp\{C_3|\eta|/R\}. \tag{1.51}$$

Last, for $j = 1, \ldots, n$, we have

$$\left(\frac{\partial}{\partial\xi_j} + i\frac{\partial}{\partial\eta_j}\right) g^R(\zeta) = \tag{1.52}$$

$$\sum_\alpha \frac{1}{\alpha!}(i\eta)^\alpha \left\{\partial_\xi^{\alpha+\langle j\rangle} g^R(\xi)(\chi_\alpha - \chi_{\alpha+\langle j\rangle})(\xi) + \partial_\xi^\alpha g^R(\xi)\frac{\partial}{\partial\xi_j}\chi_\alpha(\xi)\right\},$$

where we have denoted by $\langle j\rangle$ the n-tuple with components all zero except the jth one, equal to one. In the support of the function of ξ that multiplies $(i\eta)^\alpha/\alpha!$ in the sums at the right in (1.52), we have

$$2|\alpha| < R^{-1}|\xi| < 3(|\alpha| + 1). \tag{1.53}$$

From (1.49) we derive

$$\left|\left(\frac{\partial}{\partial\xi_j} + i\frac{\partial}{\partial\eta_j}\right) g^R(\zeta)\right| \leq C\left(1 + \frac{C}{R}\right)\sum_\alpha \frac{|\eta^\alpha|}{\alpha!}(C/R)^{|\alpha|+1}\psi_\alpha(\xi), \tag{1.54}$$

where ψ_α is the characteristic function of the shell (1.53). We see that $(C\exp\{-3(|\alpha|+1)\})^{|\alpha|+1} \leq e^{-|\xi|/R}$ in the region (1.53), and therefore, by virtue of (1.54), $|\bar\partial g^R| \leq C' \exp\{(C|\eta| - |\xi|)/R\}$. After suitable redefinitions of C and R we see that the conclusions in Lemma 1.4 have been reached.

COROLLARY 1.2. *The inverse Fourier transform $G^R(x)$ of $g^R(\xi)$ is analytic in the set*

$$x \in \mathbb{R}^n; \qquad |x| > C/R. \tag{1.55}$$

PROOF. The function g^R has all the properties required of g in Lemma 1.3, where we take $r = C/R$. □

One can also construct partitions of unity made up of functions like g^R: let $\Gamma_1, \ldots, \Gamma_\nu$ be a finite number of open cones whose union is equal to $\mathbb{R}_n\backslash\{0\}$. We define ν functions $g_j'^R$ by induction on $j = 1, \ldots, \nu$ as follows: $g_1'^R$ has all the properties in Lemma 1.4 when one takes $\Gamma^* = \Gamma_1$ and $\Gamma = (\mathbb{R}_n\backslash\{0\})\backslash(\Gamma_2 \cup \cdots \cup \Gamma_\nu)$; and if $j \geq 2$, then $g_j'^R$ has the same properties when one takes $\Gamma^* = \Gamma_j$ and Γ a suitable open cone containing

$$(1.56)\quad [(\mathbb{R}_n\backslash\{0\})\backslash(\Gamma_{j+1} \cup \cdots \cup \Gamma_\nu)] \cap \operatorname{supp}[(1 - g_1'^R) \cdots (1 - g_{j-1}'^R)].$$

It is understood that, when $j = \nu$, the union $\Gamma_{j+1} \cup \cdots \cup \Gamma_\nu$ is empty. Observe that the set (1.56) is contained in Γ_j, because

$$(1.57)\qquad \operatorname{supp}[(1 - g_1'^R) \cdots (1 - g_{j-1}'^R)] \subset \Gamma_j \cup \cdots \cup \Gamma_\nu.$$

If we assume that $\nu \geq 2$, then

$$(1.58)\quad 1 = g_1'^R + (1 - g_1'^R)g_2'^R + \cdots + (1 - g_1'^R) \cdots (1 - g_{j-1}'^R)g_j'^R + \cdots + (1 - g_1'^R) \cdots (1 - g_{\nu-1}'^R)g_\nu'^R,$$

since, by (1.57), $(1 - g_1'^R) \cdots (1 - g_\nu'^R) = 0$.

Finally it suffices to observe that if g^R has the properties (1.29), (1.30), (1.32), and (1.33) in Lemma 1.4 so does $1 - g^R$, and if g'^R also has those same properties, so does $g^R g'^R$ (with possibly increased constant C).

The next statement shows that, regardless of the nature of the cutoff functions with respect to ξ, the convolution operators that they define are analytic regularizing (in the microlocal sense) in any open cone in which the functions in question vanish.

The next lemma will be used frequently in the sequel; for this reason we state and prove it in somewhat greater generality than immediately needed.

LEMMA 1.5. *Let $k(x, y, \xi)$ be a complex-valued C^∞ function in $\Omega \times \Omega \times (\mathbb{R}_n\backslash\{0\})$ having the following property, for some real number m:*

(1.59) *To every compact subset $\mathcal{K}$ of $\Omega \times \Omega$ there is a constant $C > 0$ such that for all $\alpha \in \mathbb{Z}_+^n$, (x, y) in $\mathcal{K}$, ξ in $\mathbb{R}_n$,*

$$|D_x^\alpha k(x, y, \xi)| \leq C^{|\alpha|+1}\alpha!(1 + |\xi|)^m.$$

For any $u \in \mathscr{E}'(\Omega)$, set

$$(1.60)\qquad Ku(x) = (2\pi)^{-n} \iint e^{i(x-y)\cdot\xi} k(x, y, \xi)u(y)\, dy\, d\xi$$

(which makes sense by standard considerations, cf. Chapter I, Section 2).

Suppose that $k(x, y, \xi)$ *vanishes identically if* ξ *belongs to a certain open cone* $\Gamma \subset \mathbb{R}_n\backslash\{0\}$; *let* Γ_* *be a closed cone contained in* Γ, *and* $\{h_N\}_{N=1,2,\ldots}$ *a sequence of* C^∞ *functions having their supports contained in a compact subset of* Ω *independent of* N *and such that, for each* N, $|D^\alpha h_N| \le (CN)^{|\alpha|}$, $|\alpha| \le N$, *with a constant* $C > 0$ *independent of* N.

Given any $u \in \mathscr{E}'(\Omega)$ *there is an integer* ν *and a constant* $C_* > 0$ *such that for all* $N \ge \nu$ *and writing* $M = N - \nu$, *we have*

$$|\widehat{(h_N K u)}(\xi)| < C_*^{M+1} M! |\xi|^{-M}, \qquad \forall \xi \in \Gamma_*, |\xi| \ge 1. \tag{1.61}$$

PROOF. We may write $u = \sum_{|\beta| \le s} D^\beta u_\beta$, with u_β continuous and compactly supported in Ω. Thus $Ku = \sum_{|\beta| \le s} K_\beta u_\beta$, with K_β defined like K, except that the "amplitude" $k(x, y, \xi)$ must be replaced by $e^{iy\cdot\xi}(-D_y)^\beta[e^{-iy\cdot\xi}k(x, y, \xi)]$. We may as well assume that u is a compactly supported continuous function in Ω, and replace m by $m + s$. Let us select an integer j such that $N \le 2j \le N + 1$, and assume that $m + n + s + 1 \le 2j$.

There is a number $c > 0$ such that $|\xi - \eta| \ge c(|\xi| + |\eta|)$ for all ξ in Γ_* and all η such that $k(x, y, \eta) \ne 0$ for some x, y in Ω. From now on we restrict the variation of ξ to Γ_*. We have

$$\widehat{(h_N K u)}(\xi) =$$

$$(2\pi)^{-n} \iiint \exp\{-ix\cdot(\xi - \eta) - iy\cdot\eta\} h_N(x) k(x, y, \eta) u(y)\, dx\, dy\, d\eta =$$

$$(2\pi)^{-n} \iiint \exp\{-ix\cdot(\xi - \eta) - iy\cdot\eta\} |\xi - \eta|^{-2j}$$

$$\times\ \Delta_x^j[h_N(x) k(x, y, \eta)] u(y)\, dx\, dy\, d\eta.$$

We derive from this

$$|\widehat{(h_N K u)}(\xi)| \le c^{-2j} |\xi|^{m+n+s+1-2j} C^{2j+1} (2j)! \int (1 + |\eta|)^{-n-1}\, d\eta \int |u(y)|\, dy,$$

for all $\xi \in \Gamma_*$, $|\xi| \ge 1$, assuming that $C > 0$ is large enough. □

COROLLARY 1.3. *Under the hypotheses of Lemma* 1.5, *whatever* $u \in \mathscr{E}'(\Omega)$, *the analytic wave-front set of* Ku *does not intersect* $\Omega \times \Gamma$.

LEMMA 1.6. *Let* Ω, Ω' *be two open subsets of* $\mathbb{R}^n$ *such that the closure of* Ω' *is compact and contained in* Ω. *Let* $\Gamma^\#$ *be an open cone in* $\mathbb{R}_n\backslash\{0\}$ *containing the closure of the cone* Γ^* *in Lemma* 1.4. *For* $R > 0$ *we denote by* g^R *the cutoff function in Lemma* 1.4.

If R is large enough, then given any distribution $u \in \mathscr{E}'(\mathbb{R}^n)$ whose analytic wave-front set does not intersect $\Omega \times \Gamma^{\#}$, $g^R(D)u$ is analytic in Ω'.

PROOF. Let x_o be an arbitrary point in Ω. Given any point ξ^o in $\Gamma^{\#}$, we can find a cone Γ^o containing ξ^o and an open neighborhood U_o of x_o such that there is a sequence $\{h_N\}$ of C^∞ functions with support in a compact set independent of N, all equal to one in U_o and such that

$$|(\widehat{h_N u})(\xi)| \le C_1^{N+1} \alpha! |\xi|^{-|\alpha|}, \qquad |\alpha| \le N,\ \xi \in \Gamma^o,\ |\xi| \ge 1.$$

Let ξ^o vary over the closure of the cone Γ^* of Lemma 1.4. We may find a finite covering $\Gamma_1, \dots, \Gamma_r$ of this closure by cones like Γ^o. Let g_j^R, $j = 1, \dots, r$, be elements of a partition of unity subordinate to this covering, of the kind (1.58). For each j we associate an open neighborhood U_j of x_o and a sequence of functions h_N^j like U_o and h_N. Since $\operatorname{supp} g_j^R \subset \Gamma_j$, inverse Fourier transformation yields, if we choose $N \ge n + 1$,

$$\text{(1.62)} \quad |D^\alpha g_j^R(D)(h_N^j u)| \le C_2^{N+1} \alpha! \quad \textit{in } \mathbb{R}^n, \qquad |\alpha| \le N - n - 1.$$

Note that these estimates are valid for any $R > 0$. Let U be an open neighborhood of x_o whose closure is compact and contained in $U_1 \cap \cdots \cap U_r$. In U consider

$$\text{(1.63)} \qquad g_j^R(D)(u - h_N^j u) = G_j^R * (u - h_N^j u)$$

where $G_j^R(x)$ is the inverse Fourier transform of $g_j^R(\xi)$. By Corollary 1.2, we know that G_j^R is analytic in the set (1.55).

If we combine this information with the inclusion relation (1.13) we conclude that for R large enough, (1.63) is analytic in U. If we combine the latter property with (1.62), we see that

$$|D^\alpha g_j^R(D)u(x)| \le C_3^{N+1} \alpha! \quad \textit{in } U, \qquad |\alpha| \le N - n - 1.$$

But since the left-hand side is independent of N, it means that $g_j(D)u$ is analytic in U, and so is $v = g_1^R(D)u + \cdots + g_r^R(D)u$. Note that regardless of the value of R,

$$g^R(D)u = g^R(D)v = G^R * v.$$

The latter convolution is well defined, say via Fourier transformation, even though v is not compactly supported; and (1.13) applies to it, if we put $w = G^R$. Therefore we have the right to conclude, possibly after some increasing of R and shrinking of U, that $g^R(D)u$ is analytic in U. By means of a finite covering of Ω' made up of neighborhoods of the kind U and by taking R appropriately large, we reach the desired conclusion. □

In the following corollaries X will denote an arbitrary open subset of $\mathbb{R}^n$.

COROLLARY 1.4. *Suppose that* $(x_o, \xi^o) \in T^*X\backslash 0$ *does not belong to the analytic wave-front set of* $u \in \mathscr{D}'(X)$. *There is an integer* s, *an open neighborhood* U_o *of* x_o, *an open cone* Γ^o *containing* ξ^o *such that the following is true: Given any sequence of functions* $\{h_N\}$ $(N = 1, 2, \ldots)$ *in* $C_c^\infty(U_o)$ *such that* $|D^\alpha h_N| \leq (CN)^{|\alpha|}$, $|\alpha| \leq N$, *with* $C > 0$ *independent of* N, *we have, for every* $N \geq s$,

$$|\widehat{(h_N u)}(\xi)| < C'^{N+1}(N-s)!(1+|\xi|)^{s-N}, \qquad \forall \xi \in \Gamma^o, \tag{1.64}$$

with $C' > 0$ *independent of* N.

PROOF. Obviously we may assume u compactly supported. Let U, Γ be as in Definition 1.1. Then let Γ^o, Γ', Γ^* be three open cones in $\mathbb{R}_n\backslash\{0\}$ such that $\xi^o \in \Gamma^o$, the closure of Γ^o is contained in Γ, that of Γ' in Γ^* and that of Γ^* in Γ. We select g^R as in Lemma 1.4, except that here $g^R = 1$ in Γ'; let U_o be an open neighborhood of x_o having compact closure contained in U. We apply Lemma 1.6 and conclude that if R is sufficiently large, then $g^R(D)u$ is analytic in U_o. Of course, if h_N is as in the statement of Corollary 1.4, we have (see proof of Lemma 1.2)

$$|\widehat{h_N g^R(D)u}(\xi)| \leq C_1^{N+1} N!(1+|\xi|)^{-N}, \qquad \forall \xi. \tag{1.65}$$

Last we apply Lemma 1.5 to $k = 1 - g^R$. From (1.61) we derive, for $N \geq \nu$,

$$|\{\widehat{h_N[1-g^R(D)]u}\}(\xi)| \leq C_2^{N-1}(N-\nu)!(1+|\xi|)^{\nu-N}, \qquad \forall \xi \in \Gamma^o. \tag{1.66}$$

By combining (1.65) and (1.66) we obtain (1.64) (with $s = \nu$). □

Among other things Corollary 1.4 shows that analytic wave-front sets behave as they ought to under restriction; this is not immediately obvious in Definition 1.1.

COROLLARY 1.5. *Let* Y *be an open subset of* X, u *a distribution in* X. *The analytic wave-front set of the restriction of* u *to* Y *is equal to the intersection of* $\mathrm{WF}_a(u)$ *with* $T^*X|_Y$.

COROLLARY 1.6. *The base projection of the analytic wave-front set of a distribution* u *in* X *is exactly equal to the analytic singular support of* u.

Proof. If u is analytic in an open subset Ω of X, certainly its analytic wave-front set does not intersect $T^*X|_\Omega$. Conversely, suppose this statement to be true and take u compactly supported; apply Lemma 1.6 with $\Gamma^\# = \mathbb{R}_n\backslash\{0\}$ and g^R identically one. □

Corollary 1.7. *Let X be an open subset of $\mathbb{R}^n$, Ω an open subset of X, and $\Gamma^\#$ an open cone in $\mathbb{R}_n\backslash\{0\}$. In order that the analytic wave-front set of a distribution u in X not intersect $\Omega \times \Gamma^\#$, it is necessary and sufficient that given any relatively compact open subset Ω' of Ω, there be a function g^R like the one in Lemma* 1.6 *such that $g^R(D)u$ is analytic in Ω'.*

Proof. We may assume u to be compactly supported. The condition is necessary, by Lemma 1.6. If $g^R(D)u$ is analytic in Ω', the wave-front set of $u = g^R(D)u + [u - g^R(D)u]$ does not intersect $\Omega' \times \Gamma$, by Corollary 1.3. As Ω' "grows" to Ω and Γ to $\Gamma^\#$, we get the desired result. □

Needless to say, the wave-front set, in the C^∞ sense, of a distribution is contained in its analytic wave-front set. In general it is distinct from the latter. It is easy to construct a distribution whose C^∞ wave-front set is empty, i.e., which is a C^∞ function, and whose analytic wave-front set is equal to a single, arbitrarily chosen ray:

Example 1.1. Let ξ^o be a unit vector in $\mathbb{R}_n$. If $0 < \varepsilon < 1$, the analytic wave-front set of the function in $\mathbb{R}^n$,

$$\int \exp(ix \cdot \xi - |\xi|^{1-\varepsilon} - |x|^2|\xi| - |\xi - |\xi|\xi^o|)\, d\xi \tag{1.67}$$

is the single ray $\{(x, \xi); x = 0, \xi = \rho\xi^o, \rho > 0\}$, whereas its C^∞ wave-front set is empty.

2. Pseudoanalytic and Analytic Amplitudes

Pseudodifferential operators lead us to study distribution kernels in $\Omega \times \Omega \times \mathbb{R}^n$ of the kind

$$K_o(x, y, q) = (2\pi)^{-n} \int e^{iq\cdot\xi} k(x, y, \xi)\, d\xi. \tag{2.1}$$

In this chapter our hypotheses concerning the amplitudes $k(x, y, \xi)$ will have the following implications on the kernel $K_o(x, y, q)$:

(2.2) *Given any compact subset* $\mathcal{K}$ *of* Ω *there is an integer* $m \geq 0$ *and a constant* $C > 0$ *such that for all* $f \in C_c^\infty(\mathcal{K})$, *all* $\alpha, \beta \in \mathbb{Z}_+^n$, *and all* x *in* $\mathcal{K}$,

$$\left|\int [D_x^\alpha D_y^\beta K_o(x, y, q)|_{q=x-y}] f(y)\, dy\right| \leq C^{|\alpha+\beta|+1} \alpha!\beta! \sup \sum_{|\gamma|\leq m} |D^\gamma f|; \tag{2.3}$$

(2.4) $K_o(x, y, q)$ *is an analytic function of* (x, y, q) *in* $\Omega \times \Omega \times (\mathbb{R}^n \setminus \{0\})$.

We are going to use the notation (for any u in $C_c^\infty(\Omega)$)

$$Ku(x) = \int K_o(x, y, x - y) u(y)\, dy. \tag{2.5}$$

The reader can easily check that property (2.2) implies that the kernel distribution $K_o(x, y, x - y)$ is separately regular in x and y and that property (2.4) then implies that it is very regular (see Chapter I, Section 2). It follows from these two facts that K maps $C_c^\infty(\Omega)$ into $C^\infty(\Omega)$ and extends as a continuous linear map $\mathcal{E}'(\Omega) \mapsto \mathcal{D}'(\Omega)$ and that this operator is pseudolocal (Chapter I, Lemma 2.1). What is of interest to us here, then, is the following theorem.

THEOREM 2.1. *Under hypotheses* (2.2) *and* (2.4) K *is analytic pseudolocal, i.e., given any* $u \in \mathcal{E}'(\Omega)$, Ku *is an analytic function in every open set in which this is true of* u.

PROOF. Let $u \in \mathcal{E}'(\Omega)$, analytic in an open neighborhood V of x_o. Since K is very regular Ku is C^∞ in V. Let U be a relatively compact open subset of V containing x_o and a number d such that $0 < d < \text{dist}(U, \mathbb{R}^n \setminus V)$. Let $\rho \in C_c^\infty(\mathbb{R}^n)$ have its support in the open ball $|x| < d$, $\rho = 1$ in the ball $|x| < d/2$, and $g \in C_c^\infty(V)$, $g(x) = 1$ if $\text{dist}(x, U) < d$. It is convenient (although not necessary) to impose additional conditions on g: select arbitrarily an integer $N \geq 0$ and take g to have properties analogous to those of the function g_N in Lemma 1.1: namely, $|D^\alpha g| \leq C^N \alpha!$ if $|\alpha| \leq N$ (cf. Remark 1.1). We recall that for each N such g can be found, with a constant C independent of N.

We then write, for x in U,

$$\begin{aligned} D^\alpha Ku(x) = &\int D_x^\alpha [K_o(x, y, x - y)](gu)(y)\, dy \\ &+ \int [1 - \rho(x - y)] D_x^\alpha [K_o(x, y, x - y)][1 - g(y)] u(y)\, dy \end{aligned} \tag{2.6}$$

since $g(y)+\rho(x-y)[1-g(y)]+[1-\rho(x-y)][1-g(y)]=1$ and

(2.7) *the supports of* $1-g(y)$ *and* $\rho(x-y)$ *(as a function of* y*) are disjoint.*

Let us deal first with the second integral on the right-hand side of (2.6). Since u is a compactly supported distribution, $(1-g)u$ is a finite sum of derivatives of compactly supported continuous functions: $(1-g)u = \sum_{|\beta|\le m'} D^\beta u_\beta$. We have

$$\int [1-\rho(x-y)]D_x^\alpha[K_o(x,y,x-y)]D^\beta u_\beta(y)\,dy = \tag{2.8}$$
$$(-1)^\beta \sum_{\beta'\le\beta}\binom{\beta}{\beta'}\int \{D^{\beta-\beta'}[1-\rho(x-y)]\}D_x^\alpha D_y^{\beta'}[K_o(x,y,x-y)]u_\beta(y)\,dy.$$

By virtue of the fact that $K_o(x,y,x-y)$ is an analytic function of (x,y) in a neighborhood of the support of $1-\rho(x-y)$, we reach the conclusion that

$$\left|\int [1-\rho(x-y)]D_x^\alpha[K_o(x,y,x-y)][1-g(y)]u(y)\,dy\right| \le C^{|\alpha|+1}\alpha!. \tag{2.9}$$

In dealing with the first integral on the right-hand side of (2.6), we write

$$D_x^\alpha[K_o(x,y,x-y)] = \sum_\beta \binom{\alpha}{\beta}[D_x^{\alpha-\beta}D_y^{\beta-\beta'}K_o(x,y,z)|_{z=x-y}]$$
$$= \sum_{\beta,\beta'} \pm \binom{\alpha}{\beta}\binom{\beta}{\beta'}(-D_y)^{\beta'}[D_x^{\alpha-\beta}D_y^{\beta-\beta'}K_o(x,y,z)|_{z=x-y}],$$

whence

$$\int [D_x^\alpha K_o(x,y,x-y)](gu)\,dy = \tag{2.10}$$
$$\sum_{\beta,\beta'} \pm \binom{\alpha}{\beta}\binom{\beta}{\beta'}\int [D_x^{\alpha-\beta}D_y^{\beta-\beta'}K_o(x,y,z)|_{z=x-y}]D_y^{\beta'}(gu)(y)\,dy.$$

We apply (2.2):

$$\left|\int D_x^\alpha[K_o(x,y,x-y)](gu)(y)\,dy\right| \tag{2.11}$$
$$\le \sum_{\beta'\le\alpha} C_1^{|\alpha-\beta'|+1}(\alpha-\beta')!\,\sup \sum_{|\gamma|\le m}\frac{1}{\beta'!}|D^{\beta'+\gamma}(gu)|.$$

Let us then require $|\alpha| \le N - m$. Since u is analytic in an open neighborhood of supp g and by the estimates on the derivatives of order $\le N$ of g, we derive from (2.11)

$$\left| \int D_x^\alpha [K_o(x, y, x-y)](gu)(y)\, dy \right| \le C_2^{N-m+1} \alpha!. \tag{2.12}$$

If we combine (2.9) and (2.12) we get $|D^\alpha(Ku)| \le C_3^{N-m+1}\alpha!$ in U, for all $\alpha \in \mathbb{Z}_+^n$, $|\alpha| \le N - m$. Applying this with $|\alpha| = N - m$ and then letting N run over $\mathbb{Z}_+$ yields exactly what we wanted. □

When K_o is defined as in (2.1) we shall denote by Op k the operator K defined in (2.5). This agrees with the notation in Chapter I. The question we must now address is how to select the amplitude k so that (2.2) and (2.4) hold. Our approach is based on the following result:

Lemma 2.1. *Suppose that $k(x, y, \xi)$ is a C^∞ function in $\Omega \times \Omega \times \mathbb{R}_n$ which can be extended as a C^∞ function of (z, w, ζ), holomorphic with respect to (z, w), in an open set*

$$(z, w) \in \Omega^{\mathbb{C}} \times \Omega^{\mathbb{C}}, \qquad \zeta \in \mathbb{C}_n, \qquad |\text{Im}\, \zeta| < \delta_o |\text{Re}\, \zeta|, \tag{2.13}$$

where δ_o is a number >0 and $\Omega^{\mathbb{C}}$ an open neighborhood of Ω in $\mathbb{C}^n$, such that the following properties hold:

To every compact subset $\mathcal{K}$ of $\Omega \times \Omega$ and to every $\varepsilon > 0$ there are numbers $r > 0$, δ, $0 < \delta < \delta_o$, and constants $C, C' > 0$ such that if (z, w, ζ) belongs to the set (2.13) *and satisfies*

$$\text{dist}((z, w), \mathcal{K}) < r, \qquad |\text{Im}\, \zeta| < \delta |\text{Re}\, \zeta|, \tag{2.14}$$

then

$$|k(z, w, \xi)| \le C(1 + |\xi|)^m, \qquad \forall \xi \in \mathbb{R}_n; \tag{2.15}$$

$$|k(z, w, \zeta)| \le C' \exp\{\varepsilon\delta |\text{Re}\, \zeta|\}; \tag{2.16}$$

$$|\overline{\partial}_\zeta k(z, w, \zeta)| \le C \exp\{-|\text{Re}\, \zeta|/C\}. \tag{2.17}$$

Under these conditions the kernel $K_o(x, y, q)$ defined in (2.1) *has the properties* (2.2) *and* (2.4), *and* Op k *is analytic pseudolocal.*

Remark 2.1. Since (2.15) and (2.17) do not involve ε, we may assume that these estimates hold for a choice of δ, r, and C depending solely on $\mathcal{K}$ and not at all on ε.

REMARK 2.2. By virtue of Cauchy's inequalities, (2.15) is equivalent to inequalities of the kind

(2.18) $|D_x^\alpha D_y^\beta k(x, y, \xi)| \le C^{|\alpha+\beta|+1}\alpha!\beta!(1+|\xi|)^m, \forall(x, y) \in \mathscr{K}, \xi \in \mathbb{R}_n.$

PROOF. We have, for all $f \in C_c^\infty(\Omega)$,

$$\iint e^{i(x-y)\cdot\xi} D_x^\alpha D_y^\beta k(x, y, \xi) f(y)\, dy\, d\xi$$
$$= \iint (1-\Delta_y)^N (e^{i\xi\cdot(x-y)})(1+|\xi|^2)^{-N} D_x^\alpha D_y^\beta k(x, y, \xi) f(y)\, dy\, d\xi$$
$$= \iint e^{i\xi\cdot(x-y)}(1+|\xi|^2)^{-N}(1-\Delta_y)^N [f(y) D_x^\alpha D_y^\beta k(x, y, \xi)]\, dy\, d\xi.$$

We take $N > \frac{1}{2}(m+n)$ (but N independent of $\alpha, \beta \in \mathbb{Z}_+^n$). We have

$$\left|\iint e^{i(x-y)\cdot\xi} D_x^\alpha D_y^\beta k(x, y, \xi) f(y)\, dy\, d\xi\right| \le \left\{\int (1+|\xi|^2)^{-N+m/2}\, d\xi\right\}$$
$$\sup_\xi \left\{\int |(1-\Delta_y)^N [f(y) D_x^\alpha D_y^\beta k(x, y, \xi)]|\, dy \Big/ (1+|\xi|^2)^{m/2}\right\}.$$

To obtain (2.2), with N replacing m, it suffices to use the Leibniz formula and take (2.18) into account.

That the properties in Lemma 2.1 imply (2.4) is a restatement of Lemma 1.3, or rather a version with parameters (here z, w or, if one prefers, x, y) of that lemma. □

We say that a continuous linear operator $\mathscr{E}'(\Omega) \to \mathscr{D}'(\Omega)$ is *analytic regularizing* if its range is contained in $\mathscr{A}(\Omega)$. From now on, we shall often reason modulo analytic-regularizing operators. It will therefore be convenient to have a criterion ensuring that a given amplitude defines an analytic regularizing operator.

LEMMA 2.2. *Let $a(z, w, \xi)$ be a C^∞ function in $\Omega^{\mathbb{C}} \times \Omega^{\mathbb{C}} \times \mathbb{R}_n$, holomorphic with respect to (z, w), having the following property:*

(2.19) *To every compact subset $\mathscr{K}^{\mathbb{C}}$ of $\Omega^{\mathbb{C}} \times \Omega^{\mathbb{C}}$ there is a constant $C > 0$ such that*

(2.20) $|a(z, w, \xi)| \le C e^{-|\xi|/C}, \qquad \forall(z, w) \in \mathscr{K}^{\mathbb{C}}, \xi \in \mathbb{R}_n.$

Then Op *a is analytic regularizing.*

If an amplitude $a(z, w, \xi)$ satisfies (2.19), it is natural to say that it is *exponentially decaying at infinity* (with respect to ξ).

PROOF. It is obvious that, thanks to (2.20), the integral

$$\int e^{i(x-y)\cdot\xi} a(x, y, \xi)\, d\xi$$

can be extended holomorphically to an open neighborhood of $\Omega \times \Omega$ in $\mathbb{C}^{2n}$. It is also clear that if a distribution kernel $K(x, y)$ in $\Omega \times \Omega$ is analytic with respect to (x, y), the associated operator K is analytic regularizing. □

COROLLARY 2.1. *If $a(z, w, \xi)$ is holomorphic with respect to (z, w) and compactly supported with respect to ξ, then* Op *a is analytic regularizing.*

We now come to the central definitions of the present chapter:

DEFINITION 2.1. *A C^∞ function $k(x, y, \xi)$ in $\Omega \times \Omega \times \mathbb{R}_n$ will be called a pseudoanalytic amplitude of degree m in $\Omega \times \Omega$ if it can be extended as a C^∞ function of (z, w, ξ) in $\Omega^{\mathbb{C}} \times \Omega^{\mathbb{C}} \times \mathbb{R}_n$, with $\Omega^{\mathbb{C}}$ an open neighborhood of Ω in $\mathbb{C}^n$, holomorphic with respect to (z, w), and having the following property:*

To every compact subset $\mathcal{K}$ of $\Omega^{\mathbb{C}} \times \Omega^{\mathbb{C}}$ there are two numbers $C, R_o > 0$ such that, for all (z, w) in $\mathcal{K}$, all ξ in $\mathbb{R}_n$, all α in $\mathbb{Z}^n_+$,

$$|D^\alpha_\xi k(z, w, \xi)| \leq C^{|\alpha|+1} \alpha! |\xi|^{m-|\alpha|} \qquad \text{if } |\xi| \geq R_o \sup(|\alpha|, 1). \tag{2.21}$$

Much of the motivation for Definition 2.1 lies in the next result:

LEMMA 2.3. *If $k(x, y, \xi)$ is a pseudoanalytic amplitude of degree m in $\Omega \times \Omega$, it can be extended to a set* (2.13) *as a function $k(z, w, \zeta)$ having all the properties listed in Lemma* 2.1.

PROOF. We refer the reader to the second part of the proof of Lemma 1.4. We use the same cutoff functions χ_α as those used there (cf. (1.47)–(1.49)); we take $R > 0$ and set, for $\zeta = \xi + \sqrt{-1}\,\eta$,

$$k(z, w, \zeta) = \sum_\alpha \frac{1}{\alpha!} (-\eta)^\alpha D^\alpha_\xi k(z, w, \xi) \chi_\alpha(\xi). \tag{2.22}$$

Property (2.15) is an immediate consequence of (2.21); (2.16) follows by requiring $|\eta| < \delta|\xi|$ and $C\delta < 1$, since (2.22) yields

$$|k(z, w, \zeta)| \leq C|\xi|^m \sum_\alpha (C|\eta|/|\xi|)^\alpha.$$

Actually we get (for $\delta > 0$)

$$|k(z, w, \zeta)| \le C_1|\xi|^m \qquad \textit{if } |\eta| < \delta|\xi|. \tag{2.23}$$

Last, the derivation of (2.17) is based on the analogue of (1.52) and proceeds exactly like that of (1.33). We leave the details to the reader. □

In practice one often deals with *formal pseudoanalytic amplitudes*

$$\sum_{j=0}^{+\infty} k_j(z, w, \xi). \tag{2.24}$$

These formal amplitudes are submitted to much stricter conditions than in the C^∞ case:

There is an open neighborhood $\Omega^{\mathbb{C}}$ of Ω in $\mathbb{C}^n$ and a continuous function $R_o(z, w) > 0$ in $\Omega^{\mathbb{C}} \times \Omega^{\mathbb{C}}$ such that the following are true:

(2.25) *For every j, k_j is a C^∞ function of (z, w, ξ), holomorphic with respect to (z, w), in the set*

$$\{(z, w, \xi) \in \Omega^{\mathbb{C}} \times \Omega^{\mathbb{C}} \times R_n;\ |\xi| > R_o(z, w)\sup(j, 1)\}.$$

(2.26) *There is a continuous function $C_o(z, w) > 0$ in $\Omega^{\mathbb{C}} \times \Omega^{\mathbb{C}}$ such that, for all $j = 0, 1, \ldots$, all α in $\mathbb{Z}_+^n$, all z, w in $\Omega^{\mathbb{C}}$ and all ξ in $\mathbb{R}_n$, $|\xi| > R_o(z, w)\sup(j + |\alpha|, 1)$,*

$$|D_\xi^\alpha k_j(z, w, \xi)| \le C_o(z, w)^{j+|\alpha|+1} j!\alpha!|\xi|^{m-j-|\alpha|}. \tag{2.27}$$

Now we show how to construct a true pseudoanalytic amplitude from the formal one, (2.24). This will only be done when x and y remain in a relatively compact open subset of Ω, $\tilde{\Omega}$ (which can be chosen arbitrarily), or, if one prefers, under the additional hypothesis that the positive functions R_o, C_o in (2.25), (2.26) can be taken to be constant.

We use a sequence of cutoff functions $\phi_j(\xi)$ having the following properties:

(2.28) $0 \le \phi_j(\xi) \le 1$ *for all ξ, and* $\phi_j(\xi) = 0$ *if* $|\xi| < 2R\sup(j, 1)$, $\phi_j(\xi) = 1$ *if* $|\xi| > 3R\sup(j, 1)$;

$$|D^\alpha \phi_j| \le (C/R)^{|\alpha|} \quad \textit{if } |\alpha| \le 2j \qquad \text{(cf. (1.36))}. \tag{2.29}$$

We take R to be a number $> R_o$ and set

$$k(z, w, \xi) = \sum_{j=0}^{+\infty} \phi_j(\xi) k_j(z, w, \xi). \tag{2.30}$$

According to (2.25) each term is now a C^∞ function of (z, w, ξ) in $\Omega^{\mathbb{C}} \times \Omega^{\mathbb{C}} \times \mathbb{R}_n$, holomorphic with respect to (z, w). By the Leibniz formula,

$$D_\xi^\alpha k = \sum_{j=0}^{+\infty} \sum_{\beta \leq \alpha} \binom{\alpha}{\beta} D_\xi^\beta \phi_j D_\xi^{\alpha-\beta} k_j.$$

Suppose $|\xi| \geq 2R \sup(|\alpha|, 1)$. Then, for ξ in supp ϕ_j we have $j + |\alpha| \leq |\xi|/R$, and therefore (2.27) is applicable. Suppose furthermore that $\beta \neq 0$; then on supp $D^\beta \phi_j$ we have $R^{-1}|\xi| \leq 3j$ which demands, by what was just said, $|\alpha| \leq 2j$. At this point we may apply (2.29). From this we derive the following:

(2.31) *if* $|\xi| \geq 2R \sup(|\alpha|, 1)$, *then*

$$|D_\xi^\alpha k| \leq C_o^{|\alpha|+1} \alpha! \left\{ \sum_{j=0}^{+\infty} \sum_{\beta \leq \alpha} \frac{1}{\beta!} (C/R)^{|\beta|} [C_o/R(j + |\alpha|)]^{j+|\alpha-\beta|} \right\} |\xi|^{m-|\alpha|}.$$

By choosing R suitably large, in comparison to C_o and to C, we obtain inequalities of the kind (2.21).

Modifying the cutoff functions ϕ_j, if reasonably done, adds to the resulting amplitude (2.30) one that is exponentially decaying at infinity, i.e., one that has property (2.19). Indeed, consider an amplitude

$$r(z, w, \xi) = \sum_{j=0}^{+\infty} \psi_j(\xi) k_j(z, w, \xi), \tag{2.32}$$

where the cutoffs ψ_j have the following two properties: (1) $\sup_\xi |\psi_j(\xi)|$ is bounded independently of j; (2) for each j, the support of ψ_j is contained in a shell $j \leq |\xi|/R \leq Mj$, with $M > 1$ independent of j. The amplitude (2.32) typifies the difference between the choice of (2.30) and the choice of another, similar amplitude.

If we avail ourselves of (2.27) (actually for $\alpha = 0$) we see that

$$|r(z, w, \xi)| \leq C_1 |\xi|^m \sum_{j=0}^{+\infty} (C_1/R)^j e^{-j} |\psi_j(\xi)| \leq C_2 \left\{ \sum_j (C_1/R)^j \right\} |\xi|^m e^{-|\xi|/MR},$$

which proves our assertion if we assume that $R > C_1$. (The constant C_1 depends only on the k_j.)

Dealing with pseudoanalytic amplitudes has the advantage that we can rather freely use cutoff functions like χ_α, ϕ_j. However, in many applications one needs amplitudes that are analytic with respect to ξ, often for purposes of substitution when ξ itself is made to depend, analytically, on other variables.

Definition 2.2. *A pseudoanalytic amplitude* $k(x, y, \xi)$ *in* $\Omega \times \Omega$ *will be called an analytic amplitude if the open neighborhood* $\Omega^{\mathbb{C}}$ *of* Ω *in* $\mathbb{C}^n$ *and the number* $\delta_o > 0$ *can be so chosen that k can be extended as a holomorphic function of* (z, w, ζ) *in the open set*

$$(2.33) \qquad (z, w) \in \Omega^{\mathbb{C}} \times \Omega^{\mathbb{C}}, \qquad \zeta \in \mathbb{C}_n, \qquad 1 + |\mathrm{Im}\, \zeta| < \delta_o |\mathrm{Re}\, \zeta|.$$

Remark 2.3. In order for a holomorphic function $k(z, w, \zeta)$ in the set (2.33) to qualify as an analytic amplitude, it suffices that to every compact subset $\mathcal{K}^{\mathbb{C}}$ of $\Omega^{\mathbb{C}} \times \Omega^{\mathbb{C}}$, there be two positive constants C, R such that

$$(2.34) \qquad |k(z, w, \zeta)| \le C|\zeta|^m, \quad \forall (z, w) \in \mathcal{K}^{\mathbb{C}},$$
$$\forall \zeta \in \mathbb{C}_n, \quad 1 + |\mathrm{Im}\, \zeta| < R^{-1}|\mathrm{Re}\, \zeta|,$$

with $m \in \mathbb{R}$ independent of $\mathcal{K}^{\mathbb{C}}$. Indeed, Cauchy's inequalities applied in the ball $|\zeta - \xi| \le \frac{1}{2}R^{-1}|\xi|$ imply at once estimates of the kind (2.21).

The meaning of the next result is that, at least semiglobally, the classes (modulo analytic-regularizing operators) of pseudodifferential operators that we define by means of pseudoanalytic amplitudes are the same as those we would define by means of analytic amplitudes.

Lemma 2.4. *Let* $k(x, y, \xi)$ *and* $\Omega^{\mathbb{C}}$ *be as in Definition* 2.1 (*possibly after replacing* Ω *by any one of its relatively compact open subsets*).

There is a number $\delta_o > 0$ *and a holomorphic function of* (z, w, ζ) *in the set* (2.33), $k_{\#}(z, w, \zeta)$, *such that* $a = k - k_{\#}$ *has the property analogous to* (2.19):

There is a constant $C > 0$ *such that*

$$(2.35) \quad |a(z, w, \xi)| \le Ce^{-|\xi|/C}, \quad \forall (z, w) \in \Omega^{\mathbb{C}} + \Omega^{\mathbb{C}}, \quad \xi \in \mathbb{R}_n, |\xi| > \delta_o^{-1}.$$

Proof. Let s denote the variable in $\mathbb{C}_1$, and Γ_κ^+ the open sector $|\mathrm{Im}\, s| < \kappa\, \mathrm{Re}\, s$. In the product $\Gamma_\kappa^+ \times \Gamma_\kappa^+ \subset \mathbb{C}_2$ we define the function

$$(2.36) \qquad E(s, s') = \frac{1}{\pi} \frac{1}{s - s'} \exp[-(s - s')^2/s'].$$

We note for later reference that $E(s, s')$ is a two-sided fundamental kernel of $\partial/\partial\bar{s}$ in $\Gamma_\kappa^+ \times \Gamma_\kappa^+$, i.e.,

$$(2.37) \quad \frac{\partial}{\partial \bar{s}} E(s, s') = -\frac{\partial}{\partial \bar{s}'} E(s, s') = \delta(s - s') \qquad \text{(Dirac measure)}.$$

We switch to polar coordinates in $\mathbb{R}_n \backslash \{0\}$ and write $\rho = |\xi|$, $\theta = \xi/|\xi|$. We extend both ρ and θ to complex values; we let ρ vary in Γ_κ^+ for $\kappa > 0$ suitably small and let θ vary in a Stein open subset of $\mathbb{C}_n \backslash \{0\}$, $\mathcal{O}$, which contains the intersection of Γ with the unit sphere S_{n-1} in $\mathbb{R}_n$. These sets are chosen so that when ρ and θ range all over them, the point $\zeta = \rho\theta$ remains in a sector $|\text{Im } \zeta| < \delta |\text{Re } \zeta|$ for suitably small $\delta > 0$. Furthermore we can select δ_1, $0 < \delta_1 < \delta$, such that if

$$|\text{Im } \zeta| < \delta_1 |\text{Re } \zeta|, \tag{2.38}$$

then $\rho = (\zeta_1^2 + \cdots + \zeta_n^2)^{1/2}$ will stay in Γ_κ^+ and $\theta = \zeta/\rho$ will stay in $\mathcal{O}$. For ζ in (2.38) we define

$$\tilde{a}(z, w, \zeta) = \iint_{\Gamma_\kappa^+} E(\rho, \rho') \frac{\partial k}{\partial \bar{\rho}}(z, w, \rho'\theta)\, d\bar{\rho}' \wedge d\rho'/2\sqrt{-1}, \tag{2.39}$$

where $k(z, w, \zeta)$ is the extension in Lemma 2.3. Of course we have, in (2.38),

$$\frac{\partial}{\partial \bar{\rho}}(\tilde{a} - k) = 0. \tag{2.40}$$

In (2.40) and in what now follows we are regarding ζ as a function of (ρ, θ); of course $\zeta = \rho\theta$. We have

$$\frac{\partial}{\partial \theta_j} \tilde{a}(z, w, \zeta) = \iint_{\Gamma_\kappa^+} E(\rho, \rho') \frac{\partial}{\partial \bar{\rho}'} \left[\frac{\partial k}{\partial \theta_j}(z, w, \rho'\theta) \right] d\bar{\rho}' \wedge d\rho'/2\sqrt{-1}. \tag{2.41}$$

We integrate by parts with respect to ρ' in the right-hand side in (2.41). We take (2.37) into account and also decrease $\delta_1 > 0$ sufficiently, so that when ρ' remains on the *boundary* of Γ_κ^+ (defined by $\text{Im } s = \pm\kappa \text{ Re } s$) and ζ remains in (2.38) we have, for some constant $c > 0$,

$$\text{Re}[(\rho - \rho')^2/\rho'] \geq c(|\zeta| + |\rho'|). \tag{2.42}$$

Thus we obtain

$$\bar{\partial}_\theta(\tilde{a} - k) = b, \tag{2.43}$$

with b holomorphic with respect to ρ in Γ_κ^+ and such that

$$|b(z, w, \zeta)| \leq C' e^{-c|\zeta|} \tag{2.44}$$

in the set (2.38). Since the open set $\mathcal{O}$ in which θ varies is Stein, we can solve the equation

$$\bar{\partial}_\theta a_1 = b, \tag{2.45}$$

with a_1 holomorphic with respect to ρ in Γ_κ^+ and satisfying, in the set (2.38),

$$|a_1(z, w, \zeta)| \le C'' e^{-c'|\zeta|} \qquad (c' > 0). \tag{2.46}$$

We shall take the amplitude a in Lemma 2.4 to be $\tilde{a} - a_1$. One can show that holomorphy with respect to (z, w) is preserved throughout the current reasoning; the proof of Lemma 2.4 will be completed if we establish that $\tilde{a}$, like a_1, satisfies an inequality of the kind (2.35). This will follow from (2.39) in which we take (2.17) into account. Indeed, if both ρ and ρ' belong to Γ_κ^+ we have for a suitable $c > 0$,

$$\mathrm{Re}[(\rho - \rho')^2/\rho'] + \mathrm{Re}\,\rho' \ge c(|\rho| + |\rho'|),$$

and our assertion follows easily. □

REMARK 2.4. In the preceding proof we have used the fact that any compact subset of $\mathbb{R}_n$ (such as S_{n-1}) has a basis of neighborhoods in $\mathbb{C}_n$ consisting of Stein open sets, and then the fact that, in such an open set, one can solve the Cauchy–Riemann equations with appropriate estimates. On the latter subject we refer the reader to Chapter V of L. Hörmander, *An Introduction to Complex Analysis in Several Variables* (Van Nostrand, Princeton, New Jersey, 1966).

REMARK 2.5. When dealing with a pseudoanalytic amplitude $k(z, w, \xi)$ which, even if only for large values of $|\xi|$, is positive-homogeneous of degree m, the analytic amplitude $k_\#$ equivalent to k can be taken to be equal to $k(z, w, \xi)$ for large $|\xi|$ and to be positive-homogeneous of degree m for all $\xi \neq 0$. The inequalities (2.21) ensure that $k_\#$ extends as a holomorphic function of (z, w, ζ) in a set of the kind (2.13). A similar remark applies to formal pseudoanalytic amplitudes such as (2.24), in which each individual term k_j is positive-homogeneous of degree $m - j$ with respect to ξ for large $|\xi|$.

3. Analytic Pseudodifferential Operators

DEFINITION 3.1. *A pseudodifferential operator A in Ω will be called an analytic pseudodifferential operator if, given any relatively compact open subset $\tilde{\Omega}$ of Ω, there is a pseudoanalytic amplitude $\tilde{a}$ in $\tilde{\Omega} \times \tilde{\Omega}$ such that $A - \mathrm{Op}\,\tilde{a}$ is analytic regularizing in $\tilde{\Omega}$.*

From Lemmas 2.1 and 2.3 it follows at once that *an analytic pseudodifferential operator is analytic pseudolocal.* For a more precise result see Theorem 3.2 below.

3.1. Symbolic Calculus

Consider a pseudoanalytic amplitude of degree m in $\Omega \times \Omega$, $k(x, y, \xi)$. Just as in the C^∞ case, it defines a *formal symbol*, namely,

$$\sum_{\alpha \in \mathbf{Z}_+^n} \frac{1}{\alpha!} \partial_\xi^\alpha D_y^\alpha k(x, y, \xi)|_{y=x}. \tag{3.1}$$

If we take advantage of the inequalities (2.21) we see that, given any compact subset $\mathcal{K}_o$ of $\Omega^{\mathbf{C}}$, for suitable constants C, $R_o > 0$, we have

$$\frac{1}{\alpha!} |\partial_\xi^\alpha D_w^\alpha k(z, z, \xi)| \leq C^{|\alpha|+1} \alpha! |\xi|^{m-|\alpha|} \tag{3.2}$$

$$\textit{if } z \in \mathcal{K}_o \textit{ and } |\xi| \geq R_o \sup(|\alpha|, 1).$$

We have tacitly used Cauchy's inequalities with respect to w. The preceding means that if we write

$$k_j(x, \xi) = \sum_{|\alpha|=j} \frac{1}{\alpha!} \partial_\xi^\alpha D_y^\alpha k(x, x, \xi), \tag{3.3}$$

then the formal symbol

$$\sum_{j=0}^{+\infty} k_j(x, \xi) \tag{3.4}$$

satisfies the analogues of properties (2.25), (2.26). By means of cutoff functions ϕ_j as in (2.28), (2.29), we may form a true symbol $\tilde{k}$ in some arbitrary relatively compact open subset $\tilde{\Omega}$ of Ω. Any change in the choice of the cutoffs ϕ_j only modifies the corresponding operator, Op $\tilde{k}$, by an analytic-regularizing operator, as shown by the argument about (2.32). As a consequence this construction defines *a class $\dot{K}$ of analytic pseudodifferential operators modulo analytic-regularizing operators:* A pseudodifferential operator in Ω belongs to the class $\dot{K}$ if, given any $\tilde{\Omega} \subset\subset \Omega$ and restricted to $\mathcal{E}'(\tilde{\Omega})$, it differs from Op $\tilde{k}$ by an analytic-regularizing operator in $\tilde{\Omega}$.

THEOREM 3.1. *Op k belongs to the class $\dot{K}$.*

PROOF. Since both Op k and Op $\tilde{k}$ are analytic pseudolocal, it suffices to show that if Ω' is an arbitrarily small open subset of $\tilde{\Omega}$, for any $u \in \mathcal{E}'(\Omega')$ we have $(\text{Op } k)u - (\text{Op } \tilde{k})u \in \mathcal{A}(\Omega')$. We choose for Ω' an open ball centered at some point x_o and select an integer $s \geq 0$ such that u can be represented as a finite sum of derivatives of order $\leq s$ of continuous functions with compact support in Ω'.

We make systematic use of the cutoff functions ϕ_j in (2.28), (2.29) (the number R is suitably large). In particular, we assume that the (true) symbol $\tilde{k}$ is defined by means of those cutoffs. In what follows N is an arbitrary positive integer. If we set $K_N = \mathrm{Op}[\phi_{N+1}(\xi)k(x, y, \xi)]$, there is a constant $C_1 > 0$, independent of R and of N, such that, in Ω',

$$|D^\alpha(\mathrm{Op}\, k - K_N)u| \le C_1(C_1RN)^{|\alpha|+m+s} \qquad \text{if } |\alpha| \le N, \tag{3.5}$$

since $|\xi| \le 3R(N+1)$ on the support of $1 - \phi_{N+1}$. By Taylor expansion we have

$$k(x, y, \xi) = \sum_{|\alpha| \le N} \frac{(y-x)^\alpha}{\alpha!} \partial_y^\alpha k(x, x, \xi) + \sum_{|\alpha| = N+1} \frac{(y-x)^\alpha}{\alpha!} k_\alpha(x, y, \xi), \tag{3.6}$$

where

$$k_\alpha(x, y, \xi) = (N+1) \int_0^1 \partial_y^\alpha k\big(x, x + t(y-x), \xi\big)(1-t)^N \, dt. \tag{3.7}$$

We derive

$$K_N = K_{(N)} + R_N + S_N + T_N, \tag{3.8}$$

with

$$K_{(N)} = \mathrm{Op}\Big\{\sum_{j \le N} \phi_j(\xi) k_j(x, \xi)\Big\} \qquad (k_j \text{ defined in (3.3)}), \tag{3.9}$$

$$R_N = \mathrm{Op}\Big\{\sum_{j \le N} [\phi_{N+1}(\xi) - \phi_j(\xi)] k_j(x, \xi)\Big\}, \tag{3.10}$$

$$S_N = \mathrm{Op}\Big\{\phi_{N+1}(\xi) \sum_{|\alpha| = N+1} \frac{1}{\alpha!} D_\xi^\alpha k_\alpha(x, y, \xi)\Big\}, \tag{3.11}$$

$$\mathrm{T_N} = \mathrm{Op}\Big(\sum_{|\alpha| \le N+1} \frac{1}{\alpha!} \{D_\xi^\alpha[\phi_{N+1}(\xi)k_\alpha] - \phi_{N+1}(\xi) D_\xi^\alpha k_\alpha\}\Big), \tag{3.12}$$

where we agree that $k_\alpha = \partial_y^\alpha k(x, x, \xi)$ when $|\alpha| \le N$; when $|\alpha| = N+1$, k_α is defined by (3.7). We note that

$$\mathrm{Op}\, \tilde{k} - K_{(N)} = \mathrm{Op}\Big\{\sum_{j > N} \phi_j(\xi) k_j(x, \xi)\Big\}. \tag{3.13}$$

We shall of course avail ourselves of (2.21) and of the subsequent properties (2.25) and (2.26), in which we take k_j according to (3.3).

As usual we take R large in comparison with C; here $\mathcal{K}_o$ will be a suitably large compact subset of Ω^{C} (e.g., its interior contains the closure of

Ω'). For a suitably large constant C_2, independent of N, and all z in $\mathscr{K}_o$,

$$(3.14) \qquad \Big|\sum_{j>N} \phi_j(\xi)k_j(z,\xi)\Big| \leq C_2^{N+1}N!|\xi|^{m-N}\chi_N(\xi),$$

where χ_N is the characteristic function of the set $|\xi| \geq 2R(N+1)$. Consequently, for $C_3 > 0$ suitably large, independent of N, we have, in Ω',

$$(3.15) \qquad |D^\alpha[\mathrm{Op}\,\tilde{k} - K_{(N)}]u| \leq C_3^{N+1}\alpha! \qquad \textit{if } |\alpha| \leq N - m - s.$$

Essentially the same argument applies to S_N; the presence of the variable y, which varies in the support of u, does not change anything. We have, in Ω',

$$(3.16) \qquad |D^\alpha S_N u| \leq C_3^{N+1}\alpha! \qquad \textit{if } |\alpha| \leq N - m - s,$$

and if C_3 is suitably large.

In order to obtain useful estimates for R_N and T_N we take advantage of the fact that $|\xi| \leq 3R(N+1)$ on the support of $\phi_{N+1} - \phi_j$ $(j \leq N)$ and on that of $D^\beta\phi_{N+1}$ if $\beta \neq 0$. Thus we obtain for R_N an estimate analogous to (3.5). In Ω',

$$(3.17) \qquad |D^\alpha R_N u| \leq C_1(C_1RN)^{|\alpha|+m+s} \qquad \textit{if } |\alpha| \leq N,$$

possibly after increasing C_1.

We rewrite (3.12) in the fashion

$$(3.18) \qquad T_N = \mathrm{Op}\Big\{\sum_{|\alpha|\leq N+1}\ \sum_{0\neq\beta\leq\alpha} \frac{1}{\beta!}D_\xi^\beta\phi_{N+1}(\xi)\frac{1}{(\alpha-\beta)!}D_\xi^{\alpha-\beta}k_\alpha\Big\}.$$

Since $|\alpha| \leq N+1$ we may use the fact that $|D^\beta\phi_{N+1}| \leq C^{N+1}\beta!$. From (2.21) and Cauchy's inequalities we derive (cf. (3.7))

$$(3.19) \qquad |D_\xi^{\alpha-\beta}k_\alpha(z,y,\xi)| \leq C^{|\alpha|+1}\alpha!(\alpha-\beta)!|\xi|^{m-|\alpha-\beta|}$$

$$\textit{if } z \in \mathscr{K}_o,\ y \in \Omega',\ |\xi| \geq R_o \sup(|\alpha-\beta|, 1).$$

The requirement on $|\xi|$ is certainly satisfied if ξ belongs to the support of ϕ_{N+1}. Since, as already noted, we have $|\xi| \leq 3R(N+1)$ on the support of $D^\beta\phi_{N+1}$ when $\beta \neq 0$, we derive easily that, for $C_4 > 0$ suitably large, independent of N, we have, in Ω',

$$(3.20) \qquad |D^\alpha T_N u| \leq C_4^{N+1}N! \qquad \textit{if } |\alpha| \leq N.$$

We combine the estimates (3.5), (3.15)–(3.17), and (3.20) with the equality (3.8). We see that, for $C_5 > 0$ large enough and, as always, independent of N, the following holds in Ω',

$$(3.21) \qquad |D^\alpha(\mathrm{Op}\,k - \mathrm{Op}\,\tilde{k})u| \leq C_5^{N+1}N! \qquad \textit{if } |\alpha| \leq N - m - s.$$

Take $|\alpha|$ exactly equal to $N - m - s$ (supposed to be nonnegative). We at once reach the conclusion that $(\mathrm{Op}\, k - \mathrm{Op}\, \tilde{k})u$ is analytic in Ω'. □

Thus if we are willing to limit our attention to *classes* of analytic pseudodifferential operators modulo analytic-regularizing ones and to relatively compact open subsets of the basic open set Ω, it is enough to consider representatives of the kind

$$(\mathrm{Op}\, a)u(x) = (2\pi)^{-n} \int e^{ix\cdot\xi} a(x, \xi)\hat{u}(\xi)\, d\xi, \tag{3.22}$$

where $a(x, \xi)$ is a *pseudoanalytic symbol*, that is, a pseudoanalytic amplitude independent of the second variable y.

Actually if we wish to talk about the *symbol* of an analytic pseudodifferential operator, we must establish an equivalence relation between the various "true" symbols that can be constructed from the formal one, (3.4). Even the latter is not unambiguously defined, since we may certainly modify somewhat the terms k_j and still get a formal symbol giving rise to the same equivalence class $\dot{K}$. The definition of the equivalence relation is obvious:

DEFINITION 3.2. *Let $k(x, \xi)$ be a pseudoanalytic symbol in Ω, U an open subset of Ω, Γ^o an open cone in $\mathbb{R}_n \backslash \{0\}$. We shall say that k is equivalent to zero in $U \times \Gamma^o$ if it can be extended as a C^∞ function of (z, ξ) in $U^{\mathbb{C}} \times \Gamma^o$, where $U^{\mathbb{C}}$ is an open neighborhood of U in $\mathbb{C}^n$, holomorphic with respect to z and such that the following is true:*

(3.23) *Given any conically compact subset $\mathscr{C}$ of $U^{\mathbb{C}} \times \Gamma^o$, there is a constant $C > 0$ such that*

$$|k(z, \xi)| \leq C e^{-|\xi|/C}, \qquad \forall (z, \xi) \in \mathscr{C}. \tag{3.24}$$

We say that k is equivalent to zero near a point (x_o, ξ^o) of $\Omega \times (\mathbb{R}_n \backslash \{0\})$ if it is equivalent to zero in a product set $U \times \Gamma^o$ of the preceding kind, with U an open neighborhood of x_o in Ω and Γ^o an open cone containing ξ^o.

We say that k is equivalent to zero in a conic open subset Γ of $\Omega \times (\mathbb{R}_n \backslash \{0\})$ if k is equivalent to zero near every point of Γ.

We say that two pseudoanalytic symbols a, b in Ω are equivalent in Γ if $a - b$ is equivalent to zero in Γ.

REMARK 3.1. The proof of Lemma 2.4 shows that over each relatively compact open subset of Ω every pseudoanalytic symbol is equivalent to an *analytic* one.

All the true symbols constructed from (3.4) are equivalent, in the entirety of their common domain of definition. But there might be in the equivalence class of operators $\dot{K}$ elements of the kind Op $k^{\#}$ with $k^{\#} \neq k$. It is not *a priori* obvious that the symbol class defined by means of the formal symbol originating with $k^{\#}$ is the same as that originating with (3.4). We shall not prove this fact but shall admit it as being true; a proof would be obtained along the same lines as in the C^∞ case, based on a more precise version, under suitable analyticity hypotheses, of Theorem 3.1 of Chapter VI. The equivalence class of symbols defined by all those amplitudes is called the *symbol* (or the symbol class) of the (operator) class $\dot{K}$.

DEFINITION 3.3. *Let A be an analytic pseudodifferential operator in* Ω, $\dot{a}(x, \xi)$ *its symbol. We shall use the term analytic microsupport of A and denote by* μ $\text{supp}_a A$ *the smallest closed conic subset of* $\Omega \times (\mathbb{R}_n \backslash \{0\})$ *in whose complement* $\dot{a}$ *vanishes. (That is, all the representatives of* $\dot{a}$ *that are defined in a conic open neighborhood of some point of that complement are equivalent to zero near such a point).*

The analytic microsupport of A is, in general, different from its (C^∞) microsupport (Definition 6.4, Chapter I).

THEOREM 3.2. *Let A be an analytic pseudodifferential operator, u a compactly supported distribution in* Ω. *We have*

$$\mathrm{WF}_a(Au) \subset \mathrm{WF}_a(u) \cap \mu\ \mathrm{supp}_a A. \tag{3.25}$$

PROOF. We may limit ourselves to the case $A = \text{Op}\, a$ as in (3.22).

Consider a product set $\tilde{\Omega} \times \Gamma^{\#}$ with $\tilde{\Omega} \subset \Omega$ open and $\Gamma^{\#}$ an open cone in $\mathbb{R}_n \backslash \{0\}$, in which either a is equivalent to zero or else u is analytic, i.e., or else the analytic wave-front set of u does not intersect it. Let Ω' be an arbitrary relatively compact open subset of $\tilde{\Omega}$ and Γ^* an open cone whose closure is contained in $\Gamma^{\#}$. Let Γ be another open cone whose closure is contained in Γ^*, and let g^R be a function like the one in Lemma 1.4.

If a is equivalent to zero in $\tilde{\Omega} \times \Gamma^{\#}$ the product $a(z, \xi)g^R(\xi)$ has the property (2.19). By Lemma 2.2, $Ag^R(D)$ is analytic regularizing.

If u is analytic in $\tilde{\Omega} \times \Gamma^{\#}$ and if R is suitably large, it follows from Lemma 1.6 that $g^R(D)u$ is analytic in Ω', and so is $Ag^R(D)u$. It suffices therefore to observe that according to Corollary 1.3, the analytic wave-front set of $v = A[1 - g^R(D)]u$ does not intersect $\Omega' \times \Gamma$. □

We shall use the following "microlocal" terminology: A, a continuous linear operator $\mathscr{E}'(\Omega) \to \mathscr{D}'(\Omega)$, is *analytic regularizing in a conic open subset of* $\Omega \times (\mathbb{R}_n \backslash \{0\})$, $\mathscr{C}$, if for any compactly supported distribution u in Ω, $\mathrm{WF}_a(Au) \cap \mathscr{C} = \varnothing$.

COROLLARY 3.1. *Any analytic pseudodifferential operator in* Ω *is analytic regularizing in the complement of its analytic microsupport.*

REMARK 3.2. We have *not* proved that if A is analytic regularizing in a conic open subset $\mathscr{C}$ of $\Omega \times (\mathbb{R}_n \backslash \{0\})$, then

$$\mathscr{C} \cap \mu \operatorname{supp}_a A = \varnothing. \tag{3.26}$$

This is equivalent to the "uniqueness" of the symbol, which we have not proved.

REMARK 3.3. The argument in the proof of Theorem 3.2 shows that microlocally an analytic pseudodifferential operator is equivalent to an operator of the kind $(\mathrm{Op}\, a)g^R(D)$, with $\mathrm{Op}\, a$ as in (3.22) and a suitable choice of g^R.

Theorem 3.1 enables us to develop the symbolic calculus for analytic pseudodifferential operators in the same manner as for standard (C^∞) ones. The formulas are the same, and we content ourselves with refering to Chapter I, Section 4, in particular, (4.21)–(4.24). The composition of two analytic pseudodifferential operators A and B calls for some comment, since we cannot make use of properly supported amplitudes. Our amplitudes, here, must be analytic with respect to (x, y). Suppose that

$$Au(x) = (2\pi)^{-n} \iint e^{i(x-y)\cdot\xi} a(x, y, \xi) u(y)\, dy\, d\xi, \tag{3.27}$$

$$Bu(x) = (2\pi)^{-n} \iint e^{i(x-y)\cdot\xi} b(x, y, \xi) u(y)\, dy\, d\xi. \tag{3.28}$$

Let $\tilde{\Omega}$ be an arbitrary relatively compact open subset of Ω, $h \in C_c^\infty(\Omega)$, $h = 1$ in $\tilde{\Omega}$. The restriction to $\tilde{\Omega}$ of $u \mapsto A(hBu)$ is an analytic pseudodifferential operator equal to $\mathrm{Op}\, k$, with

$$k(x, y, \xi) = (2\pi)^{-n} \iint e^{i(y-z)\cdot(\xi-\eta)} a(x, z, \xi) h(z) b(z, y, \eta)\, dz\, d\eta \tag{3.29}$$

(cf. Chapter I, (3.11)). It is enough to give the proof of this assertion when a and b are pseudoanalytic *symbols*; we leave it to the reader. If we modify h (outside $\tilde{\Omega}$), we merely add to AhB an operator that is analytic regularizing in $\tilde{\Omega}$; this simply follows from the fact that A is analytic pseudolocal. This defines the *class* $\dot{A} \circ \dot{B}$ in $\tilde{\Omega}$, and therefore everywhere in Ω. Formula (4.24) of Chapter I shows what the symbol of this class is; the formal symbol (4.24) enables us to construct a true symbol $\tilde{k}$ in $\tilde{\Omega} \subset\subset \Omega$. If h is the above cutoff function, we see right away that Op $\tilde{k} - AhB$ is analytic regularizing in $\tilde{\Omega}$.

3.2. Parametrices of Elliptic Analytic Pseudodifferential Operators

An *elliptic* analytic pseudodifferential operator is simply a standard pseudodifferential operator that is both analytic and elliptic (of a given order). On the latter notion we refer to the end of Section 4 of Chapter I. All results in the C^∞ case (Proposition 4.1 and its corollaries) extend to the analytic case—with the proviso that one must either deal with *classes* of analytic pseudodifferential operators modulo analytic-regularizing ones, or alternatively, restrict composes, inverses, and parametrices to relatively compact open subsets of the basic set Ω. For instance

THEOREM 3.3. *Let A be an analytic pseudodifferential operator in Ω, elliptic of order m. Given any relatively compact open subset $\tilde{\Omega}$ of Ω, there is a pseudodifferential operator B in Ω, of order $-m$, having the following properties*:

(3.30) *B extends as a continuous linear map of $\mathcal{D}'(\Omega)$ into $\mathcal{E}'(\Omega)$. (One could say that B is "compactifying".)*

(3.31) *The restriction of B to $\tilde{\Omega}$ is an analytic pseudodifferential operator (or order m) in $\tilde{\Omega}$.*

(3.32) *$AB - I$ and $BA - I$ are analytic regularizing in $\tilde{\Omega}$.*

PROOF. We may limit ourselves to the case $A = \text{Op}\, a$ as in (3.22). We define symbols $b_j(x, \xi)$ by means of the equations (4.40) or (4.41), Chapter I. We must check that

$$\sum_{j=0}^{+\infty} b_j(x, \xi) \tag{3.33}$$

is indeed a formal pseudoanalytic symbol in Ω (cf. (2.25), (2.26)). If this is so, let $b(x, \xi)$ be any true pseudoanalytic symbol constructed from (3.33), by the standard technique of cutoff functions of the kind (2.28), (2.29). Assume that $b(x, \xi)$ is a pseudoanalytic symbol (of degree $-m$) in a relatively compact open subset Ω' of Ω containing the closure of $\tilde{\Omega}$, and let $h \in C_c^\infty(\Omega')$ be equal to one in $\tilde{\Omega}$: $B = \mathrm{Op}[h(x)h(y)b(x, \xi)]$ has the properties in Theorem 3.3. We have $AB = \mathrm{Op}[k(x, y, \xi)h(y)]$ with k given by (3.29); by Theorem 3.1, $AB \sim \mathrm{Op}\,\tilde{k}$ in $\tilde{\Omega}$ with $\tilde{k}$ a true symbol constructed from the formal symbol $a \odot b$. By definition of the b_j, the latter is equal (or equivalent, if one prefers) to *one* in $\tilde{\Omega}$.

We know that the first partial derivatives of $a(x, \xi)$ with respect to x are pseudoanalytic symbols of degree m in Ω. By the mean value theorem we derive that to every $\varepsilon > 0$ there is $\delta > 0$ such that, if $z \in \mathbb{C}^n$, $|z - x| < \delta$, then

$$|a(z, \xi) - a(x, \xi)| \leq \varepsilon(1 + |\xi|)^m, \qquad \forall \xi \in \mathbb{R}_n. \tag{3.34}$$

We take advantage of the ellipticity of A, embodied in property (4.38), Chapter I. Combined with (3.34) this leads to the following result:

There is an open neighborhood $\Omega^{\mathbb{C}}$ of Ω in $\mathbb{C}^n$ and two continuous functions $R_o(z)$, $C_o(z)$ in $\Omega^{\mathbb{C}}$, both strictly positive, such that $a(z, \xi)$ is a C^∞ function of (z, ξ), holomorphic with respect to z, in the set

$$\{(z, \xi) \in \Omega^{\mathbb{C}} \times \mathbb{R}_n;\, |\xi| \geq R_o(z)\}, \tag{3.35}$$

where the following also holds:

$$|\xi|^m \leq C_o(z)|a(z, \xi)|. \tag{3.36}$$

We define, in the set (3.35),

$$b_0(z, \xi) = 1/a(z, \xi), \tag{3.37}$$

and for $j = 1, 2, \ldots,$

$$b_j = -b_0 \sum_{1 \leq |\alpha| \leq j} \frac{1}{\alpha!} \partial_\xi^\alpha a D_x^\alpha b_{j-|\alpha|}. \tag{3.38}$$

By induction on j this shows that all the b_j are C^∞ functions in the set (3.35), holomorphic with respect to z, and thus (3.33) partakes of property (2.25). We must check (2.26).

We begin by proving estimates of the kind

$$|D_\xi^\alpha b_0| \leq C_1(z)^{|\alpha|+1} \alpha! |\xi|^{-m-|\alpha|} \qquad \textit{if } |\xi| \geq R_1(z) \sup(|\alpha|, 1). \tag{3.39}$$

Here R_1, C_1 are continuous functions in $\Omega^{\mathbb{C}}$, with positive values, not less than R_o and C_o respectively. Thus (3.39) is certainly true when $\alpha = 0$, by (3.36). From there on we reason by induction on $|\alpha|$, assumed to be ≥ 1. The

Leibniz formula implies, in view of (3.37),

$$D_\xi^\alpha b_0 = -b_0 \sum_{0\neq\beta\leq\alpha} \binom{\alpha}{\beta} D_\xi^\beta a D_\xi^{\alpha-\beta} b_0. \tag{3.40}$$

Assuming that the analogue of (3.39) for a in the place of b_0 holds true when R_o replaces R_1 and C_o replaces C_1, and with m instead of $-m$, we obtain

$$|D_\xi^\alpha b_0| \leq C_o \alpha!|\xi|^{-m-|\alpha|} \sum_{0\neq\beta\leq\alpha} C_o^{|\beta|+1} C_1^{|\alpha-\beta|+1}$$

$$\leq C_1^{|\alpha|+1} \alpha!|\xi|^{-m-|\alpha|} \sum_{\beta\neq 0} (C_o/C_1)^{|\beta|} C_o^2.$$

As the summation ranges over the indices $\beta \neq 0$, the last factor can be made not to exceed one, provided that C_1 is large enough in comparison to C_o, whence (3.39).

Next we prove the analogues of (3.39) for b_j, $j = 1, 2, \ldots$.

$$|D_\xi^\alpha b_j| \leq C_2(z)^{j+|\alpha|+1} \alpha! j! |\xi|^{-m-|\alpha|-j} \quad \textit{if } |\xi| \geq R_1(z) \sup(j + |\alpha|, 1). \tag{3.41}$$

If $z_o \in \Omega^{\mathbb{C}}$ and $r > 0$ we write

$$D(z_o; r) = \{z \in \mathbb{C}^n; |z^\iota - z_o^\iota| < r, \iota = 1, \ldots, n\}.$$

We shall assume that r is so small that the closure of $D(z_o; r)$ is contained in $\Omega^{\mathbb{C}}$. For any number s, $0 < s < 1$, and any holomorphic function f in $D(z_o; r)$, we set

$$\|f\|_s = \sup_{z\in D(z_o;\, sr)} |f(z)|. \tag{3.42}$$

Cauchy's inequalities imply, if $0 < s' < s < 1$,

$$\left\| \frac{\partial f}{\partial z^\iota} \right\|_{s'} \leq \frac{r^{-1}}{s - s'} \|f\|_s. \tag{3.43}$$

Suppose then that the following holds, for some constant $M > 0$,

$$\|f\|_s \leq M\left(\frac{e}{1-s}\right)^j, \qquad \forall s, 0 < s < 1. \tag{3.44}$$

If we apply (3.43), with $s = s' + \varepsilon$, and drop the "primes", then

$$\left\| \frac{\partial f}{\partial z^\iota} \right\|_s \leq M(r\varepsilon)^{-1}\left(\frac{e}{1-s-\varepsilon}\right)^j. \tag{3.45}$$

We may choose $\varepsilon = (1-s)/(1+j)$. It yields

$$(3.46)\qquad \left\|\frac{\partial f}{\partial z^\iota}\right\|_s \le Mr^{-1}(j+1)\left(\frac{e}{1-s}\right)^{j+1}.$$

By iteration we obtain, for any n-tuple α,

$$(3.47)\qquad \|D_z^\alpha f\|_s \le Mr^{-|\alpha|}(j+1)\cdots(j+|\alpha|)\left(\frac{e}{1-s}\right)^{j+|\alpha|}.$$

This said, let us return to (3.38). Let $\beta \in \mathbb{Z}_+^n$ be arbitrary and suppose that $|\xi| \ge R_1(z)\sup(j+|\beta|, 1)$. We apply the Leibniz formula:

$$(3.48)\qquad \frac{1}{\beta!}|D_\xi^\beta b_j| \le$$

$$\sum \frac{1}{\beta'!}|D_\xi^{\beta'} b_0|\frac{1}{\alpha!}\frac{1}{(\beta-\beta'-\beta'')!}|D_\xi^{\alpha+\beta-\beta'-\beta''}a|\frac{1}{\beta''!}|D_z^\alpha D_\xi^{\beta''} b_{j-|\alpha|}|.$$

The summation ranges over α, $1 \le |\alpha| \le j$, $\beta' \le \beta$, $\beta'' \le \beta-\beta'$. Thus we have $|\xi| \ge R_1(z)\sup(|\beta'|+j, |\alpha+\beta-\beta'-\beta''|, j-|\alpha+\beta''|, 1)$. We begin by using (3.39), and the analogous estimates for a. We get

(3.49)

$$\frac{1}{\beta!}|D_\xi^\beta b_j| \le C_o C_1 \sum C_1^{|\beta'|}(2C_o)^{|\alpha+\beta-\beta'-\beta''|}|\xi|^{-|\alpha+\beta-\beta''|}|D_z^\alpha D_\xi^{\beta''} b_{j-|\alpha|}|/\beta''!.$$

From this we want to derive, by induction, that for a suitable choice of constants M, $C>0$ we have

$$(3.50)\qquad \|D_\xi^\beta b_j\|_s \le Mj!\beta!\left(\frac{e}{1-s}\right)^j C^{j+|\beta|}|\xi|^{-m-j-|\beta|}.$$

Notice that this is certainly true when $j=0$. The fact that (3.44) implies (3.47) shows that (3.50) implies

$$(3.51)\qquad \frac{1}{\beta!}\|D_z^\alpha D_\xi^\beta b_j\|_s \le Mr^{-|\alpha|}(j+|\alpha|)!\left(\frac{e}{1-s}\right)^{j+|\alpha|} C^{j+|\beta|}|\xi|^{-m-j-|\beta|}.$$

In (3.51) we substitute $j-|\alpha|$ for j, β'' for β and put the result into the right-hand side in (3.49), thus obtaining

$$(3.52)\qquad \frac{1}{j!}\frac{1}{\beta!}|\xi|^{m+j+|\alpha|}\|D_\xi^\beta b_j\|_s \le$$

$$MC_0C_1\left\{\sum r^{-|\alpha|}C_1^{|\beta'|}(2C_o)^{|\alpha+\beta-\beta'-\beta''|}C^{|\beta''|-|\alpha|}\right\}C^j\left(\frac{e}{(1-s)}\right)^j.$$

This means that the ratio of the left-hand side in (3.50) divided by the right-hand side does not exceed

$$C_o C_1 \sum (2C_o/Cr)^{|\alpha|}(C_1/C)^{|\beta'|}(2C_o/C)^{|\beta-\beta'-\beta''|},$$

a quantity that (because the summation ranges over $\alpha \neq 0$) can be made not to exceed one, provided that C is large enough in comparison with all the other constants. By letting s go to zero in (3.50) we obtain an estimate of the kind (3.41) at $z = z_o$. □

REMARK 3.4. The formal pseudoanalytic symbol (3.33), with the b_j defined by (3.37), (3.38), defines the symbol class of the inverse $\dot{A}^{-1}$ of the class $\dot{A}$ of A.

COROLLARY 3.2. *Let A be an elliptic analytic pseudodifferential operator in Ω. Given any distribution $u \in \mathscr{E}'(\Omega)$ we have*

$$\mathrm{WF}_a(Au) = \mathrm{WF}_a(u). \tag{3.53}$$

3.3. Analytic Pseudodifferential Operators on a Real Analytic Manifold

Let ϕ be an analytic diffeomorphism of the open subset Ω of $\mathbb{R}^n$ onto another open subset of $\mathbb{R}^n$, Ω'. If A is any analytic pseudodifferential operator in Ω, its transfer A^ϕ to Ω' (see Chapter I, (3.14)) is a pseudodifferential operator (of the same order) in Ω' (Chapter I, Theorem 3.3). It is also obvious that A^ϕ is analytic pseudolocal. Actually

THEOREM 3.4. *The transfer A^ϕ of A to Ω' via ϕ is an analytic pseudodifferential operator in Ω'.*

PROOF. We shall content ourselves with sketching it. It suffices to reason in relatively compact open subsets of Ω' (or, by pullback via ϕ, in Ω) and assume that $A = \mathrm{Op}\ a$, as in (3.22), with a an analytic amplitude (Lemma 2.4). If we then use the formula (3.20) we see that it defines an amplitude in a suitable open neighborhood of the diagonal, in $\Omega' \times \Omega'$, specifically

$$a^\phi(y, y', \eta) = |\det[\mathscr{J}(y)\mathscr{J}_o(y, y')^{-1}]|a(\overset{-1}{\phi}(y), {}^t\mathscr{J}_o^{-1}(y, y')\eta), \tag{3.54}$$

where $\mathscr{J}(y)$ is the Jacobian matrix of $\overset{-1}{\phi}$ at the point y, and

$$\mathscr{J}_o(y, y') = \int_0^1 \mathscr{J}(ty + (1-t)y')\, dt. \tag{3.55}$$

In fact, if Ω'' is any suitably small open subset of Ω', the restriction to Ω'' of A^ϕ is equal to that of Op a^ϕ.

On the other hand, equation (3.54), which is obviously "analytic" in its domain of definition, defines a formal analytic symbol,

$$\exp\{\partial_\eta D_{y'}\}a^\phi(y, y', \eta)|_{y'=\eta}, \tag{3.56}$$

which in turn defines an equivalence class of analytic pseudodifferential operators in Ω'. Let $\tilde{\Omega}'$ be any relatively compact open subset of Ω'; we may form a true pseudoanalytic symbol $\tilde{a}^\phi$ in $\tilde{\Omega}'$ out of (3.56). If the small open set Ω'' is contained in $\tilde{\Omega}'$ we know by Theorem 3.1 that $A^\phi = \text{Op}\, a^\phi$ differs from Op $\tilde{a}^\phi$ merely by an analytic-regularizing operator in Ω''. But since both A^ϕ and Op $\tilde{a}^\phi$ are analytic pseudolocal in $\tilde{\Omega}'$, this means that given any $u \in \mathscr{E}'(\Omega'')$, $A^\phi u - (\text{Op}\, \tilde{a}^\phi)u \in \mathscr{A}(\tilde{\Omega}')$. This must therefore also be true for all u in $\mathscr{E}'(\tilde{\Omega}')$. □

Theorem 3.4 is the base for the definition of analytic pseudodifferential operators on a real analytic manifold X. A pseudodifferential operator A on X is an analytic pseudodifferential operator if its restriction to any analytic local chart gives rise (by transfer to an open set in $\mathbb{R}^n$; $n = \dim X$) to an analytic pseudodifferential operator in that open set. All the theory develops in the analytic case in perfect parallelism with the C^∞ case, and we shall not dwell on it.

In the C^∞ case the "invariance" of the wave-front set could be derived from the analogues of Theorem 3.2 and Corollary 3.2; because of the property that if Γ and $\Gamma^\#$ are two conic open subsets of $T^*X\backslash 0$ and if the closure of Γ is conically compact and contained in $\Gamma^\#$, there exists a standard (C^∞) pseudodifferential operator that is elliptic in Γ and regularizing in the complement of $\Gamma^\#$. We do not have at our disposal such a property in the analytic case. We shall therefore prove the result directly:

THEOREM 3.5. *Let u be a compactly supported distribution in Ω, $\phi_* u$ its direct image via the analytic diffeomorphism $\phi : \Omega \to \Omega'$. We have*

$$\text{WF}_a(\phi_* u) = \phi_+ \text{WF}_a(u), \tag{3.57}$$

where ϕ_+ is the bundle diffeomorphism $T^*\Omega \to T^*\Omega'$ defined by ϕ (see Chapter I, Section 5).

PROOF. We suppose that u is analytic in $U \times \Gamma^\#$ with $U \subset \Omega$ open and $\Gamma^\#$ an open cone in $\mathbb{R}_n\backslash\{0\}$. Let us assume that the compact support of u is contained in an open neighborhood $U^\#$ of the closure of U, and select arbitrarily a relatively compact open subset U^* of U and a cone Γ^* whose closure is contained in $\Gamma^{\#}$. By Lemma 1.6 we know that if R is large enough, $g^R(D)u$ will be an analytic function in U^*.

For $N = 1, 2, \dots$, let us select a C^∞ function ψ_N in $\mathbb{R}_n$ such that $\psi_N(\xi) = 0$ if $|\xi| < RN$, $\psi_N(\xi) = 1$ if $|\xi| > 2RN$. We note that, whatever the distribution $w \in \mathscr{E}'(\mathbb{R}^n)$, we have, for suitable constants C, $s \geq 0$,

$$(3.58)\qquad |D^\alpha\{[1 - \psi_N(D)]w\}| \leq C(RN)^{|\alpha|+s}, \qquad \forall \alpha \in \mathbb{Z}^n_+.$$

This remains valid in cases where w is not compactly supported, such as $w = g^R(D)u$. From the analyticity of $g^R(D)u$ in U^* (and possibly after some decreasing of the latter) we see that in U^*,

$$(3.59)\qquad |D^\alpha[\psi_N(D)g^R(D)u]| \leq C(RN)^{|\alpha|+s} \qquad \textit{if } |\alpha| \leq N - s.$$

We may also apply (3.58) to $w = [1 - g^R(D)]u$. In view of these three estimates (that is, (3.58) with w equal either to $g^R(D)u$ or to $[1 - g^R(D)]u$, and (3.59)) we must now obtain suitable estimates for

$$u_N = \psi_N(D)[1 - g^R(D)]u.$$

It suffices to consider the case where $U^{\#}$ and therefore U and U^* are very small. Then if y stays in $\phi(U)$ we have

$$(3.60)\quad v_N(y) = (\phi_* u_N)(y) - (2\pi)^{-n} \iint e^{i(y-y')\cdot\eta} c(y, y', \eta)(\phi_* u)(y')\, dy'\, d\eta,$$

where

$$(3.61)\qquad c(y, y', \eta) = \psi_N({}^t\mathscr{J}_o^{-1}(y, y')\eta)[1 - g^R({}^t\mathscr{J}_o^{-1}(y, y')\eta)]|\det[\mathscr{J}(y)\mathscr{J}_o^{-1}(y, y')]|$$

(cf. (3.54)). We observe that if $U^{\#}$ is sufficiently small, and if y remains in $\phi(U)$ while y' remains in $\phi(U^{\#})$ and ${}^t\mathscr{J}^{-1}(y)\eta \in \Gamma^*$, then $c(y, y', \eta) = 0$.

Though not needed so far we must now impose some requirements on ψ_N, namely that, for a suitable constant $C_o > 0$,

$$(3.62)\qquad |D^\alpha \psi_N| \leq (C_o/R)^{|\alpha|} \qquad \textit{if } |\alpha| \leq N.$$

They can easily be satisfied (cf. Lemma 1.1). Furthermore we make use of a sequence $\{h_N\}$ ($N = 1, 2, \dots$) of functions belonging to $C_c^\infty(U^*)$ such that

$$(3.63)\qquad |D^\alpha h_N| \leq (C_1 N)^{|\alpha|} \qquad \textit{if } |\alpha| \leq N.$$

We set $h_N^\phi(y) = h_N(\overset{-1}{\phi}(y))$. It is clear the h_N^ϕ satisfy estimates similar to (3.63).

We shall avail ourselves of the property (1.30) of g^R, which implies

(3.64) $$|D^\alpha g^R(\xi)| \leq (C/R)^{|\alpha|} \qquad \textit{if } |\alpha| \leq N \textit{ and } \xi \in \operatorname{supp} \psi_N.$$

We adapt to the present situation the argument used in the proof of Lemma 1.5. We write

(3.65)
$$(\widehat{h_N^\phi v_N})(\xi) = (2\pi)^{-n} \iiint e^{-iy\cdot(\xi-\eta)-iy'\cdot\eta} h_N^\phi(y) c(y, y', \eta)(\phi_* u)(y')\, dy\, dy'\, d\eta.$$

Actually we represent $\phi_* u$ as a finite sum of derivatives of order $\leq s$ of continuous functions supported in $\phi(U^\#)$. Also we assume that there is an open cone $\Gamma^\varnothing$ such that for some $c_o > 0$

(3.66) $$\Gamma^\varnothing \subset {}^t\mathscr{J}^{-1}(y)\Gamma^* \qquad \textit{for all } y \in \phi(U);$$

(3.67) $$|\xi - \eta| \geq c_o(|\xi| + |\eta|) \qquad \textit{if } \xi \in \Gamma^\varnothing, y \in \phi(U), y' \in \phi(U^\#) \textit{ and } c(y, y', \eta) \neq 0.$$

Selecting an integer j such that $N \leq 2j \leq N + 1$, we may write, for ξ in $\Gamma^\varnothing$,

(3.68) $$|(\widehat{h_N^\phi v_N})(\xi)| \leq C_2^{j+1} |\xi|^{-2j+s+n+1} \sup_{y'|U^\#} \iint |\Delta_y^j [h_N^\phi(y) c(y, y', \eta)]|(1 + |\eta|)^{-n-1}\, dy\, d\eta.$$

We take advantage of (3.62)–(3.64) and of the analyticity of the diffeomorphism ϕ, and we conclude that

(3.69) $$(\widehat{h_N^\phi v_N})(\xi)| \leq C_3^{N+1} N! |\xi|^{-N+s+n+1}, \qquad \xi \in \Gamma^\varnothing.$$

Since analogous inequalities can be derived when v_N is replaced by $\phi_*\{[1 - \psi_N(D)]u\}$ and by $\phi_*[\psi_N(D) g^R(D) u]$, according to what was said at the beginning, we reach the conclusion that the analytic wave-front set of $\phi_* u$ does not intersect $\phi(U^*) \times \Gamma^\varnothing$. This implies at once what we were seeking. □

4. Microlocalization All the Way. The Holmgren Theorem

In this section we show that all the concepts and results of the preceding section which are not microlocal can be microlocalized by making judicious use of the cutoff functions g^R of Lemma 1.4. Since most of the arguments are merely refinements of those used in the proofs of the earlier results, we often content ourselves with sketching them.

DEFINITION 4.1. *Let* (x_o, ξ^o) *be an arbitrary point in* $\mathbb{R}^n \times (\mathbb{R}_n \backslash \{0\})$. *By an analytic amplitude of degree m near* (x_o, ξ^o), *we mean a holomorphic function* $k(z, w, \zeta)$ *in a set*

$$(4.1) \qquad (z, w) \in U^{\mathbb{C}} \times U^{\mathbb{C}}, \qquad \zeta \in \mathbb{C}^n, \quad \operatorname{Re} \zeta \in \Gamma^o, \quad 1 + |\operatorname{Im} \zeta| \leq \delta_o |\operatorname{Re} \zeta|,$$

where $U^{\mathbb{C}}$ *is an open neighborhood of* x_o *in* $\mathbb{C}^n$, Γ^o *an open cone in* $\mathbb{R}_n \backslash \{0\}$ *containing* ξ^o, δ_o *a number* >0, *such that for a suitable constant* $C > 0$, *the following holds in the set* (4.1):

$$(4.2) \qquad |k(z, w, \zeta)| \leq C|\zeta|^m.$$

Write $U = \mathbb{R}^n \cap U^{\mathbb{C}}$ and let Γ, Γ^* be two open cones containing ξ^o such that the closure of Γ^* is contained in Γ^o and that of Γ in Γ^*. Let g^R denote a function like the one in Lemma 1.4. Given any distribution $u \in \mathscr{E}'(U)$ we may consider the distribution in U,

$$(4.3) \qquad K^R u(x) = (2\pi)^{-n} \iint e^{i(x-y)\cdot\xi} k(x, y, \xi) g^R(\xi) u(y)\, dy\, d\xi.$$

LEMMA 4.1. *Let* $g'^{R'}$ *denote another function with properties similar to those of* g^R *relative to* Γ *and* Γ^o; *and let* $K'^{R'}$ *denote the analogue of* (4.3). *Then* $K^R - K'^{R'}$ *is analytic regularizing near* (x_o, ξ^o).

PROOF. Follows at once from Lemma 1.5. □

LEMMA 4.2. *Suppose that the analytic wave-front set of* $u \in \mathscr{E}'(U)$ *does not intersect* $V \times \Gamma^o$, *with* $V \subset U$ *open. Let* V' *be any relatively compact open subset of* V. *If* R *is large enough, then* $K^R u$ *is an analytic function in* V'.

PROOF. For x, y in U and q in $\mathbb{R}^n$, set

$$(4.4) \qquad K_o^R(x, y, q) = (2\pi)^{-n} \int e^{iq\cdot\xi} k(x, y, \xi) g^R(\xi)\, d\xi.$$

By suitably adapting the proof of Lemma 2.1 and using the "natural" extension $g^R(\zeta)$ to $\mathbb{C}_n$, one can prove that there is an integer $M \geq 0$ and a constant $C > 0$ such that, for all $\alpha, \beta \in \mathbb{Z}_+^n$, all $f \in C_c^\infty(U)$ and all x, y in U,

$$(4.5) \qquad \left| \int [D_x^\alpha D_y^\beta K_o^R(x, t, q)|_{q=x-y}] f(y)\, dy \right| \leq C^{|\alpha+\beta|+1} \alpha! \beta! \sup \sum_{|\gamma| \leq M} |D^\gamma f|$$

(cf. (2.2)). On the other hand, by reasoning as in the proof of Lemma 1.3, one sees that

(4.6) *$K_o^R(x, y, q)$ is an analytic function of (x, y, q) in the set $\Omega \times \Omega \times \{q \in \mathbb{R}^n;\ |q| \geq C/R\}$.*

for a suitable choice of $C > 0$. Finally, adaptation of the proof of Theorem 2.1 implies the conclusion in Lemma 4.2. □

Let us then denote by $\mathscr{D}^{@}_{(x_0,\xi^o)}$ the quotient of $\mathscr{D}'(\mathbb{R}^n)$ modulo the following equivalence relation:
x_o. We may define $K^R(hu)$, and from Lemma 4.2 we see that modifying h only adds to $K^R(hu)$ a distribution that is an analytic function in a full neighborhood of x_o, provided that R is large enough.

Let us then denote by $\mathscr{D}^{@}_{(x_o,\xi^o)}$ the quotient of $\mathscr{D}'(\mathbb{R}^n)$ modulo the following equivalence relation:

(4.7) *The analytic wave-front set of $u - v$ does not contain (x_o, ξ^o).*

The elements of $\mathscr{D}^{@}_{(x_o,\xi^o)}$ will be called *germs of microdistributions* at (x_o, ξ^o) (*in the analytic sense*, whenever the additional precision is needed). As (x_o, ξ^o) varies in $\mathbb{R}^n \times (\mathbb{R}_n \backslash \{0\})$ these stalks $\mathscr{D}^{@}_{(x_o,\xi^o)}$ make up the *sheaf* $\mathscr{D}^{@}$ *of microdistributions* in $\mathbb{R}^n$. A distribution u in an open subset Ω of $\mathbb{R}^n$ defines a section over Ω of the sheaf $\mathscr{D}^{@}$, $u^{\#}$, whose support is exactly the analytic wave-front set of u. The relation (4.7) is obviously "conic," i.e., invariant under dilations in the ξ variables, and $\mathscr{D}^{@}$ can be regarded as a sheaf over $\mathbb{R}^n \times S_{n-1}$. But the elements of $\mathscr{D}^{@}_{(x_o,\xi^o)}$ are characterized by the behavior of "their Fourier transforms" at infinity, in cones around the ray through ξ^o.

We may also define germs of analytic amplitudes at (x_o, ξ^o) by means of "microlocal" amplitudes like k in Definition 4.1. Via these germs of analytic amplitudes we may in turn define germs of operators, by means of the corresponding K^R of (4.3). Lemma 4.2 tells us that K^R defines an endomorphism of $\mathscr{D}^{@}_{(x_o,\xi^o)}$, provided that R is large enough, and Lemma 4.1 says that this endomorphism does not depend on R, that in a sense we have the right to go to the limit as $R \to +\infty$. For this reason we denote by $K_{(x_o,\xi^o)}$ the endomorphism in question. We define *the sheaf* $\Psi^{@}$ *of* (germs of) *analytic pseudodifferential operators* as the sheaf whose stalk at (x_o, ξ^o) consists of the "operators" $K_{(x_o,\xi^o)}$. Let A be an analytic pseudodifferential operator in Ω, in the sense of Definition 3.1, and let $\tilde{a}$ be a pseudoanalytic amplitude in a relatively compact open neighborhood $\tilde{\Omega}$ of x_o in Ω such that $A - \mathrm{Op}\,\tilde{a}$ is analytic regularizing in $\tilde{\Omega}$. By Lemma 2.4 we know that there is an analytic amplitude $a_{\#}$ in $\tilde{\Omega}$ such that $\mathrm{Op}\,\tilde{a} - \mathrm{Op}\,a_{\#}$ is analytic regularizing in an open neighborhood of x_o. Let $A^R_\#$ be the analogue of (4.3) when we substitute $a_\#$ for k; $A - A^R_\#$ is analytic regularizing in a conic open neighborhood of (x_o, ξ^o), and thus A and $A^R_\#$ define the same endomorphism of $\mathcal{D}^@_{(x_o,\xi^o)}$. Thus A defines naturally a section $A^\#$ over Ω of the sheaf $\Psi^@$. We also see that there would have been no gain in generality had we used pseudoanalytic amplitudes in Definition 4.1 instead of analytic ones. That the section $A^\#$ vanishes in a conic open subset $\mathscr{C}$ of $\Omega \times (\mathbb{R}_n\backslash\{0\})$ means that A is analytic regularizing in $\mathscr{C}$. The support of $A^\#$ is equal to the analytic microsupport of A is we admit the "uniqueness" of the symbol (Remark 3.1).

In all this, $\mathbb{R}^n$ can be replaced by an arbitrary analytic manifold X. The role of $\mathbb{R}^n \times (\mathbb{R}_n\backslash\{0\})$ is then played by $T^*X\backslash 0$, that of $\mathbb{R}^n \times S_{n-1}$ by the cosphere bundle over X, S^*X.

The microlocal symbolic calculus can be developed essentially in the same manner as the "global" one. An analytic symbol near (x_o, ξ^o) is simply an analytic amplitude near the point (x_o, x_o, ξ^o) independent of the second base variable, w. If $k(z, w, \zeta)$ is any analytic amplitude near the same point, we may form a formal analytic symbol near (x_0, ξ^o), namely the series

$$\sum_{\alpha \in \mathbb{Z}^n_+} \frac{1}{\alpha!} D^\alpha_\zeta \partial^\alpha_w k(z, w, \zeta)|_{w=z}. \tag{4.8}$$

By restricting the variation of ζ to an open cone in real space $\mathbb{R}^n$, containing ξ^o, and using the standard cutoff functions, we can form from (4.8) a true pseudoanalytic symbol. Then by applying Lemma 2.4 we can construct an analytic symbol "equivalent" to that pseudoanalytic one.

If $\tilde{k}(z, \zeta)$ is an analytic symbol in a conic neighborhood of (x_o, ξ^o) in $\mathbb{C}^n \times (\mathbb{C}_n\backslash\{0\})$ of the kind (cf. (4.1))

$$z \in U^{\mathbb{C}}, \quad \zeta \in \mathbb{C}^n, \quad \operatorname{Re}\zeta \in \Gamma^o, \quad 1 + |\operatorname{Im}\zeta| \le \delta_o|\operatorname{Re}\zeta|, \tag{4.9}$$

and if we select the cones $\Gamma \subset \Gamma^* \subset \Gamma^o$ appropriately and associate with them the "cutoff function" g^R, we may form the operator $\tilde{K}^R$ in analogy to (4.3):

$$\tilde{K}^R u(x) = (2\pi)^{-n} \int e^{ix\cdot\xi} \tilde{k}(x, \xi) g^R(\xi) \hat{u}(\xi)\, d\xi. \tag{4.10}$$

Observe in passing that $\tilde{K}^R$ acts on arbitrary distributions $u \in \mathscr{E}'(\mathbb{R}^n)$, not just on those whose support lies in U. But $\tilde{K}^R u$ is a distribution in U, possibly not extendable beyond the boundary of U.

Let K^R be defined by (4.3) and suppose that the symbol $\tilde{k}(x, \xi)$ is derived from the amplitude $k(x, y, \xi)$ by the procedure just described. The analogue of Theorem 3.1 is here

LEMMA 4.3. *$K^R - \tilde{K}^R$ is analytic regularizing in a conic open neighborhood of (x_o, ξ^o).*

PROOF. It suffices to show that $K^R - K^R_{\#}$ is analytic regularizing near (x_o, ξ^o) where

$$K^R_{\#}u(x) = (2\pi)^{-n} \int e^{ix\cdot\xi} k_{\#}(x, \xi) g^R(\xi) \hat{u}(\xi)\, d\xi, \tag{4.11}$$

$$k_{\#}(x, \xi) = \sum_{j=0}^{\infty} \phi_j(\xi) k_j(x, \xi), \tag{4.12}$$

$$k_j(x, \xi) = \sum_{|\alpha|=j} \frac{1}{\alpha!} D_\xi^\alpha \partial_y^\alpha k(x, y, \xi)|_{y=x},$$

with cutoffs ϕ_j as in (2.28), (2.29).

Let us introduce the symbol

$$k^R_*(x, \xi) = \sum_{j=0}^{+\infty} \phi_j(\xi) k^R_{*j}(x, \xi) \tag{4.13}$$

$$k^R_{*j}(x, \xi) = \sum_{|\alpha|=j} \frac{1}{\alpha!} D_\xi^\alpha [g^R(\xi) \partial_y^\alpha k(x, y, \xi)]_{y=x},$$

and let us set $K^R_* = \operatorname{Op} k^R_*$. We begin by showing that $K^R_{\#} - K^R_*$ is analytic regularizing in a conic open neighborhood of (x_o, ξ^o). Indeed, there, the symbol of $K^R_{\#} - K^R_*$ is equal to

$$\sum_{j=0}^{+\infty} \phi_j(\xi) \sum_{|\alpha|=j} \sum_{\beta\le\alpha, \beta\neq 0} \frac{1}{\alpha!} [D_\xi^\beta g^R(\xi)] \frac{1}{(\alpha-\beta)!} D_\xi^{\alpha-\beta} \partial_y^\alpha k(x, y, \xi)|_{y=x}. \tag{4.14}$$

We apply (1.30). Since $\phi_j(\xi) = 0$ if $|\xi| \le 2Rj$ $(j \ge 1)$, and since $|\beta| \le j$, we have $|D_\xi^\beta g^R(\xi)| \le (C/R)^{|\beta|}$ when $\phi_j(\xi) \neq 0$. On the other hand, for those same ξ (and for x, y in a suitable neighborhood of x_o),

$$\begin{aligned} \frac{1}{(\alpha-\beta)!} |D_\xi^{\alpha-\beta} \partial_y^\alpha k(x, y, \xi)| &\le C^{|\alpha|+1} \alpha! |\xi|^{m-|\alpha-\beta|} \\ &\le C^{j+1} j! (1+|\xi|)^m (2Rj)^{|\beta|-j}, \end{aligned}$$

which shows that the absolute value of (4.14) does not exceed

$$C_1(1+|\xi|)^m \sum_{j=0}^{+\infty} \phi_j(\xi)(C/R)^j \sum_{j'=0}^{j} \binom{j}{j'} C^{j'} 2^{-(j-j')}, \tag{4.15}$$

which converges as soon as $R > C(\frac{1}{2}+C)$. By using the analyticity of $k(x, y, \xi)$ with respect to (x, y), we obtain a similar estimate after having applied D_x^γ to (4.14); the only difference is that the constant C_1 in (4.15) must be replaced by $C_1^{|\gamma|+1}\gamma!$. This means that (4.14) satisfies the hypotheses of Lemma 1.5. Notice that (4.14) vanishes identically for $\xi \in \Gamma$, where $g^R = 1$. It follows that, given any $u \in \mathscr{E}'(U)$, the analytic wave-front set of $K_{\#}^R u - K_*^R u$ does not contain (x_o, ξ^o).

From there on we adapt the proof of Theorem 3.1. We substitute K^R for Op k, $K_N^R = \mathrm{Op}[\phi_{N+1}(\xi)k(x, y, \xi)g^R(\xi)]$ for K_N, K_*^R for Op $\tilde{k}$, k_{*j}^R for k_j, the latter in (3.9), (3.10). With these substitutions the analogues of the estimates (3.5), (3.15), (3.17) are easily derived.

We substitute $g^R(\xi)k_\alpha(x, y, \xi)$ for $k_\alpha(x, y, \xi)$ in (3.11), (3.12). It is then easy to check that the estimate (3.20) remains valid. The proof of the analogue of (3.16) is slightly more delicate. Let us set

$$S_{*N} = \mathrm{Op}\left\{\phi_{N+1}(\xi)g^R(\xi) \sum_{|\alpha|=N+1} \frac{1}{\alpha!} D_\xi^\alpha k_\alpha(x, y, \xi)\right\}.$$

Exactly the same argument that proved (3.16) proves

$$|D^\alpha S_{*N}u| \le C_2^{N+1}\alpha! \qquad \textit{if } |\alpha| \le N - m - \nu, \tag{4.16}$$

for a suitable choice of $C_2 > 0$ and of the integer ν (depending on $u \in \mathscr{E}'$). We must look at $S_N^\# = \mathrm{Op}\, s_N^\#$, with

$$s_N^\#(x, y, \xi) = \phi_{N+1}(\xi) \sum_{\substack{|\alpha|=N+1 \\ 0\neq\beta\le\alpha}} \frac{1}{\beta!}[D_\xi^\beta g^R(\xi)] \frac{1}{(\alpha-\beta)!} D_\xi^{\alpha-\beta} k_\alpha(x, y, \xi).$$

We repeat the argument that led us to the conclusion that the absolute value of (4.14) does not exceed (4.15). We find here that

$$|D_x^\gamma s_N^\#(x, y, \xi)| \le C_3^{|\gamma|}\gamma!(C_3/R)^{-N}, \tag{4.17}$$

with $C_3 > 0$ independent of N and of γ. Once again we use Lemma 1.5, but with added precision, namely that $S_N^\#$ is analytic regularizing in a conic open neighborhood of (x_o, ξ^o) *uniformly with respect to N*. This means that the conic neighborhood in question as well as the relevant constants are independent of N. These facts enable us to conclude the proof of Lemma 4.3. □

The upshot of Lemma 4.3 is that the amplitude $k(z, w, \zeta)$ and any analytic symbol such as $k(z, \zeta)$ defined by means of the formal symbol (4.8) lead to the same section of the sheaf of analytic pseudodifferential operators in a neighborhood of (x_o, ξ^o). Thus we could have started with analytic symbols rather than analytic amplitudes. But in order to handle easily transposes, adjoints, composes, and images via analytic diffeomorphisms, it is always more convenient to deal with amplitudes. For instance, if $k_i(z, w, \zeta)$ $(i = 1, 2)$ are two analytic amplitudes in the set (4.1), we may define their compose near (x_o, ξ^o) by selecting a function $h \in C_c^\infty(U)$ $(U = \mathbb{R}^n \cap U^C)$ equal to one in an open neighborhood $V \subset U$ of x_o and by forming (cf. (3.29))

$$
\begin{aligned}
& k_3^R(z, w, \zeta) = \\
(4.18) \qquad & (2\pi)^{-n} \iint e^{i(w-y)\cdot(\zeta-\eta)} k_1(z, y, \zeta) h(y) k_2(y, w, \eta) g^R(\eta)\, dy\, d\eta.
\end{aligned}
$$

The element $\Psi^{@}_{(x_o, \xi^o)}$ defined by k_3^R is the compose of those defined by k_1 and k_2 (in that order), and is independent of R, as seen by applying a variant of Lemma 1.5. Actually we may assume that k_1 and k_2 are symbols, that is, independent of the second base variable. The formal symbol naturally associated with the amplitude k_3^R has the usual expression:

$$
(4.19) \qquad (k_1 \odot k_2)(z, \zeta) = \sum_{\alpha \in \mathbb{Z}_+^n} \frac{1}{\alpha!} D_\zeta^\alpha k_1(z, \zeta)\, \partial_z^\alpha k_2(z, \zeta),
$$

etc.

We may now define microlocally elliptic analytic pseudodifferential operators, and invert them microlocally; formally everything looks exactly as it did in the global situation. For instance, if $a(z, \zeta)$ is the symbol of the operator that we seek to invert, the terms in the formal symbol of its inverse are given by equations (3.37) and (3.38). Incidentally this yields a new proof of the invariance under analytic diffeomorphism of the analytic wave-front set, if we accept this invariance for analytic pseudodifferential operators—or, rather, for sections of the sheaf $\Psi^{@}$. Indeed, let $u^{\#}$ be a microdistribution in a conic open subset $\mathscr{C}$ of $\mathbb{R}^n \times (\mathbb{R}_n \backslash \{0\})$, $A^{\#}$ a section of $\Psi^{@}$ over the same set. We have (cf. (3.25))

$$
(4.20) \qquad \operatorname{supp}(A^{\#} u^{\#}) \subset (\operatorname{supp} A^{\#}) \cap (\operatorname{supp} u^{\#}).
$$

If $A^{\#}$ is elliptic in $\mathscr{C}$, let $B^{\#}$ be its inverse. Still in $\mathscr{C}$, we must have

$$
\operatorname{supp} u^{\#} = \operatorname{supp}(B^{\#} A^{\#} u^{\#}) \subset (\operatorname{supp} B^{\#}) \cap (\operatorname{supp}(A^{\#} u^{\#})).
$$

But of course supp $A^{\#} = \mathscr{C} = \text{supp } B^{\#}$; hence

(4.21) $$\textit{If } A^{\#} \textit{ is elliptic, } \operatorname{supp}(A^{\#}u^{\#}) = \operatorname{supp} u^{\#}.$$

It follows from this that a distribution u is analytic in $\mathscr{C}$ if and only if there is an analytic pseudodifferential operator (or a germ of such an operator), elliptic in $\mathscr{C}$, A, such that Au is analytic in $\mathscr{C}$.

We draw another consequence of (4.21). We consider a *classical* analytic pseudodifferential operator P, of order m, in $\Omega \subset \mathbb{R}^n$; everything that we are going to say is valid for germs of such operators. The symbol of P is given by

(4.22) $$\sum_{j=0}^{+\infty} P_{m-j}(x, \xi),$$

in which each term P_{m-j} extends as a holomorphic function of (z, ζ) in a region

(4.23) $$z \in \Omega^{\mathbb{C}}, \qquad |\operatorname{Im} \zeta| < \delta |\operatorname{Re} \zeta|,$$

where $\Omega^{\mathbb{C}}$ is an open subset of $\mathbb{C}^n$ containing Ω, and δ is some number >0. Furthermore $P_{m-j}(z, \zeta)$ is positive-homogeneous of degree $m - j$ with respect to ζ and satisfies the standard estimates

(4.24) $$|P_{m-j}(z, \zeta)| \leq C_o(z)^{j+1} j! |\zeta|^{m-j},$$

with C_o a strictly positive continuous function in $\Omega^{\mathbb{C}}$.

The *characteristic set* of P is defined exactly as in the C^∞ case. Char P is the zero set of the principal symbol of P, $P_m(x, \xi)$. (Unless specified otherwise, characteristic sets lie in the real cotangent bundle.) In the complement of Char P, P is elliptic of order m. Thus by virtue of (4.21) we have

THEOREM 4.1. *Let $P^{\#}$ denote the section of the sheaf of analytic pseudodifferential operators defined over $\Omega \times (\mathbb{R}_n \backslash \{0\})$ by the classical operator P, $u^{\#}$ a microdistribution in the same set. We have*

(4.25) $$\operatorname{supp} u^{\#} \subset (\operatorname{Char} P) \cap \operatorname{supp}(P^{\#}u^{\#}).$$

COROLLARY 4.1. *Suppose that $P^{\#}u^{\#} = 0$. Then* supp $u^{\#} \subset$ Char P.

COROLLARY 4.2. *Let u belong to $\mathscr{E}'(\Omega)$. Then if Pu is analytic in a conic open subset $\mathscr{C}$ of $\Omega \times (\mathbb{R}_n \backslash \{0\})$, we have*

(4.26) $$\mathscr{C} \cap \operatorname{WF}_a(u) \subset \mathscr{C} \cap (\operatorname{Char} P).$$

As an application we derive a proof of the Holmgren theorem about uniqueness in the Cauchy problem (cf. Chapter II, Section 3).

We assume now that P is a differential operator of order m ($\in \mathbb{Z}_+$) in Ω, with analytic coefficients. The symbol of P is

$$P(x, \xi) = P_m(x, \xi) + P_{m-1}(x, \xi) + \cdots + P_0(x), \tag{4.27}$$

with $P_{m-j}(x, \xi)$ a homogeneous polynomial of degree $m - j$ with respect to ξ. Let us suppose that Ω is connected and that we are given an analytic hypersurface Σ in Ω such that the following is true: there are two open subsets Ω^+, Ω^- of Ω, nonempty and connected, not intersecting Σ nor each other, such that $\Omega = \Omega^+ \cup \Sigma \cup \Omega^-$ (Ω^+ and Ω^- lie on the "sides" of Σ). Furthermore we make the following assumption:

(4.28) *The hypersurface Σ is noncharacteristic with respect to P at every one of its points.*

That is, the conormal bundle to Σ, $N^*\Sigma$, does not intersect Char P. We always assume that Char P is made up of points (x, ξ) with $\xi \neq 0$. The following is one of the standard versions of the classical Holmgren theorem.

THEOREM 4.2. *Let u be a distribution in Ω such that $Pu = 0$ in Ω. If u vanishes in Ω^-, it vanishes in a full neighborhood of Σ.*

Standard proofs of Theorem 4.2 are based on a dual form of the Cauchy–Kovalevska theorem (e.g., see Treves [3], Sections 20, 21). We present a proof, due to L. Hörmander, using analytic wave-front sets:

PROOF. The property to be proved is purely local: we reason in a neighborhood (taken to be Ω) of a point of Ω, which we take to be the origin in $\mathbb{R}^n$. An analytic change of variables allows us to suppose that Ω^+ is contained in a region

$$x^n \geq |x'|^2/\varepsilon, \tag{4.29}$$

with $\varepsilon > 0$ suitably small, and writing $x' = (x^1, \ldots, x^{n-1})$. Indeed one can first effect a linear change of variables such that the normal to Σ at the origin becomes the x^n-axis, and then take $x^n + 2|x'|^2/\varepsilon$ as the new variable, while the remaining variables, x', are left unchanged. Thus supp u is contained in the region (4.29) and therefore u can be viewed as a distribution with respect to $x^n < \varepsilon_1$, for $\varepsilon_1 > 0$ suitably small, valued in the space of *compactly supported* distributions with respect to x'. Actually it is a C^∞ function of $x^n < \varepsilon_1$ valued in $\mathscr{E}'(\mathbb{R}_{x'}^{n-1})$, but we shall not use this stronger property. We can form the duality bracket of $u(x', x^n)$ with any C^∞ function $h(x')$ in $\mathbb{R}^{n-1}$. Actually we shall take h in the space $E(\mathbb{R}^{n-1})$ of L^2 functions whose Fourier transform verifies

$$(4.30) \qquad |\hat{h}(\xi')| \leq \text{const}\, e^{-|\xi'|}, \qquad \forall \xi' \in \mathbb{R}_{n-1}.$$

Consider then

$$(4.32) \quad U(x^n) = \int u(x', x^n)\overline{h(x')}\, dx' = \int \hat{u}(\xi', x^n)\overline{\hat{h}(\xi')}\, d\xi'/(2\pi)^{n-1}.$$

Let $\chi_N \in C_c^\infty(\mathbb{R}^1)$, $\chi_N(x^n) = 0$ if $x^n < \frac{1}{2}\varepsilon_1$, $|D^\alpha \chi_N| \leq (CN)^{|\alpha|}$ if $|\alpha| \leq N$. We observe that the analytic wave-front set of u is contained in Char P by Corollary 4.2; and $(0, \xi^o)$, $\xi^o = (0, \ldots, 0, 1)$, does not belong to Char P by our hypothesis (4.28), and by the fact that the normal to Σ at the origin is the x^n-axis. It follows from this that $\mathrm{WF}_a(u)$ does not contain any point of the form (x, ξ) if $|x| < \delta$, $|\xi'| < \delta|\xi_n|$ and if $\delta > 0$ is small enough. Actually, because supp u is contained in (4.29), this remains true if we merely require $x^n < \delta$. Consequently, if $\varepsilon_1 < \delta$ and if $|\xi'| < \delta|\xi_n|$,

$$(4.32) \quad \left|\int \exp\{-ix^n\xi_n\}\chi_N(x^n)\hat{u}(\xi', x^n)\, dx^n\right| \leq C^{N+1}N!(1 + |\xi|)^{-N}.$$

From this and from (4.30) we derive, for some constants $C > 0$, $M \in \mathbb{R}$,

$$(4.33) \quad \left|\int \exp\{-ix^n\xi_n\}U(x^n)\chi_N(x^n)\, dx^n\right| \leq C^{N+1}N!(1 + |\xi_n|)^{-N}\int_{|\xi'|<\delta|\xi_n|} |\hat{h}(\xi')|\, d\xi' + Ce^{-\delta|\xi_n|/2}\int_{|\xi'|>\delta|\xi_n|} (1 + |\xi'|)^M e^{-|\xi'|/2}\, d\xi.$$

Since $e^{-\delta|\xi_n|/2} \leq N!(\frac{1}{2}\delta|\xi_n|)^{-N}$ we derive from (4.33) that the analytic wave-front set of $U(x^n)$ is empty, at least over the half-line $x^n < \delta$; in other words $U(x^n)$ is analytic there. But since $U(x^n) = 0$ when $x^n < 0$ (since supp $u \subset \overline{\Omega^+}$), we have $U(x^n) = 0$ for all $x^n < \delta$. This means that for these values of x^n, as a distribution in x', $u(x', x^n)$ is orthogonal to all functions h belonging to $E(\mathbb{R}^{n-1})$. Theorem 4.2 will be proved if we show that $E(\mathbb{R}^{n-1})$ is dense in $C^\infty(\mathbb{R}^{n-1})$. Let $f \in C_c^\infty(\mathbb{R}^{n-1})$ be arbitrary. Set

$$(4.34) \qquad f_j(x') = (2\pi)^{-n+1}\int \exp\{ix'\cdot\xi' - |\xi'|^2/j\}\hat{f}_j(\xi')\, d\xi'.$$

We leave it as an exercise to the reader to show (1) that every f_j belongs to $E(\mathbb{R}^{n-1})$, (2) that f_j converges to f in $C^\infty(\mathbb{R}^{n-1})$ as $j \to +\infty$. □

5. Application to Boundary Problems for Elliptic Equations: Analyticity Up to the Boundary

We go back to the boundary problems for elliptic equations studied in Chapter III and resume that study, but now under natural hypotheses of analyticity.

We begin by assuming that the manifold X of Chapter III is analytic. Actually we shall confine ourselves to a local analysis, and we replace X by an open subset of $\mathbb{R}^n$, Ω. We assume that the operator $A(t)$ of equation (1.3), Chapter III, is congruent, modulo analytic-regularizing operators depending analytically on t in $[0, T[$, to an operator of the form $\mathrm{Op}\, a(x, t, \xi)$. For the sake of simplicity we suppose that $A(t)$ is equal to such an operator:

$$(5.1)\qquad A(t)u(x) = (2\pi)^{-n}\int e^{ix\cdot\xi}a(x, t, \xi)\hat{u}(\xi)\, d\xi.$$

Our basic assumption is that

(5.2) *$a(x, t, \xi)$ is a pseudoanalytic symbol of degree one in Ω, depending analytically on t in $[0, T[$.*

Explicitly this means the following:

(5.3) *a can be extended as a C^∞ function of (x, t, ξ), holomorphic with respect to x,† in $\Omega^{\mathbf{C}} \times [0, T[\times \mathbb{R}_n$, where $\Omega^{\mathbf{C}}$ is an open subset of $\mathbb{C}^n$ containing Ω, having the following property:*

There are two constants C, $R > 0$ such that, given any nonnegative integer j, any n-tuple α, we have

$$(5.4)\qquad \|\partial_t^j \partial_\xi^\alpha a(x, t, \xi)\| \le C^{j+|\alpha|+1} j!\,\alpha!\,|\xi|^{1-|\alpha|},$$
$$\forall x \in \Omega^{\mathbf{C}},\ t \in [0, T[,\ \xi \in \mathbb{R}_n,\ |\xi| \ge R \sup(|\alpha|, 1).$$

In (5.4), $\|\ \ \|$ denotes the norm in the Banach space of continuous linear operators on the Hilbert space H.

REMARK 5.1. In the applications $a(x, t, \xi)$ is constructed from a formal *analytic* symbol (Definition 2.2 and (2.24)) $\sum_{j=0}^{+\infty} a_j(x, t, \xi)$ with a_j positive-

† In this section z always denotes the variable in the complex plane, and we shall use the notation x to denote the extension to complex values (in $\mathbb{C}^n$) of the variable in $\mathbb{R}^n$.

homogeneous of degree $1-j$ with respect to ξ (in other words, a classical analytic symbol).

We shall always reason under the "negativity" hypothesis (1.6) of Chapter III, which we state here in the following fashion:

(5.5) *To every compact subset $\mathcal{K}$ of $\Omega^{\mathbb{C}}$ there is a compact subset $\mathcal{K}'$ of the open half-plane $\mathbb{C}_- = \{z \in \mathbb{C};\ \operatorname{Re} z < 0\}$ such that*

$$zI - a(x, t, \xi)/(1 + |\xi|^2)^{1/2} : H \to H$$

is a bijection, for all $x \in \mathcal{K}$, $0 \le t < T$, $z \in \mathbb{C}\backslash\mathcal{K}'$, ξ in $\mathbb{R}_n$.

For the sake of simplicity, in the sequel we use the notation

$$\rho = (1 + |\xi|^2)^{1/2}. \tag{5.6}$$

We seek a solution $U(t)$ to the initial value problem:

$$\frac{dU}{dt} - A(t)U \sim 0, \qquad 0 \le t < T, \tag{5.7}$$

$$U(0) = I, \tag{5.8}$$

where $\sim$ indicates equivalence modulo analytic-regularizing operators which depend analytically on t in $[0, T[$. If $R(t)$ is such an operator, given any compactly supported distribution u in Ω, then $R(t)u$ is an analytic function of (x, t) in $\Omega \times [0, T[$. We seek $U(t)$ in the form of an analytic pseudodifferential operator (of order zero) depending analytically on t. Formally it is determined as in the C^∞ case (see proof of Theorem 1.1, Chapter III, part A, existence of the parametrix $U(t)$), but we need more precise estimates of the terms $\mathcal{U}_j$ in (1.12), Chapter III.

5.1. Construction and Estimates of the Local Parametrix $U(t)$

We use the notation $E = [zI - a(x, t, \xi)/\rho]^{-1}$. We try to solve equation (1.22) of Chapter III, which we rewrite here

$$k = E\left[I - \rho^{-1}\left(\frac{\partial k}{\partial t} - a \odot k + ak\right)\right]. \tag{5.9}$$

We interpret k as a formal symbol $\sum_{j=0}^{+\infty} k_j$ and we determine the successive terms k_j by the equations (1.24), Chapter III. Here, since the number m of

Chapter III is now equal to one, we shall have deg $k_j \leq -j$. We take then, as in the C^∞ case,

$$(5.10) \qquad \mathcal{U}_j(x, t, \xi) = (2\pi i)^{-1} \oint_\gamma e^{\rho t z} k_j(x, t, \xi; z)\, dz,$$

with the same meaning for the contour γ as in Chapter III.

In order to estimate the k_j we shall use the norms analogous to (3.42). Let x_o be any point in $\Omega^{\mathbb{C}}$, $D(x_o; r)$ the polydisk $|x^\iota - x_o^\iota| < r$, $\iota = 1, \ldots, n$. We assume that r is so small that the closure of $D(x_o; r)$ is contained in $\Omega^{\mathbb{C}}$. Let f be a holomorphic function in $D(x_o; r)$ valued in $L(H)$ and s any number, $0 < s < 1$. We set here

$$(5.11) \qquad \|f\|_s = \sup_{x \in D(x_o; sr)} \|f(x)\|.$$

We shall make use of the fact, proved in Section 3, that if for some constants M, $C > 0$ and some integer $j \geq 0$, (3.44) holds, then so does (3.47). We recall that the variable denoted by z in Section 3 is now denoted by x.

Our starting point will be (5.4) and similar estimates for $\|E\|$. We select the compact set $\mathcal{K}$ in (5.5) large enough that it contains the polydisk $D(x_o; r)$, and the compact set $\mathcal{K}' \subset \mathbb{C}_-$ correspondingly. Then let $\mathcal{K}''$ be any compact subset of $\mathbb{C}\backslash\mathcal{K}'$. We have, for some $C_o > 0$, all integers $l \geq 0$, n-tuples α,

$$(5.12) \qquad \|D_t^l D_\xi^\alpha E(\cdot, t, \xi; z)\|_1 \leq C_o^{l+|\alpha|+1} l!\, \alpha!\, \rho^{-|\alpha|},$$

$$\forall t \in [0, T[,\ z \in \mathcal{K}'',\ \xi \in \mathbb{R}_n,\ |\xi| \geq R \sup(|\alpha|, 1).$$

We derive from this (cf. (3.47)), whatever s, $0 < s < 1$,

$$(5.13) \qquad \|D_t^l D_\xi^\alpha D_x^\beta E(\cdot, t, \xi; z)\|_s \leq C_o^{l+|\alpha|+1} l!\, \alpha!\, \rho^{-|\alpha|} r^{-|\beta|} |\beta|! \left(\frac{e}{1-s}\right)^{|\beta|},$$

for the same t, z, ξ as before. We want to use the recursive relations (1.24), Chapter III, to prove, by induction on j and for a suitable choice of the constants C_1, C_2 (both larger than C_o), the following estimates:

$$(5.14) \qquad \|D_t^l D_\xi^\alpha k_j(\cdot, t, \xi; z)\|_s \leq C_1^{l+|\alpha|} C_2^{j+1} j!\, l!\, \alpha! \left(\frac{e}{1-s}\right)^j \rho^{-|\alpha|-j},$$

$$\forall t \in [0, T[,\ z \in \mathcal{K}'',\ \xi \in \mathbb{R}_n,\ |\xi| \geq R \sup(j + |\alpha|, 1).$$

According to what was said before, we derive from (5.14), for the same s, t, z, ξ,

$$(5.15) \qquad \|D_t^l D_\xi^\alpha D_x^\beta k_j(\cdot, t, \xi; z)\|_s \leq C_1^{l+|\alpha|} C_2^{j-1} j!\, l!\, \alpha!\, |\beta|!\, r^{-|\beta|} \left(\frac{e}{1-s}\right)^{j+|\beta|}.$$

Since $k_0 = E$, (5.14) (resp., (5.15)) reduces to (5.12) (resp., (5.13)) when $j = 0$. For $j > 0$ we use the relation (1.24), Chapter III, which we rewrite here:

$$k_j = -E\rho^{-1}\left[\frac{\partial k_{j-1}}{\partial t} - \sum_{1\le|\mu|\le j}\frac{1}{\mu!}\partial_\xi^\mu a D_x^\mu k_{j-|\mu|}\right]. \tag{5.16}$$

We have, by the Leibniz formula,

$$D_t^l D_\xi^\alpha k_j = \tag{5.17}$$

$$-\sum_{l'\le l,\alpha'\le\alpha}\binom{l}{l'}\binom{\alpha}{\alpha'}[D_t^{l-l'}D_\xi^{\alpha-\alpha'}(E/\rho)]D_t^{l'}D_\xi^{\alpha'}\left[\frac{\partial k_{j-1}}{\partial t}\right.$$

$$\left.-\sum_{1-|\mu|\le j}\frac{1}{\mu!}\partial_\xi^\mu a D_x^\mu k_{j-|\mu|}\right].$$

We avail ourselves of the following estimates:

$$\|D_t^{l-l'}D_\xi^{\alpha-\alpha'}(E/\rho)\|_s \le C_o^{l-l'+|\alpha-\alpha'|+1}(l-l')!(\alpha-\alpha')!\rho^{-1-|\alpha-\alpha'|}; \tag{5.18}$$

$$\left\|D_t^{l'}D_\xi^{\alpha'}\left(\frac{\partial k_{j-1}}{\partial t}\right)\right\|_s \le C_1^{l'+|\alpha'|+1}C_2^j(j-1)!\,l'!\,\alpha'!\,\rho^{-|\alpha'|-j+1}\left(\frac{e}{1-s}\right)^{j-1}; \tag{5.19}$$

$$\frac{1}{\mu!}\|D_t^{l'}D_\xi^{\alpha'}(\partial_\xi^\mu a D_x^\mu k_{j-|\mu|})\|_s \tag{5.20}$$

$$\le l'!\alpha'!\sum_{l''\le l',\,\alpha''\le\alpha'}\frac{1}{\mu!}\frac{1}{l''!}\frac{1}{\alpha''!}\|D_t^{l''}D_\xi^{\mu+\alpha''}a\|_s$$

$$\times\frac{1}{(l'-l'')!(\alpha'-\alpha'')!}\|D_t^{l'-l''}D_\xi^{\alpha'-\alpha''}D_x^\mu k_{j-|\mu|}\|_s$$

$$\le l'!\alpha!\sum_{l''\le l',\,\alpha''\le\alpha'}C_o^{l''+|\mu+\alpha''|+1}\rho^{1-|\mu+\alpha''|}C_1^{l'-l''+|\alpha'-\alpha''|}C_2^{j-|\mu|+1}$$

$$\times(j-|\mu|)!\,|\mu|!\,r^{-|\mu|}\left(\frac{e}{1-s}\right)^j\rho^{-j-|\mu|-|\alpha'-\alpha''|}.$$

We have used the fact that $|\xi|\ge R(j+|\alpha|)$ implies $|\xi|\ge R|\mu+\alpha''|$. If we put all this together in an estimate of the s-norm of the left-hand side in (5.17), we get

$$\left[j!\,l!\,\alpha!\,C_2^{l+|\alpha|}\left(\frac{e}{1-s}\right)^j\right]^{-1}\rho^{j+|\alpha|}\|D_t^lD_\xi^\alpha k_j\|_s \tag{5.21}$$

$$\le\frac{C_oC_1}{C_2}\sum_{l'\le l,\alpha'\le\alpha}\left(\frac{C_o}{C_1}\right)^{l-l'+|\alpha-\alpha'|}$$

$$+\sum_{1\le|\mu|\le j}\sum_{\substack{l''\le l'\le l\\ \alpha''\le\alpha'\le\alpha}}\left(\frac{C_o}{C_1}\right)^{l-(l'-l'')+|\alpha-(\alpha'-\alpha'')|}\left(\frac{C_o}{rC_2}\right)^{|\mu|}C_o.$$

The first sum, on the right-hand side of (5.21), can be made arbitrarily small by choosing C_1 large in comparison to C_o, and C_2 large in comparison to C_oC_1. The second sum can be made arbitrarily small by choosing $C_1 \gg C_o$ and $C_2 \gg C_o/r$, this time also due to the fact that the summation is effected over $|\mu| \geq 1$. By selecting C_1 and C_2 in such a way that the right-hand side in (5.21) is ≤ 1, we have proved (5.14).

By the Borel–Lebesgue lemma we can cover the compact subset $\mathcal{K}$ of $\Omega^{\mathbb{C}}$ with finitely many disks $D(x_o; \frac{1}{2}r) \subset\subset \Omega^{\mathbb{C}}$. We select $\mathcal{K}''$ to be the contour γ of the integral in (5.10). For a suitable choice of $M > 0$ we obtain

$$\|D_t^l D_\xi^\alpha k_j(x, t, \xi; z)\| \leq M^{j+l+|\alpha|+1} j!\, l!\, \alpha!\, \rho^{-|\alpha|-j}, \tag{5.22}$$

$$\forall x \in \mathcal{K}, t \in [0, T[, z \in \gamma, \xi \in \mathbb{R}_n, |\xi| \geq R \sup(j + |\alpha|, 1).$$

We apply (5.22) in conjunction with the integral representation (5.10). By virtue of the fact that $\operatorname{Re} z < 0$ on the contour γ, we obtain, after a suitable increase of M, for some constant $c > 0$,

$$\|D_t^l D_\xi^\alpha \mathcal{U}_j(x, t, \xi)\| \leq M^{j+l+|\alpha|+1} j!\, l!\, \alpha!\, \rho^{-|\alpha|-j} \left(\sum_{l'=0}^{l} \frac{\rho^{l'}}{l'!} \right) e^{-ct\rho}, \tag{5.23}$$

for all $x \in \mathcal{K}, t \in [0, T[, \xi \in \mathbb{R}_n, |\xi| \geq R \sup(j + |\alpha|, 1)$.

This means that

$$\sum_{j=0}^{\infty} \mathcal{U}_j(x, t, \xi) \tag{5.24}$$

is a formal pseudoanalytic symbol of degree zero in Ω, depending analytically on $t \in [0, T[$. The precise meaning of this analytic dependence is that made exact by (5.23).

Let $\tilde{\Omega}$ be any relatively compact open subset of Ω. We may select the sequence of cutoff functions ϕ_j ($j = 0, 1, \ldots$) according to the rules (2.28), (2.29), so that

$$\mathcal{U}(x, t, \xi) = \sum_{j=0}^{\infty} \phi_j(\xi) \mathcal{U}_j(x, t, \xi) \tag{5.25}$$

is a true pseudoanalytic symbol (of degree zero) in $\tilde{\Omega}$ (depending analytically on t, $0 \leq t < T$), and set $U(t) = \operatorname{Op} \mathcal{U}(x, t, \xi)$, that is,

$$U(t)u(x) = (2\pi)^{-n} \int e^{ix \cdot \xi} \mathcal{U}(x, t, \xi) \hat{u}(\xi)\, d\xi, \tag{5.26}$$

for all $u \in C_c^\infty(\tilde{\Omega}; H)$ and $x \in \tilde{\Omega}$.

Before proceeding we must check that equations (5.7)–(5.8) are indeed verified. Property (5.8), which must be valid in $\tilde{\Omega}$, has been established in Chapter III (see the proof of (1.21)). Concerning (5.7) there is the usual difficulty in composing analytic pseudodifferential operators. We might as well suppose that $U(t)$ is defined, by means of (5.26), in an open neighborhood Ω' of the closure of $\tilde{\Omega}$. Select $h \in C_c^\infty(\Omega')$ to be equal to one in $\tilde{\Omega}$ and define $A(t)U(t)$ in $\tilde{\Omega}$ as the restriction to this set of the compose $A(t)hU(t)$ where, as is customary, h stands for the operator of multiplication by h. What we need then is a slightly stronger version of Theorem 3.1, a version in which the operators are allowed to depend analytically on $t \in [0, T[$ and classes are defined modulo analytic-regularizing operators with respect to (x, t), and not just with respect to x alone—precisely, operators that act on (compactly supported) distributions in x and transform them into analytic functions of (x, t). The proof of such a version is analogous to that of Theorem 3.1; we shall skip it.

In what follows we omit the tildes and assume that $\mathcal{U}(x, t, \xi)$ has all required properties and satisfies the appropriate estimates (see (5.29)) "globally" in Ω, that $U(t)$ is defined by (5.26) in Ω, and that equations (5.7), (5.8) are valid, also in the whole of Ω.

5.2. The Operator $U(t)$ Is Analytic Pseudolocal in a Strong Sense

Let us set (cf. (2.1), (2.5))

$$U_o(x, t, q) = (2\pi)^{-n} \int e^{iq\cdot\xi}\mathcal{U}(x, t, \xi)\, d\xi; \tag{5.27}$$

we have

$$U(t)u(x) = \int U_o(x, t, x-y)u(y)\, dy, \qquad u \in C_c^\infty(\Omega). \tag{5.28}$$

If, as we tacitly assume, one chooses the number R large in relation to M, we derive from the estimates (5.23) for the $\mathcal{U}_j$ the following estimates for $\mathcal{U}$ itself:

$$\begin{aligned} \|D_t^l D_\xi^\alpha \mathcal{U}(x, t, \xi)\| &\le M^{l+|\alpha|+1} l!\,\alpha!\,\rho^{-|\alpha|}\left(\sum_{l'=0}^{l} \frac{\rho^{l'}}{l'!}\right) e^{-ct\rho} \\ &\le M_1^{l+|\alpha|+1} l!\,\alpha!\,(1+1/t)^l \rho^{-|\alpha|}. \end{aligned} \tag{5.29}$$

Recalling that earlier we shrank Ω, and observing that $R^{-1}|\xi| \ge 2\sup(j, |\alpha|)$ implies $R^{-1}|\xi| \ge j + |\alpha|$, we may assume that (5.29) is valid for

$$x \in \Omega^{\mathrm{C}}, \qquad 0 \le t < T, \qquad |\xi| \ge 2R\sup(|\alpha|, 1). \tag{5.30}$$

By essentially duplicating the proof of Lemma 2.1 one can prove the following:

LEMMA 5.1. *The distribution* U_o, *defined in* (3.27), *is an analytic function of* (x, t, q), *holomorphic with respect to* x, *in the region*

$$x \in \Omega^C, \quad 0 < t < T, \qquad q \in \mathbb{R}^n, \quad q \neq 0. \tag{5.31}$$

There is a constant $C > 0$ *such that for all* $f \in C_c^\infty(\Omega; H)$, *all integers* $l \geq 0$, *all* $\alpha \in \mathbb{Z}_+^n$ *and all* x *in* Ω,

$$\left| \int [D_t^l D_x^\alpha U_o(x, t, q)|_{q=x-y}] f(y)\, dy \right| \leq C^{l+|\alpha|+1} l!\, \alpha! \sum_{l'=0}^{l} \|f\|_{l'}/l'!. \tag{5.32}$$

In (5.32), $\| \ \|_{l'}$ stands for the usual Sobolev norm (here, of H-valued functions) of order l'.

THEOREM 5.1. *Let* $\mathcal{O}$ *be any open subset of* Ω, f *any compactly supported, H-valued distribution in* Ω.

If f *is an analytic function* (*valued in* H) *in* $\mathcal{O}$, *then* $U(t)f$ *and* $U(t)^*f$ *are analytic functions in* $\mathcal{O} \times [0, T[$.

We have denoted by $U(t)^*$ the *adjoint* of $U(t)$, as a pseudodifferential operator in Ω, valued in $L(H)$ (where the adjoint is defined by the Hilbert space structure of H).

The proof of Theorem 5.1 is a straightforward extension of that of Theorem 2.1, with differentiations with respect to t thrown in, and kept under control by means of (5.32). We leave the details to the reader.

REMARK 5.1. Suppose that $f \in \mathscr{E}'(\Omega; H)$ is analytic in a conic open neighborhood of a point $(x_o, \xi^o) \in \Omega \times (\mathbb{R}_n \backslash \{0\})$. We may select two open cones in $\mathbb{R}_n \backslash \{0\}$, Γ, Γ^*, with the closure of Γ contained in Γ^*, and $\xi^o \in \Gamma$, and a function g^R associated with these cones as in Lemma 1.4, such that $g^R(D)f$ is analytic in some open neighborhood $\mathcal{O}$ of x_o. By Theorem 5.1 we know then that $U(t)[h\{g^R(D)f\}]$ is analytic in $\mathcal{O} \times [0, T[$—if we take $h \in C_c^\infty(\Omega)$ to be equal to one in some open subset of Ω containing the support of f (and $\mathcal{O}$). An obvious modification of the proof of Lemma 1.5 shows that $v(x, t) = U(t)[h\{f - g^R(D)f\}]$ is analytic with respect to (x, t) in $\Omega \times \Gamma \times [0, T[$. The latter means the following:

(5.33) *Let Γ_* be any closed cone contained in Γ, $\{h_N\}_{N=1,2,\ldots}$ any sequence of C^∞ functions having their supports contained in a compact subset of Ω independent of N and such that, for each N, $|D^\alpha h_N| \le (CN)^{|\alpha|}$ if $|\alpha| \le N$, with $C > 0$ independent of N.*

There is an integer ν and a constant $C_* > 0$ such that for all $N \ge \nu$ and all $l = 0, 1, \ldots$, we have, writing $M = N - \nu$,

$$|D_t^l[\widehat{h_N v(\cdot, t)}](\xi)| \le C_*^{l+M+1} l!\, M!\, |\xi|^{-M}, \qquad \forall \xi \in \Gamma_*, |\xi| \ge 1. \tag{5.34}$$

This means that $U(t)$ is microlocally analytical pseudolocal, in the strong sense that we also have analyticity with respect to t in $[0, T[$, up to and including at $t = 0$.

5.3. Analyticity in the Cauchy Problem

Instead of dealing with the equations (5.7)–(5.8) we could have dealt with the *backward* Cauchy problem when $A(t)$ is replaced by $-A(t)^*$:

$$\frac{\partial V}{\partial t}(t, t') + A(t)^* V(t, t') \sim 0, \qquad 0 \le t \le t', \tag{5.35}$$

$$V(t', t') = I, \tag{5.36}$$

where t' is any number such that $0 < t' < T$. (Cf. part B, uniqueness of the parametrix, in the proof of Theorem 1.1, Chapter III). We would have reached conclusions analogous to those of the preceding paragraphs. We can even obtain that the equivalence in (5.35) is valid modulo analytic-regularizing operators with respect to (x, t, t'), that is, modulo linear operators that transform distributions in x alone into analytic functions of (x, t, t') in $\Omega \times \{(t, t'); 0 \le t \le t' < T\}$. The solution $V(t, t')$ is an analytic pseudodifferential operator of type analogous to $U(t)$:

$$V(t, t')u(x) = (2\pi)^{-n} \int e^{ix\cdot\xi} \mathscr{V}(x, t, t', \xi)\hat{u}(\xi)\, d\xi. \tag{5.37}$$

The operator $V(t, t')$ is analytic pseudolocal in the strong sense that given any open subset $\mathcal{O}$ of Ω, any compactly supported distribution f in Ω, if f is analytic in $\mathcal{O}$, then $V(t, t')f$ is an analytic function of (x, t, t') in $\mathcal{O} \times \{(t, t');$ $0 \le t \le t' < T\}$.

Actually we are interested in the backward Cauchy problem for the equation adjoint to (5.35):

$$\frac{\partial}{\partial t}V(t,t')^* + V(t,t')^*A(t) \sim 0, \qquad 0 \le t \le t', \tag{5.38}$$

$$V(t',t') = I. \tag{5.39}$$

THEOREM 5.2. *Let $\mathcal{O}$ be an open subset of Ω, u any C^∞ function of t in $[0, T[$ valued in $\mathscr{E}'(\Omega; H)$. Suppose that the restriction to $\mathcal{O}$ of $u(x, 0)$ is an analytic function, while the restriction to $\mathcal{O} \times [0, T[$ of $\partial u/\partial t - A(t)u$ is analytic with respect to (x, t) (both valued in H). Then the latter is also true of the restriction of u.*

PROOF. It suffices to observe that according to (5.38)–(5.39), we have

$$u(t) \equiv V(0,t)^*u(0) + \int_0^t V(t',t)^*\left[\frac{\partial u}{\partial t}(t') - A(t')u(t')\right] dt' \tag{5.40}$$

modulo analytic functions of (x, t) in $\Omega \times [0, T[$. In (5.40), $u(t) = u(\cdot, t)$. □

REMARK 5.3. There is a microlocal version of Theorem 5.2 along the lines of Remark 5.2.

5.4. Application to Elliptic Boundary Problems

We want to apply Theorem 5.2 to obtain the analyticity up to the boundary (which is the manifold X, now assumed to be analytic) of the solutions to the boundary problem (3.23)–(3.39) of Chapter III (see also problem (∗) in Section 3 of Chapter III). For convenience, we recall the equations under study:

$$P(x, t, D_x, \partial_t)u = f \qquad \text{in } X \times [0, T[, \tag{5.41}$$

$$B_j(x, D_x, \partial_t)u|_{t=0} = h_j \qquad \text{in } X,\ j = 1, \dots, \nu. \tag{5.42}$$

We assume that

$$P(x, t, D_x, \partial_t) = I\,\partial_t^m + \sum_{j=1}^m P_j(x, t, D_x)\,\partial_t^{m-j}, \tag{5.43}$$

$$B_j(x, D_x, \partial_t) = \sum_{j'=0}^{r_j} B_{j,j'}(x, D_x)\,\partial_t^{j'}, \tag{5.44}$$

with P_j and $B_{j,j'}$ *classical analytic* pseudodifferential operators in X valued in $L(H)$. We assume that the P_j depend analytically on t in $[0, T[$ and have order $\le j$. Let us be more specific and, for the sake of simplicity, localize the analysis, in order to be able to replace X by an open subset Ω of $\mathbb{R}^n$. We assume that the (formal) symbol of P_j is of the form

$$\sum_{i=0}^{+\infty} P_{j,i}(x, t, \xi), \tag{5.45}$$

where for each i, $P_{j,i}$ is an analytic function of (x, t, ξ), holomorphic with respect to (x, ξ), positive-homogeneous of degree $j - i$ with respect to ξ, in the set

$$x \in \Omega^{\mathbb{C}}, \quad 0 \le t < T, \quad \xi \in \mathbb{C}_n, \quad |\mathrm{Im}\, \xi| < \delta |\mathrm{Re}\, \xi|, \tag{5.46}$$

with $\Omega^{\mathbb{C}}$ an open subset of $\mathbb{C}_n$ containing Ω and δ a number >0, both independent of i, j. Moreover there is a constant $C_o > 0$ such that for all (x, t, ξ) in (5.46),

$$|D_t^l P_{j,i}(x, t, \xi)| \le C_o^{l+1} l! |\xi|^{j-1}. \tag{5.47}$$

A similar hypothesis is made about $B_{j,j'}$.

In addition to these analyticity hypotheses we make the same other hypotheses as in Chapter III. We assume that the *principal symbol of P is scalar*, equivalent to saying that for each $j = 1, \dots, m$, each homogeneous symbol $P_{j,0}(x, t, \xi)$ is a scalar multiple of the identity of H, $P_j^0(x, t, \xi)I$. We also assume the *ellipticity* hypothesis (3.6).

We avail ourselves of the same factorization as in Chapter III, (3.22), and also now of the fact that the factors M^+, M^- and the remainder R have the appropriate analyticity. This is derived from the following remark.

Let $\sigma(M^{\pm 0})$ have the same meaning as in Chapter III, (3.7). Since these polynomials in z are coprimes, their coefficients are analytic functions of (x, t, ξ), holomorphic with respect to (x, ξ), in the set (5.46). Let $\mathcal{P}^d$ denote the vector space of polynomials of degree $\le d$ with respect to z, with complex coefficients; it is a complex vector space of dimension $d + 1$. We regard

$$(p, q) \mapsto \sigma(M^{+0})p + \sigma(M^{-0})q$$

as a linear map of $\mathcal{P}^{m^- - 1} \times \mathcal{P}^{m^+ - 1}$ into $\mathcal{P}^{m-1}$, recalling that $m^+ = \deg \sigma(M^{+0})$, $m^- = \deg \sigma(M^{-0})$ and $m = m^+ + m^-$. Since $\sigma(M^{+0})$ and $\sigma(M^{-0})$ are coprime this map is injective. Since the source and target spaces have the same dimension, it is bijective. Let us set for any polynomial $p(z)$

$$N(p) = \sum_{j=0}^{+\infty} \left| \frac{d^j p}{dz^j}(0) \right| \Big/ j!. \tag{5.48}$$

We regard $N(p)$ as the canonical norm on any vector space $\mathcal{P}^d$.

Possibly after some shrinking of Ω and decreasing of $T > 0$, we see that there is a constant $C > 0$ such that the following holds:

Whatever $r(z) \in \mathcal{P}^{m-1}$, if $p(z) \in \mathcal{P}^{m^- -1}$ and $q(z) \in \mathcal{P}^{m^+ -1}$ satisfy

$$\sigma(M^{+0})p + \sigma(M^{-0})q = r, \tag{5.49}$$

then

$$N(p) + N(q) \leq CN(r). \tag{5.50}$$

Suppose furthermore that the coefficients of r depend on t, analytically in the closed interval $[0, T]$. By differentiating with respect to t both sides in (5.49) and reasoning by induction on l, we obtain easily the estimates

$$\frac{1}{l!}[N(D_t^l p) + N(D_t^l q)] \leq C'^{l+1} \sum_{l' \leq l} N(D_t^{l'} r)/l'!. \tag{5.51}$$

We determine the successive homogeneous terms in the formal symbols of M^+ and M^- exactly as in the C^∞ case (see Chapter III, (3.12)–(3.19)), except that here we complement the solution of (3.19), Chapter III, with the concomitant estimates of the kind (5.51). We thus obtain that $M^\pm$ have properties analogous to those of P:

$$M^\pm = M^\pm(x, t, D_x, \partial_t) = I\,\partial_t^{m^\pm} + \sum_{j=1}^{m^\pm} M_j^\pm(x, t, D_x)\,\partial_t^{m^\pm - j}, \tag{5.52}$$

the symbol of $M_j^\pm$ being of the form

$$\sum_{i=0}^{+\infty} M_{j,i}^\pm(x, t, \xi), \tag{5.53}$$

with $M_{j,i}^\pm$ having properties analogous to those of $P_{j,i}$ in (5.45), in particular verifying inequalities similar to (5.47).

As for $R = P - M^+M^-$ it has the form

$$R = \sum_{j=1}^{m^-} R_j(t)\,\partial_t^{m^- - j}, \tag{5.54}$$

where the $R_j(t)$ are analytic regularizing, in the sense now familiar to us: they transform any compactly supported distribution of x alone (in Ω) into an analytic function of (x, t).

From there on the argument proceeds very much as in the C^∞ case. We make exactly the same transformations as in Section 3 of Chapter III, transforming first the equation (5.41) into the coupled equations (3.32)–(3.36), Chapter III, and finally transforming the boundary conditions (5.42) into the relation (3.46) of Chapter III. This is the passage from (*) to (**) in

Section 3 of Chapter III. As we have done there we replace (**) by the approximate equations (3.50)–(3.52), Chapter III. Lemma 4.1 has an analytic analogue: replace C^∞ by $\mathcal{A}$ (analytic functions) and $\mathcal{D}'(X; H \otimes \mathbb{C}^{m^+})$ by $\mathcal{E}'(X; H \otimes \mathbb{C}^{m^+})$. Finally we reach in this way the analytic versions of Definition 4.1 and Theorem 4.1 of Chapter III:

DEFINITION 5.1. *We say that the problem* (5.41)–(5.42) *is analytic hypoelliptic if the following holds*:

(5.55) *Let Y be any open subset of X, $f \in C^\infty([0, T[; \mathcal{D}'(X; H))$, $h_j \in \mathcal{D}'(X; H)$, $j = 1, \dots, \nu$, be any set of data. Suppose that the restrictions of f and of the h_j are analytic H-valued functions, in $Y \times [0, T[$ and in Y respectively. Then every solution u of* (5.41)–(5.42) *is an analytic function of (x, t) in $Y \times [0, T[$ with values in H.*

THEOREM 5.3. *Under the preceding analyticity hypotheses the problem* (5.41)–(5.42) *is analytic hypoelliptic if and only if the Calderon operator $\mathcal{B}$ is analytic hypoelliptic.*

An analytic pseudodifferential operator (or an equivalence class of such operators, like $\mathcal{B}$) is analytic hypoelliptic if it preserves the analytic singular supports.

The proof of Theorem 5.3 is analogous to that of Theorem 4.1, Chapter III; the proof of the "sufficiency" of the condition is based on Theorem 5.2.

In the *coercive* case (Chapter III, Section 6) the Calderon operator $\mathcal{B}$ is *elliptic*; we recall that this property characterizes the coercive boundary problems. Since any elliptic analytic pseudodifferential operator on an analytic manifold is analytic hypoelliptic (by (4.21)) we get the following result of Morrey and Nirenberg [1]:

THEOREM 5.4. *If the problem* (5.41)–(5.42) *is coercive* (*and analytic*), *it is analytic hypoelliptic.*

All the preceding concepts and results can and ought to be stated *microlocally*, in terms of microdistributions and the sheaf of analytic pseudodifferential operators in the analytic manifold X (see Section 4). The reader interested in stating them in this manner will have no difficulty in doing so.

References

AKUMOVIC, V. G.

[1] Über die Eigenfunktionen auf geschlossenen Riemannschen Mannigfaltigkeiten, *Math. Z.* **65**, 327–344 (1956).

ANDERSSON, K. G.

[1] Propagation of analyticity of solutions of partial differential equations with constant coefficients, *Ark. Mat.* **8**, 277–302 (1970).

ATIYAH, M., and SINGER, I. M.

[1] The index of elliptic operators, *Ann. Math.*, Part I, **87**, 484–530 (1968); Part III, **87**, 546–604 (1968); Part IV, **93**, 119–138 (1971); Part V, **93**, 139–149 (1971).

BONY, J.-M.

[1] Equivalence des diverses notions de spectre singulier analytique, *Séminaire Goulaouic–Schwartz*, Ec. Polytechn., Exp. no. 3 (1976–77).

BEALS, R.

[1] A general calculus of pseudodifferential operators, *Duke Math. J.* **42**, no. 1, 1–42 (1975).

[2] Spatially inhomogeneous pseudodifferential operators, II, *Comm. Pure Appl. Math.* **27**, 161–205 (1974).

[3] Square roots of nonnegative systems and the sharp Gårding inequality (mimeographed).

BEALS, R. and FEFFERMAN, CH.

[1] Classes of spatially inhomogeneous pseudodifferential operators, *Proc. Nat. Acad. Sci. USA* **70**, 1500–1501 (1973).

[2] On local solvability of linear partial differential equations, *Ann. Math.* **97**, 482–498 (1973).

[3] Spatially inhomogeneous pseudodifferential operators, I, *Comm. Pure Appl. Math.* **27**, 1–24 (1974).

BOUTET DE MONVEL, L.

[1] Boundary problems for pseudodifferential operators, *Acta Math.* **126**, 11–51 (1971).

[2] Opérateurs pseudo-différentiels analytiques et opérateurs d'ordre infini, *Ann. Inst. Fourier Grenoble* **22**, 229–268 (1972).

[3] Hypoelliptic operators with double characteristics and related pseudo-differential operators, *Comm. Pure Appl. Math.* **27**, 585–639 (1974).

[4] Propagation des singularités des solutions d'équations analogues à l'équation de Schrödinger, *Fourier Integral Operators and Partial Differential Equations*, Springer Lecture Notes **459**, 1–14 (1974).

BOUTET DE MONVEL, L., and KREE, P.

[1] Pseudo-differential operators and Gevrey classes, *Ann. Inst. Fourier Grenoble* **27**, 295–323 (1967).

BOUTET DE MONVEL, L., and SJÖSTRAND, J.
[1] Sur la singularité des noyaux de Bergman et de Szegö. *Astérisque* **34–35**, 123–164 (1976).
CALDERON, A. P.
[1] Uniqueness in the Cauchy problem of partial differential equations, *Amer. J. Math.* **80**, 16–36 (1958).
[2] Existence and uniqueness theorems for systems of partial differential equations, *Symposium on Fluid Dynamics, University of Maryland*, College Park, Maryland (1961).
[3] Boundary value problems for elliptic equations, *Proceedings of the Joint Soviet–American Symposium on Partial Differential Equations.* Novosibirsk Acad. Sci. USSR 1–4, (1963).
[4] Singular integrals, *Bull. Amer. Math. Soc.* **72**, 427–465 (1966).
CALDERON, A. P., and VAILLANCOURT, R.
[1] On the boundedness of pseudo-differential operators, *J. Math. Soc. Japan* **23**, 374–378 (1971).
[2] A class of bounded pseudodifferential operators, *Proc. Nat. Acad. Sci. USA* **69**, 1185–1187 (1972).
CALDERON, A. P. and ZYGMUND, A.
[1] Singular integral operators and differential equations, *Amer. J. Math.* **79**, 901–921 (1957).
CHAZARAIN, J.
[1] Formule de Poisson pour les variétés riemanniennes, *Inventiones Math.* **24**, 65–82 (1974).
CHEVALLEY, CL.
[1] *Theory of Lie groups*, Princeton University Press, Princeton, New Jersey, 1946.
DANILOV, V. G., and MASLOV, V. P.
[1] Quasi-invertibility of functions of ordered operators in the theory of pseudodifferential operators, *J. Sov. Math.* **7**, no. 5, 695–794 (1977).
DE RHAM, G.
[1] *Variétés Différentiables*, Hermann, Paris, 1955.
DIEUDONNÉ, J.
[1] *Eléments d'Analyse*, Vols. 7, 8, Gauthier-Villars, Paris, 1978.
DUISTERMAAT, J. J.
[1] *Fourier integral operators*, Lecture Notes, Courant Institute of Mathematical Sciences, New York, 1973.
DUISTERMAAT, J. J. and HÖRMANDER, L.
[1] Fourier integral operators. II, *Acta Math.* **128**, 183–269 (1972).
DUISTERMAAT, J. J., and SJÖSTRAND, J.
[1] A global construction for pseudodifferential operators with non-involutive characteristics. *Inventiones Math.* **20**, 209–225 (1973).
EGOROV, YU, V.
[1] On canonical transformations of pseudodifferential operators. *Uspehi Mat. Nauk* **25**, 235–236 (1969).
[2] Subelliptic operators, *Uspehi Mat. Nauk* **30**, no. 2, 57–114 (1975); *Uspehi Mat. Nauk* **30**, no. 3, 57–104 (1975); *Russian Math. Surveys* **30**, no. 2, 59–118 (1975); **30**, no. 3, 55–105 (1975).
EGOROV, YU. V., and KONDRAT'EV, V. A.
[1] The oblique derivative problem, *Mat. Sbornik* **78**, 148–176 (1969); *Math. USSR Sbornik* **7**, 368–370 (1969).
ESKIN, I. G.
[1] *Boundary Value Problems for Elliptic Pseudodifferential Equations*, Nauka, Moscow, 1973 (in Russian; English translation to appear in *Transl. Amer. Math. Soc.*).
FARRIS, M.
[1] A generalization of Egorov's theorem, to be published.
FEFFERMAN, CH., and PHONG, D.
[1] On positivity of pseudodifferential operators, *Proc. Nat. Acad. Sci.* **75**, 4673–4674 (1978).

FOLLAND, G. B., and KOHN, J. J.
[1] *The Neumann problem for the Cauchy–Riemann complex*, Ann. Math. Studies, Princeton University Press, Princeton, New Jersey, 1972.
GUILLEMIN, V. W.
[1] Clean intersection theory and Fourier integrals, in *Fourier Integral Operators and Partial Differential Equations*, Springer Lecture Notes in Math. No. 459, 23–34 (1974).
[2] Symplectic spinors and partial differential equations, *Proc. C.N.R.S., Colloque Geometrie Symplectique*, Aix-en-Provence (June 1974).
GUILLEMIN, V. W., and STERNBERG, S.
[1] *Geometric Asymptotics*, American Mathematical Society, Providence, R.I., 1978.
GODEMENT, R.
[1] *Topologie Algébrique et Théorie des Faisceaux*, Hermann, Paris, 1958.
HOCHSCHILD, G.
[1] *The Structure of Lie Groups*, Holden–Day, San Francisco, 1965.
HÖRMANDER, L.
[1] Hypoelliptic differential operators, *Ann. Inst. Fourier Grenoble* **11**, 477–492 (1961).
[2] *Linear Partial Differential Operators*, Grundl. Math. Wiss., Band 116, Springer-Verlag, Berlin–Heidelberg–New York, 1963.
[3] Pseudo-differential operators, *Comm. Pure Appl. Math.* **18**, 501–517 (1965).
[4] Pseudo-differential operators and hypoelliptic equations, *Proc. Symp. Pure Math.* **10**, 138–183 (1966).
[5] Pseudo-differential operators and non-elliptic boundary problems, *Ann. Math.* **83**, 129–209 (1966).
[6] Hypoelliptic second-order differential equations, *Acta Math.* **119**, 147–171 (1967).
[7] The spectral function of an elliptic operator, *Acta Math.* **121**, 193–218 (1968).
[8] On the index of pseudodifferential operators. *Koll. Ell. Diff. Gl. II*, 127–146. Academie-Verlag, Berlin, 1969.
[9] Linear differential operators, *Actes Congr. Int. Math.* **1**, 121–133 (1970).
[10] On the singularities of solutions of partial differential equations, *Comm. Pure Appl. Math.* **23**, 329–358 (1970).
[11] Fourier integral operators, I, *Acta Math.* **127**, 79–183 (1971).
[12] L^2 estimates for pseudodifferential operators, *Comm. Pure Appl. Math.* **24**, 529–536 (1971).
[13] Uniqueness theorems and wave front sets for solutions of linear differential equations with analytic coefficients, *Comm. Pure Appl. Math.* **24**, 671–704 (1971).
[14] A remark on Holmgren's uniqueness theorem, *J. Diff. Geom.* **5**, 129–134 (1971).
[15] On the existence and the regularity of solutions of linear pseudo-differential equations, *Ens. Math.* **17**, 99–163 (1971).
[16] A class of hypoelliptic pseudodifferential operators with double characteristics, *Math. Ann.* **217**, 165–188 (1975).
[17] The Cauchy problem for differential equations with double characteristics, *J. Anal. Math.* **32**, 118–196 (1977).
[18] Propagation of singularities and semi-global existence theorems for (pseudo-)differential operators of principal type, *Ann. Math.* **108**, 569–609 (1978).
[19] The Weyl calculus of pseudo-differential operators, *Comm. Pure Appl. Math.* **32**, 359–443 (1979).
[20] Subelliptic operators, *Seminar on Singularities of Solutions of Linear Partial Differential Equations*, Ann. Math. Studies **91**, 127–207, Princeton University Press, Princeton, New Jersey, 1979.
IVRII, V. IA.
[1] Sufficient conditions for regular and completely regular hyperbolicity, *Trudy Moskow Mat. Obsc.* **33**, 1–65 (1975).

[2] Energy integrals for nonstrictly hyperbolic operators, *Uspehi Mat. Nauk.*, **30**, no. 6, 169–170 (1975).

[3] Correctness of the Cauchy problem for nonstrictly hyperbolic equations, *Trudy Moskow Mat. Obsc.* **34**, 151–170 (1977).

IVRII, V. IA, and PETKOV, V. M.

[1] Necessary conditions for the correctness of the Cauchy problem for nonstrictly hyperbolic equations, *Uspehi Mat. Nauk.* **29**, no. 5, 3–70 (1974).

KARAMATA, J.

[1] Neuer Beweis und Verallgemeinerung der Tauberschen Sätze, welche die Laplacesche- und Stieltjescher-Transformationen betreffen, *J. Reine Angew. Math.* **164**, 27–39 (1931).

KOHN, J. J.

[1] Pseudo-differential operators and non-elliptic problems, *Pseudo-differential Operators*, C.I.M.E. Stresa (Italy) 157–165 (1968).

[2] Pseudo-differential operators and hypo-ellipticity, *Proc. Symp. Pure Math.* **23**, 61–69 (1973).

[3] Subelliptic estimates, *Proc. Symp. Pure Math.* **35**, 143–152 (1979).

KOHN, J. J., and NIRENBERG, L.

[1] An algebra of pseudo-differential operators, *Comm. Pure Appl. Math.* **18**, 269–305 (1965).

KUCHERENKO, V. V.

[1] Asymptotic solutions of equations with complex characteristics, *Mat. Sbornik* **95**, 164–213 (1974); *Math. USSR Sbornik* **24**, no. 2, 159–207 (1974).

[2] Parametrix for equations with degenerate symbol, *Soviet Math. Dokl.* **17**, no. 4, 1099–1103 (1976).

KUMANO-GO, H.

[1] Algebras of pseudodifferential operators, *J. Fac. Sci. Tokyo* **17**, 31–50 (1970).

[2] Oscillatory integrals of symbols of pseudo-differential operators and the local solvability theorem of Nirenberg and Treves, *Katada Symposium on Partial Differential Equations*, 166–191 (1972).

KUMANO-GO, H., and TANIGUCHI, K.

[1] Oscillatory integrals of symbols of operators on $\mathbb{R}^n$ and operators of Fredholm type, *Proc. Japan Acad.* **49**, 397–402 (1973).

LAX, P. D.

[1] Asymptotic solutions of oscillatory initial value problems, *Duke Math. J.* **24**, 627–646 (1957).

LERAY, J.

[1] *Analyse Lagrangienne et Mécanique Quantique*, Séminaire Collège de France, Paris (1976–1977).

LIONS, J.-L., and MAGENES, E.

[1] *Problèmes aux Limites Non Homogènes et Applications*, 3 vol., Dunod, Paris, 1968. English translation: Grund. Math. Wiss., Band 181, 182, Springer-Verlag, Berlin-Heidelberg-New York, 1972.

MASLOV, V. P.

[1] *Théorie des Perturbations et Méthodes Asymptotiques*, Dunod, Paris, 1972 (French translation).

[2] *Operational Methods*, MIR Publishers, Moscow, 1973; English translation (1976).

MELIN, A.

[1] Lower bounds for pseudo-differential operators, *Ark. Mat.* **9**, 117–140 (1971).

MELIN, A., and SJÖSTRAND, J.

[1] Fourier integral operators with complex phase functions, Springer Lecture Notes No. 459, 120–223 (1974).

[2] Fourier integral operators with complex phase and application to an interior boundary problem, *Comm. Partial Diff. Eqns.*, **1**, no. 4, 313–400 (1976).

MENIKOFF, A., and SJÖSTRAND, J.
[1] On the eigenvalues of a class of hypoelliptic operators,
Part I: *Math. Ann.* **235**, 55–85 (1978).
Part II: Springer Lecture Notes, Proc. Conf. in Global Analysis, Calgary.
Part III: the non-semibounded case, to appear.

MORREY, C. B., and NIRENBERG, L.
[1] On the analyticity of the solutions of linear elliptic systems of partial differential equations, *Comm. Pure Appl. Math.* **10**, 271–290 (1957).

MOYER, R.
[1] The Nirenberg–Treves condition is necessary for local solvability, to be published.

NIRENBERG, L.
[1] Pseudo-differential operators, *Proc. Symp. Pure Math.* **16**, 147–168 (1970).
[2] A proof of the Malgrange preparation theorem, *Proc. Liverpool singularities*, Symp. I, Springer Lecture Notes in Math. No. 192, 97–104 (1971).

NIRENBERG, L., and TREVES, F.
[1] On local solvability of linear partial differential equations. I: Necessary conditions, *Comm. Pure Appl. Math.* **23**, 1–38 (1970).

OLEINIK, O. A., and RADKEVITCH, E. V.
[1] *Second-Order Equations with Nonnegative Characteristic Form*, Itogi Nauk, Moscow, 1971.

ROTHSCHILD, L., and STEIN, E. M.
[1] Hypoelliptic differential operators and nilpotent groups, *Acta Math.* **137**, 247–320 (1976).

SATO, M., KAWAI, T., and KASHIWARA, M.
[1] Microfunctions and pseudo-differential equations, *Hyperfunctions and Pseudo-Differential Equations*, Springer Lecture Notes No. 287 (1971).

SCHWARTZ, L.
[1] *Théorie des Distributions*, 2nd ed., Hermann, Paris, 1966.

SEELEY, R.
[1] Refinement of the functional calculus of Calderon and Zygmund, *Koninkl. Nederl. Akad. v. Wet. Proceedings, Ser. A*, **68**, 521–531 (1965).
[2] Singular integrals and boundary value problems, *Amer. J. Math.* **88**, 781–809 (1966).
[3] Complex powers of an elliptic operator, *Proc. Symp. Pure Math.* **10**, 288–307 (1968).
[4] Analytic extension of the trace associated with elliptic boundary problems, *Amer. J. Math.* **91**, 963–983 (1969).

SJÖSTRAND, J.
[1] A class of pseudo-differential operators with multiple characteristics, *C. R. Acad. Sci. Paris Sér. A* **275**, 817–819 (1972).
[2] Operators of principal type with interior boundary conditions, *Acta Math.* **130**, 1–51 (1973).
[3] Parametrics for pseudodifferential operators with multiple characteristics, *Ark. Math.* **12**, 85–130 (1974).
[4] Propagation of singularities for operators with multiple involutive characteristics, *Ann. Inst. Fourier Grenoble* **27**, 141–155 (1976).
[5] Propagation of analytic singularities for second-order Dirichlet problems, *Comm. Partial Diff. Eqns*, **5**, Part I: 41–94, Part II: 187–207 (1980).

SOURIAU, J.-M.
[1] *Structure des Systèmes Dynamiques*, Dunod, Paris, 1970.
[2] Construction explicite de l'indice de Maslov. Applications, Fourth International Colloquium on Group Theoretical Methods in Physics, University of Nijmegen, Netherlands (1975).

STEENROD, N.
[1] *Topology of Fibre Bundles*, Princeton University Press, Princeton, New Jersey, 1951.

TAYLOR, M.
[1] *Pseudo-differential Operators*, Springer Lecture Notes in Math. No. 416 (1974).
[2] Grazing rays and reflection of singularities of solutions to wave equations, *Comm. Pure Appl. Math.* **29**, 1–38 (1976).
TREVES, F.
[1] Opérateurs différentiels hypo-elliptiques, *Ann. Inst. Fourier* **9**, 1–73 (1959).
[2] *Topological Vector Spaces, Distributions and Kernels*, Academic Press, New York, 1967.
[3] *Basic Linear Partial Differential Equations*, Academic Press, New York, 1975.
[4] Hypoelliptic PDEs of principal type, sufficient conditions and necessary conditions, *Comm. Pure Appl. Math.* **24**, 631–670 (1971).
[5] A new method of proof of the subelliptic estimates, *Comm. Pure Appl. Math.* **24**, 71–115 (1971).
UNTERBERGER, A.
[1] Oscillator harmonique et opérateurs pseudo-différentiels, *Ann. Inst. Fourier*, **29**, 201–221 (1979).
UNTERBERGER, A., and BOKOBZA, J.
[1] Les opérateurs de Calderon–Zygmund précisés, *C. R. Acad. Sci. Paris* **259**, 1612–1614 (1965).
[2] Sur une généralisation des opérateurs de Calderon–Zygmund et des espaces H^s, *C. R. Acad. Sci. Paris* **260**, 3265–3267 (1965).
[3] Les opérateurs pseudo-différentiels d'ordre variable, *C. R. Acad. Sci. Paris* **261**, 2271–2273 (1965).
VOLEVIČ, L. R.
[1] Boundary value problems for general elliptic systems, *Mat. Sbornik* **68**, no. 110, 373–416 (1965).
YAMAMOTO, K.
[1] On the reduction of certain pseudo-differential operators with noninvolutive characteristics, *J. Differential Equations* **26**, 435–442 (1977).
WALLACH, N. R.
[1] *Symplectic Geometry and Fourier Analysis*, Mathematical Sciences Press, Brookline, Mass., 1977.
WEINSTEIN, A.
[1] On Maslov's quantization condition, *Fourier Integral Operators and Partial Differential Equations*, Springer Lecture Notes in Math. No. 459, 341–372 (1974).
[2] Fourier integral operators, quantization, and the spectra of riemannian manifolds, Colloque Intern. Geom. Sympletique et Phys. Math., C.N.R.S. Paris, (1974).
[3] The order and symbol of a distribution, *Trans. Amer. Math. Soc.* **24**, 1–54 (1958).
[4] *Symplectic Manifolds*, Regional Conference Series in Math., American Mathematical Society, Providence, Rhode Island (1977).

Index

O

P

R

S

MIX
Papier aus verantwortungsvollen Quellen
Paper from responsible sources
FSC® C105338

If you have any concerns about our products,
you can contact us on
ProductSafety@springernature.com

In case Publisher is established outside the EU,
the EU authorized representative is:
Springer Nature Customer Service Center GmbH
Europaplatz 3, 69115 Heidelberg, Germany

Printed by Libri Plureos GmbH
in Hamburg, Germany